PROTOMORPHOLOGY

THE PRINCIPLES OF
CELL AUTO-REGULATION

By

ROYAL LEE

and

WILLIAM A. HANSON
of the
Lee Foundation for Nutritional Research

WITH AN APPENDED ARTICLE ON
"CYTOTOXINES"

by

PROF. K. R. VICTOROV
Timiryazew Agricultural Academy Moscow, U.S.S.R.

1947

Published by
LEE FOUNDATION FOR NUTRITIONAL RESEARCH
MILWAUKEE 3, WISCONSIN

This book is sponsored by the Lee Foundation for Nutritional Research, a non-profit institution chartered by the State of Wisconsin, whose objectives are the investigation and dissemination of facts relating to nutrition and associated sciences. Therefore, it is freely offered as a contribution to biological research without restrictions as to reproduction if the source is properly acknowledged.

Lee Foundation for Nutritional Research

Originally Printed in the United States of America
By the W. A. KRUEGER Co.
Milwaukee, Wilconsin, U.S.A.

*   *   *

Vedder, Mark 1965−2102
        Pleomorphology / Royal Lee
        396 pp. 2 cm
ISBN 978-1-941776-52-0

2021
ISBN 978-1-941776-52-0
Publishing rights reserved
by
Mark Vedder
sufferingduckman@gmail.com

2021 EDITORS' PREFACE

THE 1947 EDITION OF *PROTOMORPHOLOGY* remains as relevant today as when it was published, albeit for different reasons. We have put together this edition in view of both its informational and historical value. The text is not altered or revised in any manner; fonts are used that simulate as closely as possible the original, and the exact same pagination is used with only the first or last sentence slipping forward or backward as suits the formatting. The entire work was painstakingly typed from a copy of the original; archaic spellings and punctuation, are preserved rather than being corrected, and the drawings were cleaned up without 'fixing' them, as the flavor of the illustrations can be considered to be as valuable as the information it carries. There were a number of errors (thirty-four) made in the original edition, and these are listed in order on the following page. The reader is thus assured that what he reads is as faithful as possible to the original publication. The editors present this, neither promulgating or dismissing its discoveries and conclusions, but with the recognition of its value in the rapidly diminishing field of foundational research.

Erratica Corrected in this Edition

Pages xix and xx, sections VII, VIII, and IX were mislabeled V, VI, and VII.

Page 19, 2nd paragraph, period missing after 0.000105.

Page 35, last line "characteristics" misspelled.

Page 36, item 5, "alcohol" and "petroleum" misspelled.

Page 54, item 4, last bibliographic entry, "Senescence" misspelled.

Page 57, paragraph 3, "accumulation" misspelled.

Page 63, 4th paragraph, "Embryo" misspelled.

Page 67, second entry in bibliography, "inhibitor" misspelled.

Page 70, title "Figure 4" misplaced.

Page 111, 7th paragraph, "proceeds" misspelled.

Page 152 1st paragraph last line, "evaluate" misspelled.

Page 155 4th paragraph, reference to 'Page 138' missing.

Page 156, first bibliography entry, first quote mark missing.

Page 157, 9th entry, dash instead of a colon.

Page 157, next to last bibliography entry, first quote mark missing.

Page 174 4th paragraph "embryo" spelled wrong.

Page 182, 2nd paragraph "superstition" spelled wrong.

Page 206, extra dash in footnote.

Page 219, third paragraph, "titre" misspelled.

Page 227. the bottom line has a correction made by the original editors which we have implemented; "bile" becomes "blood".

Page 242 17th entry, "Nucleoproteins" misspelled.

Page 242 18th entry, "Coagulation" misspelled.

Page 242 19th entry, "Cambridge" misspelled.

Page 244 4th entry, missing beginning quote mark.

Page 260, item 6, "protomorphogens" misspelled.

Page 281, 4th paragraph, "accumulated" misspelled.

Page 294, missing second quote mark after "image" at the end of the first paragraph.

Page 309, 4th entry, extra space.

Page 311, 14th entry, "Febiger" misspelled.

Page 325, 2nd paragraph, "except" misspelled.

Page 325, last paragraph, "processes" misspelled.

Page 327, 2nd paragraph, "Bogomolets" misspelled.

Page 344, under Magrou, "Croucroun" misspelled.

Page 355, under "Fibrous Proteins", "phosphatase" misspelled.

PREFACE

THAT OMNIFIC GENTLEMAN Leonardo da Vinci is reputed to have generalized: "The supreme misfortune that can befall any man is for him to embrace a theory mistaking it for a fact." ·

While engaged in compiling an exhaustive survey of the endocrine and nutritional factors affecting growth, we were struck by the similarity of certain specialized growth factors reported by various experimenters. Dr. Fenton B. Turck's quotation, "The intrusion of something which the flesh engenders ... not entering from outside" (Imhotep, 4000 B.C.) has taken on a new and significant meaning. These observations, as will be seen in the following pages, have led to speculation on the basic problem of biology and the presentation of a rationally integrated hypothesis for its explanation.

In preparing the abstract of this hypothesis published herein, we have constantly kept da Vinci's warning before us. We are fully cognizant of the inevitable danger of errors in interpretation that must, by the nature of our method, be inherent in this exposition. We realize that there is scarcely a paragraph in this volume that cannot be interpreted in many different ways other than that in which we have.

We beg the indulgence of the reader, who will realize that the work of corroboration and review cannot by its nature be established on meticulous specialized experimentation. So long as one is guided by Leonardo da Vinci's axiom and does not embrace the theory in preference to the fact there is little danger to the scientific method in its presentation. And if such an hypothesis is presented freely and with the feeling that our reward will be measured by the experimental work it stimulates, it can only be of benefit to scientific progress. Ideas and hypotheses rise and fall on their merits. It is only through critical experimental investigation that the truth can be established. Such hypotheses, therefore, should not be defended except by candid examination of factual evidence, for if founded on misconceptions, no amount of persuasion can prop them up, but if founded on basic truths, no amount of prejudice can prevent their ultimate adoption. [1]

The outline contained herein has been prepared under the most trying circumstances. Before and during the war both authors have been forced to spend all but their sparse leisure time in activities directly related to the engineering of products necessary for national defense. Our study and corroboration, therefore, is hardly begun. We have thought it expedient, however, to publish that part of our outline which has been thought through to a reasonable conclusion. It is our hope that by doing so we may stimulate experimental criticism of our hypothesis and thus add to the excellent work now being conducted in this field. In a sense this outline can be considered a report of the "Lee Foundation for Nutritional Research" on its planning of future activity in this field.

If there be any merit to the suggestions within these pages, little credit can be claimed by the authors. We have been constantly appalled by the insignificance of our difficulties in piecing together this hypothesis in comparison with the inconceivable obstacles faced by those men reviewed by us who have supplied us with the facts to work with. The list of contributors of experimental evidence and candid postulations, which comprise the true creators of this structure is substantially the entire bibliographic category of references.

We wish especially, however, to pay tribute to the three men whose work has held the key to our hypothesis. Dr. Fenton B. Turck, that practicing physician whose keen insight and monumental curiosity led him to the investigation of the specificity of ashed tissue extracts, deserves the honor of being possibly the first to demonstrate some phenomena of Protomorphology and enlarge upon its importance. Dr. T. Brailsford Robertson, the prolific thinker, whose allelocatalyst theory, also expounded by others, is, we believe, the cornerstone of the newer biology, deserves the greatest credit for original thinking in many fields of biochemistry. And Dr. Montrose T. Burrows whose investigations into the nature of "archusia" and "ergusia" have supplied, we

---

[1] Even while this volume is in process of publication we are gratified to notice that recent experimental evidence seems favorable to one hypothesis we have advanced, that toxemia of pregnancy is caused by a disorder m the metabolism of protomorphogen (thromboplastin).

See: Schneider, C. L., "The Active Principle cf Placental Toxin; Thromboplastin... " *Am. Jour. Physiol.* 149:123-129. 1947.

believe, a basic fundamental approach to the cancer problem, deserves commendation as one of the few original thinkers in American Medicine.

The list of scientific investigators who have contributed material for our project is of international scope. We pay tribute to all those whose work we have been honored to review herein.

We wish to acknowledge the financial assistance from those people whose interest in sound nutrition has led them to contribute to the support of the Lee Foundation for Nutritional Research. Their aid has played no small part in making this publication possible. We also wish to acknowledge the kindness of Dr. Otto Glaser of Amherst College reading the first four chapters and suggesting minor changes although he cannot necessarily endorse our hypotheses without more experimental substantiation. Our appreciation is also extended to Dr. Charles Packard of the Woods Hole Biological Laboratory, Woods Hole, Massachusetts, for his kindness in making possible the use of their extensive library and reprint collection, and to Dr. F. E. Chidester also at the Woods Hole Biological Laboratory for his suggestions and generous assistance. Finally our appreciation is extended to Miss Dorothy Tomsyck on the Research Staff of the Lee Foundation for her work on bibliographic material and preparation of the manuscript.

ROYAL LEE
WILLIAM A. HANSON

Milwaukee, Wisconsin February, 1947

OUTLINE AND CONTENTS
CHAPTER 1—BASIC BIOLOGICAL DETERMINANTS

CHAPTER 2—MORPHOGENS AS REGULATORS OF CELL VITALITY

I. Experimental Basis of Their Influences in Cultures

Reproductive capacity of protozoa increased by these processes. Necessity of considering the effect of these phenomena on vitality of cells when choosing transfers.
A brief review of fundamental principles basic to the allelocatalyst theory of control of mitosis.
A list of various experimental reports is presented reviewing the consensus of opinions on the allelocatalyst theory.

A factor secreted by cells accumulates in the media inhibiting the division rate in cultures of metazoan tissues.

Isolated tissue cells will not grow in a large volume of media, a fact also discussed previously pertaining to protozoa.

Various investigations are reviewed reporting the presence of a substance in the media of tissue cultures, derived from the cells themselves, which stimulates the mitotic rate in dilute amounts and inhibits it when concentrated.

The allelocatalytic theory applies to all animal cells universally. The reciprocal relationship between protoplasmic and media allelocatalyst concentration is the basic universal influence over division rate and vitality of the cell; it is primarily responsible for the degenerative processes of old age.

The morphogens secreted by the cells as a product of their metabolism are identified as the allelocatalytic growth substance. A review of the theory is presented.

CHAPTER 3—MORPHOGENS AS REGULATORS OF CELL VITALITY

II. Biodynamic Influences on Cell Metabolism

Evidence is presented concerning the above cycle by
which media protomorphogens are brought into the
nucleus for chromatin synthesis.

As a consequence of chromatin metabolism
morphogens appear in the cytoplasm under the
protection of a fatty envelope and gradually seep into
the media.

Concentrations of protomorphogens in the media
prevent further loss from the cytoplasm where they
accumulate exerting toxic effects. This prevention may
be due to the polymerization of the morphogens as they
become concentrated.

The possibility that mitogenetic rays exert their
influence by means of a control over polymerization is
discussed.

Further notes on lag period, suggesting that substances
inducing the depolymerization of morphogens shorten
the latent period preceding the first division after
transfer.

The determinant and metabolic morphogen cycles
distinguished. The metabolic cycle is under
consideration in this chapter.

CHAPTER 4—MORPHOGENS AS DETERMINANTS

CHAPTER 5—MORPHOGENS IN THE HIGHER ORGANISMS

The sex hormones are discussed as elutogenic factors
removing adsorbed protomorphogen from tissues. An
hypothesis is presented suggesting that the morphogens
thus eluted are transported to the germ cells where they
are attached to the chromosome network.

Thyroid hormone is discussed as a physiological
elutogen. It is possible that it exerts its influence over
the metabolic rate by the removal of protomorphogens
and concomitant release of pyrexin, a pyrogenic
substance.

Blood trypsin functions as an elutogen and also
enzymatically reduces protomorphogens released by
other elutogens.

Trypsin releases heparin which seems to associate with
protomorphogen forming a stable platelet structure.

A discussion of the basic differences between
elutogenic factors and those responsible for
depolymerization of protomorphogen.

Evidence that these substances are physiological
depolymerizers of protomorphogens and their
properties as growth promoters may be ascribed to this
activity. Summary of the hypothesis up to this point.

The nature of the protective lipids associated with
protomorphogens is discussed.

Vitamin A is linked with the lipid protective molecule
either as a part of or as an associated catalyst.

CHAPTER 6—MORPHOGENS AND PATHOLOGICAL PROCESSES

INDEX OF ILLUSTRATIONS

PROTOMORPHOLOGY

CHAPTER I

BASIC BIOLOGICAL DETERMINANTS

Fundamental Scientific Concepts

THE ENTIRE SCIENTIFIC WORLD, in its broader aspects, concerns itself with the laws of matter and energy.

What is matter? Our physicists define it as anything that occupies space; it is tangible; it has three dimensions. What is energy? Energy is demonstrable by its ability to produce changes in matter. In biological science life may be considered as the manifestation of highly complex and coordinated forces produced by the energy of properly organized matter. Our immediate objective is the study of the manner in which this specially organized living matter determines and reproduces its specific forms of organization.

We have been given reason to believe that all matter may be reduced to ninety odd elements. The possible number of combination of these elements is for all practical purposes infinite.

The history of the science of chemistry begins with the listing and studying of the commoner forms of matter. One has only to investigate the state of chemical science before Dalton to realize the confusion that existed before the atomic theory welded a heterogeneous mixture of facts into an integrated structure.

In many respects, the science of biology suffers from the lack of a similar theory of organization. It is true that Darwin's conception of evolution clarified and organized an otherwise disjointed collection of observations. The lack of a detailed and comprehensive theory of the dynamics of life, however, has made biological phenomena susceptible to metaphysical explanations.

"Inorganic" Evolution.—Inorganic matter assumes its form because of the inherent chemical affinities of the molecules composing it. There is usually only one combination among simpler substances which is stable in its natural environment. The fewer types of molecules and atoms composing matter, the fewer mathematically possible combinations exist. The simpler compounds of inorganic chemistry are therefore limited as to the possible combinations between the different atoms and molecules composing them.

Elements with multiple valences may produce more than one stable form of compound in a given environment. For instance, both ferrous and ferric salts are familiar compounds of iron. Ferric salts are found almost exclusively in the natural inorganic environment, however, as they are the more stable form.

In organic chemistry the innumerably different possibilities of complex structures give rise to many different stable combinations of the same atoms and molecules. The plethora of possible stable forms of these organic compounds is so great that the chemist is driven to the use of structural formulas in order to differentiate isomeric peculiarities between apparently identical combinations of the same elements. Among the higher organic compounds, the tremendous number of atoms composing these molecules results in large numbers of possible combinations which the organic chemist finds very difficult to identify or catalog. The environmental forces producing many isomeric compounds are so complex and infinitesimal that their production in the laboratory is out of reach of today's chemistry. Arachidonic acid, for instance, can have 256 different isomeric combinations of the same amount of the same elements. The attempt to synthesize biological products in the laboratory has often resulted in a product with an activity very similar to the one attempted to duplicate, but differing very slightly in its chemistry. The slight difference may be that of structure rather than of content.

The integrated idea we are trying to establish is that the various forms of matter exist as they do because there are only certain combinations allowable in a given environment, i.e., under the influence of a given pattern of forces. Secondly, the more complex these combinations become, the more complicated are their relations with their environment. We need no more explanation for the reactions of a virus structure to its environment than to state that it is the natural characteristic of this particular highly organized group of molecules to react in this manner, just as it is the natural characteristic of the simplest inorganic molecule to react to its environment, i.e., stimuli. It is now only logical to suggest that life itself is the reaction of the ultimate in molecular complexity to environmental timuli. "Life in all its complexity seems to be no more than one of the innumerable properties of the compounds of carbon."[1]

This seems, off hand, to be a very rash and ill considered statement: but it should be observed here that living processes are cyclic reactions and that, therefore, the influence of environment at any given time is not to be considered the total environmental stimulus. It is only a fragmentary part of the total environmental pattern that is involved during the entire cycle of life for a given organism. The sum total of the environment of any organism would include all the stimuli to which it *or its ancestors* had been subjected throughout its whole evolutionary development.

It is not our purpose to lose sight of the infinite detail and unquestionable magnitude of the vital phenomena. Rather we are attempting to sketch a comprehensive hypothesis, relatively simple, to correlate the phenomena of simplest chemistry with that of the most complex and inscrutable of problems, life.

With this viewpoint in mind, we shall now present certain correlations of experimental observations in order to establish a working hypothesis for one part of the machinery by which the more complicated reactions are established.

Where "Inorganic" Evolution Changes to "Organic."——It is obvious that the ability of molecules to hang together and maintain a complex intricate structure must be limited. The physical nature of these structures is such that the larger they are and the more extensive isomerism they exhibit, the more fragile and sensitive to environmental changes they become. In such an inverted relationship a point must be reached at which the sensitivity is greater than the ability to hang together. It is at this point that evolutionary development of further molecular complexity with its more profound reactions to environment would, perhaps, have been prevented were it not for the introduction of a heretofore unconsidered method for the organization and maintenance of molecular structure.

For analogy we may examine the evolution of life. The protoplasmic globule developed into multi-celled animals with more extensive environmental reactions. Until the advent of the external skeleton or shell, further progress was impossible due to the fact that, without it, it is

---

1 *Physical Chemistry of Living Tissues and Life Processes,* R. Beumer. The Williams & Wilkins Co., Baltimore, Md. 1933.

5

impractical to hold the various structures of the organism in the proper physical relationship. The external shell, too, has its limitations. The development of the rudimentary vertebrate skeleton made possible the evolutionary development of still higher forms of life.

The evolutionary development of such primary forms of organic matter is a problem of basic interest and it is subject to many different theoretical explanations. Perhaps the most easily acceptable one is that recently reviewed by Beadle (1946). He comments that the incredible probability that molecules could by chance have come together in organic forms and remain as such was infinitely more likely before life was present than today, with endless varieties of bacteria and enzymes existing ready to break them down. He postulates that the first evolutionary break into the field of life was a hypothetical "protogene." This "protogene," we shall attempt to demonstrate, is the "skeleton" of the living protein molecule, which we prefer to term "protomorphogen" for reasons outlined a few pages hence.

Drennan (1944) has also discussed this problem and emphasizes that the cell should not be considered as the fundamental unit of living matter but as a complex organization of even more minute living units. He called these units biophores. Obviously, although this field of thought has been the subject of wide speculation, some orientation of the evolutionary boundary between "living" and "non-living" substance is desirable.

We propose to advance our hypothesis that the further extensive development of the living protein molecule depends upon a "skeleton" to maintain its stable complicated forms. This hypothesis is really a molecular adaptation of the mechanism, so common in nature, of specialization of tissue for the performance of certain necessary responsibilities. The "skeleton" of the complicated protein molecule, the chromosomes of the higher organisms, the skeleton of vertebrates, even to an extent the cellulose framework of a tree or plant are specialized units whose primary responsibility is that of preserving the integrity of organization.

At this point of our discussion, such an idea is obviously pure speculation. A number of years of intensively detailed study and correlation of pertinent biological fields has preceded this conclusion. The more significant of our correlations of experimental data are discussed herewith. For clarification,however, we consider it expedient to outline the foundation of our general working hypothesis in the first chapter. In the development of an hypothesis, the correlation of experimental facts must precede the conclusion, but in the presentation of the hypothesis clarity demands that the conclusion precede the discussion of evidence.

Specificity is a property universally exhibited by proteins composing living protoplasm. This property is best demonstrated by the administration of a foreign protein into an animal organism and observing the immune reaction of the organism, which proceeds specifically to destroy this foreign protein. Proteins of apparently identical nature (as far as it is possible to determine) may be found to be specifically different when tested by means of this immune reaction.

The species specific qualities of protein can be identified only by their biological effect. Although in our search of the literature, we have reviewed articles showing that various synthetic proteins have been obtained, in no case have we found a report of the absolute synthesis of a protein exhibiting species specificity. The reactions to injection of foreign proteins have been carefully studied by many investigators. One of the most interesting examples of this study is shown by the protein of tissue of the same species at different ages. The homologous protein exhibits different antigenic characteristics at different ages.[1]

The detailed investigation of the chemistry of protein structure, responsible for the differences between specific proteins, is beyond the reach of present day methods. By means of the conventional methods of organic chemistry we are hardly able to approximate the structural formulas of relatively simple protein molecules. The identification of specificity of proteins by chemical methods is still far beyond the frontiers of protein chemistry.

Without such a detailed explanation of these complicated relationships, we must plod along with our meager information and attempt to locate avenues of research to stimulate the investigation of the future.

We arbitrarily suggest that the development of specific proteins is the dividing line between the simpler and more complicated molecules, where the necessity for a protein "skeleton" begins. The fact that the protein of living tissue is synthesized and destroyed at such a rapid rate would

1 *Chemical Embryology,* J. Needham. Vol. 111, Cambridge University Press, Cambridge, 1931.

predicate the existence of a stable "skeleton"to preserve the integrity of the molecule. The development of organized life would depend upon this "skeleton," since the integrity of the individual depends upon the integrity of its proteins. Also, the forms of life that have evolved from a common denominator, would have similar proteins, and would require protection from unwarranted changes in these proteins, causing a change in their morphology. We will later discuss experimental evidence showing chat specific proteins contain a "skeleton" or determinant component which may be separated from them.

We suggest chat the "skeleton" of the specific protein molecule has the unique function of creating and maintaining the specific character and physical form of the molecule. We postulate that the essential duty of this "skeleton" is to act as an organizer. To provide an adequate vocabulary for our discussion we are expanding Weismann's application of the word "determinant" to include this hypothetical protein organizer.

Basic Biological Determinants

Throughout this presentation, we include in the term "Biological Determinants" that group of factors whose function it seems to be co organize the structure and maintain the integrity of biological entities. The Biological Determinants, including those which are a part of our working hypothesis, may be tentatively listed as follows:

The chromosome assembly in the germ cell is the determinant for the characteristics of the species and of the individual.

The genes of the chromosome are the determinants for the separate characteristics of the individual.

We shall review evidence which has led us to suggest the hypothesis chat the genes contain an organized assemblage of smaller units which are the determinants of individual cell morphology. For these units we propose the name *cytomorphogen.* We define a cytomorphogen as an extremely complicated assemblage of molecules exhibiting some characteristics of a virus. It is a determinant for the morphology of the individual cell.

Extending the working hypothesis further, we propose that a cytomorphogen is composed in part of still simpler and quite stable

8

units composed principally of minerals which are the determinants for the specificity of the biological proteins. For these basic determinants we propose the name, *protomorphogen*.[1] We define a protomorphogen as a comparatively stable but complex group of molecules, linked together by the chemical affinities of mineral material, which by reason of its physical and chemical structure determines the exact plan or pattern by which the component parts of a specific protein are combined. It will be shown that protomorphogens exert a profound influence upon the mitotic activity and general vitality of every living cell.

In discussing further the nature of the biological determinants, we shall begin with the simplest, protomorphogen. Since we are suggesting that the specificity, or antigenic property, of proteins is due to protomorphogens we shall discuss briefly the phenomenon of protein specificity.

Wrinch (1941) has probably developed the most satisfactory and advanced concepts of the nature of the biological protein molecule. Her hypotheses are too involved to warrant a detailed discussion in these pages and we refer the reader to the bibliography for further study.

Basically, however, she has emphasized the surface pattern of a globular molecule as the factor primarily concerned in the specificity of the protein. So called "chain" or fibrous protein molecules exhibiting specificity, she envisions as globular molecules attached together much the same as beads are arranged on a string.

1 As our hypotheses are developed in this volume, the reader will begin to see that the fundamental unit which we term protomorphogen has been variously discussed, postulated and studied by numerous investigators in relation to diverse biological activities. We believe that the following terms are either descriptive of the unit we call protomorphogen or are closely related to some of the commoner molecules depending upon protomorphogen for their primary characteristics:

Allelocatalyst—T. B. Robertson Cytost—F. B.
Turck
Archusia and Ergusia—M. T. Burrows Chromidia
Plasmagene
Specialized Evocator—J. Needham
Granulum or Microzyma—Antoine Bechamp
Necrosin—V. Menkin
Proteinogen—J. H. Northrup
Protogene—G. W. Beadle
Id and Idant—A. Weismann
X-Substance—S. O. Mast and D. M. Pace
Biophores—M. R. Drennan
Heat stable growth inhibitor removed by fat solvents—A. Carrel and H. Werner

This structure still allows the important surface patterns to exert their catalytic influence.

Our hypothesis that the basic protein determinant (protomorphogen) contains mineral links of utmost importance does not basically argue against Wrinch's surface active globular molecule conception. It is altogether likely that the mineral constituents are the organized "links" for the protein molecule, and are not responsible *per se* for the antigenic activity we shall shortly review. More likely they retain their spatial relationships even after the organic parts are removed (See Turck's tissue ash experiments in the next section) and when they are administered to a living organism, supply a framework upon which an antigenically active globular protein molecule may be re-synthesized.

Dr. Otto Glaser, who has kindly consented to critically review this chapter, has cautioned us against misleading the reader into the assumption that the synthesis of the mineral pattern of protomorphogen precedes the construction of a specific protein molecule. Rather the mineral pattern, we believe, is a product of the life activity of the cell nucleus and is the means by which biological proteins are reproduced and their integrity maintained in living systems. (It is only fair to state that Dr. Glaser does not wholeheartedly subscribe to the hypotheses we present herein, preferring to await more direct experimental substantiation.)

Stern (1938) shows that the species specificity of proteins may be demonstrated either by their antigenic behavior or by their crystalline structure. The universal method of demonstrating protein specificity, however, is the experimental demonstration of the serological reaction to antigens. This is shown by the injection of a foreign protein into the blood stream of a heterologous species. This is followed by the appearance of specific antibodies whose function it is to protect the host from the deleterious effects of the foreign proteins.

This antigenic reaction is a vital protective mechanism of all animals. The morphology of a cell is dependent to a considerable extent upon the nature of its proteins. If these proteins are not protected from the possibility of dilution or replacement by foreign proteins, the nutritional action alone of such proteins would tend to change the characteristic morphology of the cell. Indeed, this actually has been experimentally accomplished in certain tissue cultures by altering the nature of the determinant morphogens present.[1]

In discussing the antigenic behavior of proteins, therefore, we are discussmg one of the fundamental phenomena of living cells and the basic factor in the organization and protection of cell morphology.

Antigenic "Non-Protein" Substances.—Having postulated the existence of this protomorphogen organizer, it will be necessary for us to look for a certain portion of the protein molecule which is singular in its ability to initiate the antigenic reaction. The work of Turck (1933) has provided us with an answer. Turck heated tissues to approximately 300 degrees C. and made a 10 per cent saline solution of the resulting ash. This solution was administered intravenously to test rats. Shock and death resulted in from five seconds to two minutes. Smaller doses caused light shock and the animal usually recovered.

A series of experiments were carried out to show the species specificity exhibited by this tissue ash. He found, in investigating the ash from guinea pigs and from cats, that both ashes when injected intraperitoneally were toxic to guinea pigs. However, considering the minimum lethal dose of the ash, the guinea pig ash was twice as potent as the cat ash in its shock producing action.

Similar experiments were carried out using the ash of beef, dog and rat tissue on guinea pigs, and in each case the homologous ash was shown to be more toxic than the heterologous ash, although the latter exhibited a lethal action in sufficiently large doses. Four modes of administration, intravenous, subcutaneous, intraspinal and intraperitoneal, were employed with conclusive results with each method.

Ten per cent ash solutions from rat, beef, horse and lion tissues were tested in cats. Homologous ash solutions caused shock, paralysis and occasionally death, while equal amounts of heterologous ash solutions only slightly inconvenienced the animal. Of ash from human, horse and lion, the latter was the most potent when injected into cats, a fact of interest as pointed out by Turck, because the lion is more closely related to the cat, phylogenetically, than the other species. In these tests, four times the amount of heterologous ash solution did not evoke the toxic reaction of homologous solution.

1 Drew, A. H.: "Growth and Differentiation in Tissue Cultures." *Brit. J. Exp. Path.,* 4:46, 1923.

This active principle of the tissue ash is remarkably thermostable, toxicity being demonstrated by tissues ashed near 300 degrees C., the lethal characteristics diminishing at higher temperatures, and finally disappearing at 700 degrees C.

Turck draws his conclusion that at 300 degrees C. this tissue ash retains some remarkably thermostable organic constituents responsible for the species specific properties. He suggests that at 700 degrees C. this organic constituent is volatilized and driven off, since he considered it improbable that any of the inorganic constituents could be destroyed or volatilized at that "low" temperature.

We present the hypothesis that these effects of tissue ash are due to the physical and chemical configuration of the mineral constituents of the original protein. This mineral microstructure we expect to establish, as a vital component of the biological determinant which we term protomorphogen. This microstructure could conceivably be destroyed at 700 degrees C. Possibly the first indication of the importance of mineral configuration in biological form was reported by Lehman (1855) who demonstrated that the ash of human sperm retains the identical microscopical form of the original sperm. We consider the idea of a complicated specific physical structure of the mineral ash of the protein molecule more plausible than Turck's postulation of an extremely thermostable organic compound which volatilizes at the comparatively high temperature of 700 degrees C. It is more likely that some inorganic component fuses or volatilizes at this degree of temperature, destroying the essential structure.

Since antigenic reactions have only in isolated special instances been reported as due to products other than proteins, we feel safe in postulating that the toxic ash reported by Turck is a protein derivative. Although antigenic activity has been reported for certain of the complex carbohydrates the universal occurrence of such activity in living protein and its specificity for organs and species makes this a sound assumption.

We further suggest that the serological reactions of protein are related to the mineral organizer (protomorphogen) in that the protomorphogen serves as a stable framework of mineral linkages or "receptors" upon which the specific spatial array of the protein molecule is organized. Since the specific serological reactions of proteins are due to their molecular geometry it follows that the antigenic properties of a protein are derived from that protein's protomorphogen.

We do not wish to infer that the antigenic reaction consequent to the injection of tissue ash is a direct response to the presence of the foreign tissue ash. Such antigenic reactions could only be due to the presence of foreign protein. Very likely in this case the tissue ash functions as a protomorphogen supplied with substrate materials, i.e., it serves as a framework upon which organic components are assembled into an antigenically active foreign protein. The serological reaction therefore is against this protein and not directly against the ash itself.

Protein specificity, or as Robertson puts it, the biological individuality of the proteins, is strikingly illustrated by the classic antigenic experiments of Nuttall (1904). He measured the amount of precipitin formed by the serum of a rabbit, immunized against human serum by repeated injections of the latter, when mixed with the serum of various species. This immunized rabbit serum contained an antibody which caused a precipitin to form when brought in contact with proteins characteristic of the human species.

Nuttall found the greatest amount of precipitin formed when this serum was mixed with human, chimpanzee, gorilla and orangutan serum in the order named. Relatively insignificant amounts were formed with dog, cat and tiger serum, and no precipitin was formed from guinea pig, rabbit and kangaroo serum. This shows, in a spectacular manner, the relative specificities for proteins of phylogenetically related species. Nuttall performed these experiments, immunizing the serum of rabbits against vertebrates other than man. He noticed the same development of a precipitin for related species and relatively no reaction with unrelated vertebrates as when human protein was used as an antigen.

Nuttall's experiments with human immunized rabbit serum showed that ox and sheep serum gave five times the amount of precipitin as tiger serum and one-fourth as much as orangutan serum. On close study of these reactions, it becomes apparent that although these effects are quite spectacular and beyond criticism in their overall entirety, yet, occasionally there appears a situation where the difference in amount of precipitin formed and the relative phylogenetic position of the species are not in the expected ratio. Turck noticed similar, but more extensive, discrepancies in his tissue ash experiments. For instance, the ash from rat was far more toxic to the cat than that from the phylogenetically related lion, although the lion tissue had a proportionately greater toxicity than that of the human or the horse.

Turck definitely refutes any suggestion that the specificity shown by his ash experiments and that shown by Nuttall's experiments have anything in common. His reasons for refuting this are: first, the discrepancies in proportionate toxicity as shown by the rat and cat tissue experiments, and second, the fact that phylogenetically unrelated tissue ash may still exhibit a lethal effect if the dosage is sufficiently high.

As we mentioned before, he suggests that tissue ash contains a universally toxic constituent of a remarkably thermostable organic nature. He states that the identical toxic potency of rat heterologous tissue ash and homologous cat tissue ash, when administered to the cat, is due to the fact that they happen to contain equal amounts of this universal toxic agent.

Without detracting from the immense importance of Turck's work, we feel that this suggestion is inaccurate in that it does not completely fit the facts. In the first place, he has shown that rat tissue ash is far more toxic to rats than cat tissue ash but exhibits the same toxicity as cat tissue ash when administered to cats. If this toxicity is due to a single universal toxic constituent, as he suggests, why should the potency be equal in one case and not in the other? (The toxicity of rat tissue ash to the cat can be due to the fact that cats are undoubtedly sensitized to rat protein by reason of it being a common constituent of the cat diet. The same "dietary" sensitization may be responsible for the greater precipitin reaction of ox and sheep over tiger serum to humans reported by Nuttall.)

In the second place, the antigenic properties exhibited by the mineral ash of a protein would not necessarily be as restricted in specificity as that of the organized cell proteins, since the mineral ash by itself contains only the more basic linkages of the determinant. Hence, the universal. toxicity of tissue ash, when administered in high enough doses, would not become an insurmountable objection to the hypothesis of the specificity of the ash.

In the third place, the hypothesis of the determinant or organizer nature of the mineral ash, such as we propose, makes the specific antigenic properties of tissue ash quite plausible, i.e., through the catalysts of the production of antigenically active protein molecules. We do not believe Turck's ash experiments alone to be extensive enough to establish this hypothesis but have correlated them with other related experiments and observations available in the literature on the subject.

In an attempt to find the "toxic agent" in tissue ash, Turck made a spectrographic analysis of the ash from several different species. Sodium, potassium, calcium, magnesium and phosphorus were present in all the tissues analyzed. He found traces of boron, zinc, aluminum and rubidium in many of the species examined. These findings were considered by Turck to have little interest in connection with the experiments on tissue ash antigenic properties. To us, however, the spectrographic analysis of various tissue ashes is quite significant as it indicates a new need and function for the "trace" mineral elements. If this mineral ash structure of tissue represents, as we postulate, the protomorphogen or determinant for the essential specific proteins of living tissue, then the integrity of tissue would depend upon the integrity of the patterned mineral structure. The absence of any or one of the individual mineral elements going into the complex physical structure of the protomorphogen would seriously impair, if not prevent, its determinant action in organizing specific proteins. Comparatively insignificant amounts of these elements would be needed to supply the necessary molecular links in the protomorphogen structure, just as "trace" amounts of certain mineral elements are needed for health and growth. More than the necessary amount of the trace mineral would not go into the building of necessary protomorphogens, and thus uncombined, could exhibit the toxic pharmacological effect attributed to overdoses of some of the trace elements.

Trace Mineral Elements and Their Importance in Morphology
—The significant effect of certain trace mineral elements on the growth and morphology of tissue seems to bear this contention out. These experiments are a key link in the chain of evidence leading to the protomorphogen hypothesis.

We must remind the reader that although we are discussing the elementary determinant, protomorphogen, which organizes protein molecules, this determinant is a part of the more complicated determinants which organize the morphology of cells. Thus, deficiencies of mineral factors resulting in change of cell morphology can be interpreted as support for our theory of the function of these minerals in the protomorphogen molecules.

Wadleigh and Shive (1939) have demonstrated that as a result of boron deficiency there is a degeneration of the protoplasm of cotton plants. Although boron affects the absorption of salts, the experiments of Rehm (1938) do not exclude the possibilities of other functions. In this respect, we recognize that there are functions of trace elements other than our suggestion of a portion of the protomorphogen molecule, such as the effect of molybdenum reported by Steinberg (1937) of acting as an enzyme activator in processes leading to synthesis of amino acids and protein.

The effects of the absence of boron are particularly significant with respect to the possibilities of a deficient determinant structure in view of the observations reported by Shive and Robbins (1939) in their review of the literature that the most pronounced effects of boron deficiency in plants are exhibited as injury to the meristems. The meristem is the new tissue at the sprouting end of a growing leaf or twig and, therefore, constitutes the botanical analogy of embryonic tissue in animals. It is principally in the newly formed cells of growing tissue that injuries resulting from disordered determinant structure should be assumed to occur: The morphology of the older cells has already been organized by the determinant, while that of newly formed cells would be influenced by the determinants formed during the starvation for the component mineral trace elements.

Additional evidence concerning the possibility that boron may be a constituent of the determinant is supplied by Bertrand and Silberstein (1938). They have demonstrated that the highest concentration of boron in the lily plant is present in the stigma, which contains six times as much as the stalk and 20 per cent more than the leaves. The Russian investigators Bobko and Tserling (1938) have also observed that the highest concentration of boron is found in the stigma and pollen. The utilization of boron by the reproductive apparatus is also suggested by Lipman (1940) in his review of plant mineral metabolism. In the higher organisms the reproductive cells have the specialized function of producing the chromosome determinant, and inasmuch as our hypothesis suggests that protomorphogens are a part of the chromosome organization, the high concentration of boron, supposedly a protomorphogen constituent, in the germ cells of the plant is not surprising.

Copper, of course, has received considerable attention as a mineral necessary to prevent anemia. This mineral may also play a part in protomorphogen formation and organization. Although Saeger (1937) was unable to prove definitely that copper is necessary for growth of Spirodela polyrrhiza, he was able to show that concentrations as minute as one-tenth part per billion of the media had a definite favorable effect, whereas one part per billion was toxic.

Bromine is another trace element that is a normal constituent of all tissues. But Winnek and Smith (1937) report that a bromine intake as low as 0.5 mgm. per kilo of dietary intake in rats results in the death of offspring.

The experiments of Bertrand (1902) suggest that arsenic is a normal constituent of tissue. Tangl (1939) has demonstrated the growth accelerating effect of arsenic in amounts less than I mgm. in the diet of test animals. Arsenic may be found to play a part in the determinant picture because its content in blood is known to increase three and four fold during the middle months of pregnancy (Guthmann and Grass, 1932). Pregnancy is a time when protomorphogen and other determinants are undergoing a greatly accelerated metabolism, for which we shall presently find reason. Daniel's review of the trace elements suggests that cobalt, aluminum, magnesium, zinc, tin, vanadium, nickel, caesium, lithium, barium, strontium, silver, germanium and titanium are all occasionally identified in tissue: Underwood (1940) has reviewed the literature on trace elements and recognizes copper, manganese, zinc, iodine and cobalt as important trace elements and suggests that arsenic, aluminum, rubidium, bromine, fluorine, silicon, barium and nickel may be found to be important trace elements since they all are found in minute amounts in animal tissues and fluids. Rusoff and Gaddum (1938), working with newborn rats, have identified by spectrographic examination aluminum, barium, copper, manganese, strontium, tin and zinc in the tissues of all rats examined; they identified lead, silver, chromium, nickel and molybdenum in several animals.

The enormous importance of trace minerals as enzyme activators in the

17

various enzyme systems of living organisms is not overlooked when we emphasize the possibilities of their entering into the protomorphogen determinant structure. We are of the opinion that many of the trace elements are important protomorphogen constituents but also engage in other activities. The possibilities of boron being necessary for the integrity of protoplasm (as in cotton plants) and also affecting the osmotic absorption of salts is a point in example.

Of the few experiments we have mentioned only boron and iron seem to be definitely classified as probable constituents of protomorphogen organizers. The absence of adequate experimental methods of accurately determining this matter must leave us uncertain in this respect, but here is a field of research anxiously awaiting the constructive activities of some biological pioneer. Other minerals are undoubtedly necessary, but their need, in much greater amounts for other vital functions, would exclude the possibilities of their deficiency in protomorphogens from being demonstrated since the organism would die from the lack of one or another of these vital functions before protomorphogen "starvation" became noticeable. We feel that potassium, calcium, magnesium and zinc may also be placed in the protomorphogen metabolism category because of their effects on mitosis and the morphology of cells.

Day and Comboni (1937) studied the effects of a potassium free nutrient solution upon the formation of starch in Pisum. The plants receiving no potassium had less than half the normal amount of starch, but the significant factor in relation to our discussion is the fact that the plants receiving no potassium showed a necrosis of leaves and did not produce as many buds as the controls. If, as we may suggest, the potassium lack resulted in a deficient protomorphogen formation, necrosis and lack of growth would be a significant observation.

Sorokin and Sommer (1940) have studied the effects of absence and deficiency of calcium in the development of the roots of Pisum sativum. They recognize the difference between the *per se* effects of calcium upon the morphology of the cell and the effects of calcium upon the cell through other means, such as the physico-chemical systems. In media containing no calcium, aberrant morphological types developed in root tips in two to five days; 0.06 ppm. Calcium, aberrant types at fifteen days; 0.125 ppm.

calcium, aberrant types inin twenty days; 0.25 ppm. calcium, normal at end of four weeks or end of experiment. Failure of plant metabolism in the media containing 0.25 ppm. calcium may be attributed therefore, to effects other than those of calcium in the protoplast. We suggest that this indicates a position of calcium in the determinant molecule, protomorphogen.

Mast and Pace (1939) have shown that the absence of either calcium or magnesium from the media of Chilomonas paramecium results in monsters due to the inhibition of cytoplasmic reproduction. They demonstrated that, to an extent, each mineral could compensate for the deficiency of the other, but minute amounts of both were absolutely necessary for normal morphology and growth. The optimum concentration of both minerals was established at 0.000105. M. Daniel (1939) comments in her review of the trace elements that at least two parts per million of magnesium is necessary in the diet to prevent the death of experimental animals. It is fitting that we should keep in mind the influence of trace minerals on enzyme activity. Manganese, in particular, has been found to activate various enzymes.[1] It is certainly true that much of the activity of the trace minerals is a consequence of this affect. This point may seem to weaken our argument but the plural influence of trace minerals must be carefully considered.

Mast and Pace (1942) have also demonstrated that phosphorus must be present in the medium to support Chilomonas. These individuals die within a few days of transfer into a medium devoid of this element. They mention that Pierce (1937) observed a decrease in the chromosome size in the root tips of phosphorus deficient violets and suggest that the influence of phosphorus over Chilomonas may be correlated with its effect on the nucleic acid of the chromosome. They found the optimum concentration of this element in the media to be 0.00109 M Na_2HPO_4.

Eltinge and Reed (1940) have observed the abnormal development of tomato root tips grown in media devoid of zinc. These roots developed a swelling with crooked root hairs and various other aberrant forms. Histologically there were abnormalities in the meristematic cells and an abnormal metabolism of these cells was indicated. In this respect zinc

1 *Chemistry and Methods of Enzymes*, J. B. Sumner and G. F. Somers. Academic Press, Inc., New York. 1943.

would seem to exert a function similar to that of boron, reviewed earlier in this chapter. This evidence seems to suggest that it is an important constituent of protomorphogen for the tomato plant. Later experiments of Reed (1942) have demonstrated that a concentration of zinc greater than 0.01 parts per million was necessary for culture growth of peas (Pisum sativum). Zinc concentrations of more than 0.05 parts per million, however, were found to be necessary for seed production. It is of special interest to note that the threshold requirement for seed and pod production is five times as high as that for growth. Seed production necessitates the construction of a complete determinant system for each seed and a deficiency of important trace minerals would hinder the synthesis of the elementary mineral constituents of the chromosome.

In the animal organism, potassium possibly has an important role as an indispensable constituent of protomorphogen. It forms a constant percentage for all ages and, according to Shohl (1939), along with sulphur, it parallels the increase in body weight during growth. Both potassium and sulphur are more extensively associated with proteins than any of the other inorganic constituents of the animal organism. Tennant and Liebow (1942) have shown that potassium is the dominant cation of the cell and sodium of the intercellular fluids. Their experiments show that although potassium is necessary within the cell, it cannot be substituted for sodium in the media without significant effects on cell growth.

Holmes (1938) has reviewed the comparative analysis of blood and tissue in respect to the distribution of cations as follows:

CATION	Mgms per 100 gms plasma	Mgms per 100 gms muscle
Sodiu8m .	158.0	60.0
Potassium	18.0	360.0
Calcium .	10.5	10.0
Magnesium	2.7	23.0

From these figures one can calculate that 95.5 per cent of the body potassium and 89 per cent of the magnesium is present in the tissues and kept there by some, as yet unexplained, mechanism.

Shohl presents a review of experiments to show that potassium in the animal organism is intracellular and increases throughout life as the intracellular fluid increases in percentage of total body fluid. He states that the chemical forces constraining potassium within the cell are as within the cell are as yet unknown. However, he concludes that the potassium within the cell must be in a combination in which it is not freely ionized since in the red blood cells osmotic relations with the surrounding fluids would not allow the high erythrocyte potassium concentration if the potassium were completely ionized. Separated muscle is freely permeable to potassium, and Fenn (1940), after a review of the literature, concludes that body cells in general are more or less permeable to potassium. He, therefore, presumes that potassium remains in the cell because no other ion can get in to take its place and because the anions with which it is combined prevent its escape. We would suggest that potassium is an irreplaceable part of the protomorphogen molecule and as such is prevented from diffusing through the cell wall because of the physical size of the latter. This explains why intact muscle with intact protomorphogens constrains its potassium in spite of the apparent permeability of the cell membrane to this element. Potassium is also found in various animal cells, particularly muscle, as a part of the phosphagen molecule and this probably accounts for much of the potassium bound within the cell.

The peculiar characteristic of potassium, its radioactivity, may be vital to the metabolism of the cell and exert its effect through protomorphogen molecules, but this possibility must be discussed later. We mention it here to bring in our suggestion that potassium is an indispensable part of the protomorphogen determinant not alone for structural reasons.

Crile (1931) has produced artificial "cells" by combining lipoids, brain proteins and a solution of brain ash. We will discuss his experiments in greater detail in a later section but it is of interest to note that if the potassium salt is omitted from his experiments the organization of the artificial "cell" is delayed. This may be interpreted to suggest an important function of potassium in the organization of specific protein molecules.

Baudisch (1943) has recently reviewed the importance of trace minerals in biological activity emphasizing their influence in enzymatic reactions. Wrinch (1941) reviews that the structure of cytoplasm is dependent upon the spatial relationships of the component protein molecules and the nature of this relationship with the trace mineral components. She has commented that the insulin molecule in particular is associated with zinc and cannot be synthesized without this trace

21

element.

We have extensively dwelled on the hypothesis that the basic biological determinant owes many, if not all, of its characteristics to the patterned configuration of a collection of mineral elements. Our reasons for forming this hypothesis may be briefly reviewed as follows: (1) experimental evidence exists tending to demonstrate an antigenic and species specific property of the ash of specific protein, and (2) in addition to other recognized biochemical influences, certain minerals are necessary in minute quantities for growth and morphological development of cells.

We specifically exclude the possibility that protomorphogen mineral ash by itself may be directly responsible for serological reactions, but rather it directs the formation of an antigenically active protein.

At this point we might raise the question of whether the basic biological determinant itself (protomorphogen) consists solely of specially organized inorganic minerals in a unique and complex pattern. Turck's tissue as experiments seem to indicate that the mineral ash pattern alone can (with proper substrate environment) construct a specific protein molecule.

This phenomena is probably due to the fact that the spatial molecular pattern of the protomorphogen of specific protein is not altered when the organic fractions are destroyed by ashing. The tissue ash therefore still contains receptive links of inorganic elements which when brought into contact again with the proper substrate can attract and bind the necessary organic fractions into a healthy and complete specific protein molecule.

The experimental evidence and the theoretical interpretation seem to be sound, and can leave us with no other conclusion but that the protomorphogen mineral array alone can *in the proper substrate environment* effectively catalyze the synthesis of an antigenically specific protein molecule. This spectacular conclusion should not, however, be interpreted as an explanation of the normal succession of events in living protoplasm. Science still hedges on the question of "which came first, the hen or the egg?" The problem of "which came first, the mineral array of the protomorphogen or the specific protein molecule?" should be relegated into the hands of the philosophical exponents of evolutionary processes. We have tentatively suggested at the beginning of this chapter that during the evolutionary development of matter, the basic protomorphogen determinate may have had a chance formation, followed much later by the development of the living protein. Such suggestions, however, are pure speculation and should be interpreted as such.

We shall review evidence later which strongly indicates that protomorphogen can only be constructed in the chromatin material of the cell nucleus. There are other properties of protomorphogen which preclude the assumption that it can be assumed to consist only of an organized array of mineral linkages. (These properties have to do with the control of growth and mitosis and are discussed in the two following chapters.) These other properties strongly suggest that in living processes protomorphogen consists of this mineral configuration associated with nucleoprotein, in fact the living protomorphogen is probably a virus. The reader will see later on that the properties of protomorphogen depend to a considerable degree on its molecular size, and there are many different physiological, chemical and physical properties which are manifest only under certain degrees of polymerization or degradation of the fundamental determinant molecule synthesized in the chromatin.

We must remember, however, that we are referring here only to the basic biological determinant, protomorphogen, whose sole function as a determinant is to establish the specificity of the biological protein molecule. Determinants to which we impute more extensive organizing influence, such as cytomorphogens and genes, will naturally be far more complex.

It is of interest to here note the hypothesis presented by Dr. John H. Northrup of the Rockefeller Institute for Medical Research at the Princeton Bicentennial Growth Conference in the fall of 1946. He postulates the existence of a primary mother substance for all proteins which he terms "proteinogen." This substance is presumably manufactured in all cells and forms the basic energized molecule from which all living proteins are synthesized. This publication will go to press before details of his theory are available, but it promises some startling possibilities in the field of protein chemistry.

Our general hypothesis proposes that the hereditary units termed genes are composed, in part, of protomorphogens, and that it is the protomorphogens which are responsible for the gene influences over specific proteins. There seems to be little doubt as to whether or not the gene exerts this influence. The studies of Cumley, Irwin and Cole (1941), have led them to dismiss any idea to the contrary. The serum antigens specific for Pearlineck as contrasted with Senegal pigeons, segregate in black crosses in accordance with genetic prediction. From this fact and other studies they have concluded that the species specific qualities of the serum complex is determined by genes. Although the Pearlineck genes are shown to to be responsible only for serum antigens, it is probable that all species specific qualities of proteins are due to gene function. Our hypothesis suggests, however, that the gene functions as a carrier and organizer of the ultimate protomorphogen units which exert the primary effect on protein specificity. The determinant action of the gene is the executive responsibility over the organization of the total serum protein complex, but we postulate that the individual proteins receive their antigenic properties through the medium of the protomorphogens.

Determinants, Viruses and Nucleoproteins.—That the genes exhibit characteristics attributed to viruses is becoming the consensus of opinion of workers in this field. The most commonly mentioned point of similarity is the property of self-reproduction only in the environment of a living cell, although each is considered to be non-living matter.

A comparison of the two on the basis of nucleoprotein activity is reviewed by Schultz (1943). Analysis of virus strains shows that the nucleic acid content of even serologically distant strains is constant. Amino acid determinations, on the other hand, show a relative similarity between closely related strains, but significantly different values in serologically distant strains. Similarly, enzymatic analysis demonstrates that the structural integrity of the chromosome depends upon protein rather than nucleic acids. Mazia (1936) and Frolowa (1939), for instance, have removed nucleic acids from chromosomes without structural disturbances. Mazia (1941) and Caspersson (1936) have each separately shown that the structural integrity of the chromosome is destroyed by trypsin.

Astbury (1941), in a review of the literature, concludes that the chromosome is composed of polypeptide chains with side chains that change shape by intramolecular folding. The tobacco mosaic virus also is composed of a repetition of small units which correspond by analogy to the genes of the chromosome.

An excellent review of chromosomes and nucleoproteins, including a discussion of virus relationship, has been supplied by Mirsky (1943).

The protomorphogen hypothesis of specificity would preclude the possibility that the antigenic properties could physiologically exist in any other part of the chromosome or gene structure than the protein. Antigenic specificity has, however, been demonstrated to exist towards certain polysaccharides of high molecular weights, freed of protein. In combination with the protein, an antibody is formed which causes the precipitation of the polysaccharide deprived of protein. These reactions are mentioned in the studies of Meyer (1942).

The similarity of virus and gene entities has a still sounder basis than that of the common property of being non-living substances which reproduce in living cells. Blakeslee and Avery (1941) have been able to produce similar effects on the shape of the flower of Datura, both by means of the gene determinant, and also by means of a specific virus.

The mode of virus reproduction is still not conclusively established. According to Meyer, the ultraviolet microscope photographs of vaccine virus and the virus of acute rheumatism made by Barnard (1939), show virus particles in the process of division. On the other hand he reports that Stanley has come to the conclusion that the virus catalyzes the process of protein synthesis so that "virus protein" is produced. Martin, Balls and McKinney (1939) have demonstrated that in a virus disease the normal protein was diminished by an amount equal to virus protein formed. Even this is not conclusive since they state that these results may be interpreted either as a conversion of normal protein constituents into virus structure or competition of two independent protein syntheses for a limited supply of protein material available for building. Meyer concludes his review by stating that if the eventual decision of science is in favor of the theory that viruses are capable of reproduction in a living environment rather than acting as a "catalyzing" substance for protein synthesis, then viruses must be compared to the genes.

We are inclined to favor this viewpoint. The viruses, we postulate, are the evolutionary hangover of that dim point where molecules began to combine in such a way as to react to stimuli in the manner that is generally termed life processes. The determinants, which make possible more complex organisms, we would place in the same evolutionary class as viruses, but this does not mean by any stretch of imagination that all, or even a large percentage of viruses are determinants, or can act as such, in spite of their close relationship. With this hypothesis, both the "vital" theory of virus reproduction and the "catalyzing" theory of virus influence over protein formation can be accepted and catalogued in its proper place in the general biological picture.

There is more evidence creeping into the literature tending to support this assumption. Bernheim and his associates (1942) report that certain types of cancer in rabbits are caused by a virus, which they suggest is a protein constituent degraded by enzymatic action of the cancer cell. This virus, in our estimation, is very likely a protomorphogen which has been affected in such a way, by injury or other means, as to form a "mutation" giving rise to a new determinant which may organize tissue foreign to its host and not controlled by the metabolic products of the latter. This suggestion finds support in the mutation theory of cancer. It is possible. to form a working hypothesis of cancer, armed with this suggestion, that seems to envision several new approaches to an understanding of this curse. We have discussed this in another chapter.

Beard and Wyckoff (1937) have reported the existence of a virus in glycerolated wart tissue. It is impossible to separate this virus from the tissue without destroying the protein itself. We suggest that the virus identified was the protomorphogen which organized and was a part of the protein molecules of the wart tissue. A separation of the protein from its determinant would, of course, destroy the protein. This information adds to that which links the determinants with the viruses.

Wyckhoff (1945) in a review of the virus problem, has noted investigations which report macromolecular particles in many tissues exhibiting the same sedimentation constant as viruses.

Much work can be profitably done in this field. Additional experimental work must be completed in order to establish the mode of "virus reproduction" so as to compare it with the gene units of the chromosomes. The "determinant" influence of viruses can be investigated by experiments involving the synthesis of protein of a specific name. Already we realize from what work has been done, the influence of viruses on production of specific nucleoproteins, but we would suggest experiments attempting to produce protein specific for the injured host of the virus. Until additional experimental work makes itself felt, we can only form a tentative hypothesis and leave many important possibilities subject to further investigation.

In concluding this chapter we leave the reader in nebulous and conflicting scientific territory. We feel that the need for the existence of protomorphogen as a part of a well organized determinant structure is a reasonable assumption. We feel that our survey has established the hypothesis that the physical configuration of specific proteins is organized by the protomorphogen determinant. The influence of genes on the specificity of protein is apparently unquestioned, leading us to the assumption that the gene is an organized group of protomorphogens. This hypothesis is enough to allow us to continue with our discussion of determinants. The information included on viruses is not necessary, but is a help in studying the effects of protomorphogens in later chapters. Let us now proceed to a discussion of the biodynamics and influence of protomorphogens on the metabolism and morphology of living cells.

BIBLIOGRAPHY

ASTBVRY., W. T.: "Protein and Virus Studies in Relation to the Problem of the Gene." *Proc. 7th Intern. Genetical Congr.,* Edinburgh, Aug. 23-30, 1939; 41-51, 1941.

BARNARD, J. E.: *J. Roy. Microscop.* Soc., 59:1, 1939.

BAUDISCH, O.: "The Importance of Trace Elements in Biological Activity." *American Scientist,* 31:211-245, 1943.

BEADLE, G. W.: "High Frequency Radiation and the Gene." *Chemical & Engineering News,* 24: 1366-1371, 1946.

BEARD, J. W., and WYCKOFF, R. W. G.: "The Isolation of a Homogenous Heavy Protein From Virus-Induced Rabbit Papillomas." *Science,* 95:201-202, 1937.

BERNHEIM, F., BERNHEIM, M. L. C., TAYLOR, A. R., BEARD, D., SHARP, D. G., and BEARD, J. W.: "A Factor in Domestic Rabbit Papilloma Tissue Hydrolyzing the Papilloma Virus Protein." *Science,* 95:230-231, 1942.

BERTRAND, G.: "Sur l'existence de l'arsenic Dans l'organisme." *Compt. rend. soc. biol.,* 134:1434-1437, 1902.

BERTRAND, G., and SILBERSTEIN, L.: "Sur la Répartition du Bore Dans Les Organes du Lis Blanc." *Ann. Agron.,* 8:443-446, 1938.

BLAKESLEE, A. F., and AVERY, A. G.: "Genes in Datura Which Induce Morphological Effects Resembling Those Due to Environment." *Science,* 93:436-437, 1941.

BOBKO, E. V., and TSERLING, V. V.: "The Effect of Boron on Reproduction m Plants." *J. Bot. USSR*, 23:3-11, 1938.

CASPERSSON, T.: *Skmd. Arch. Physiol. 73*, Suppl. No. 8, 1936.

CRILE, G. W., TELKES, M. and ROWLAND, A. F.: "The Nature of Living Cells." Arch. Surg., 23:703-714, 1931.

CUMLEY, R. W., IRWIN, M. R. and L. J. COLE: "genic Effects on Serum Proteins." Proc. Natl. Acad. Sci. U.S.A., 27:565-570, 1941.

DANIEL, E. P.: "Trace Elements." Food & Life, Yearbook of U.S. *Department of Agriculture,* 1939.

DAY, D., and COMBONI, S.: "Effects of Potassium Deficiency on the Formation of Starch in Pisum sativum." *Am. J. Bot.,* 24:594-597, 1937.

DRENNAN, M. R.: "What is the Ultimate Nature of the Cell?" *Clin. Proc.,* 3:171-181, 944.

ELTINGE, E. T. and REED, H. S.: "The Effect of Zinc Deficiency Upon the Root of Lycopersicum esculentum." *Am. J. Bot.,* 27:331-335, 1940.

FENN, W. O.: "The Role of Potassium in Physiological Processes." *Physiol. Rev.,* 20:377-415, 1940

FROLOWA, S. L.: *Compt. rend. acad. sci. U.S.S.R.,* 30:459-461, 1939.

GUTHMANN, H., and GRASS, H.: "Arsenic Content of the Blood of Women. Influence of Menstrual Cycle, Pregnancy and Carcinoma." *Arch. f. Gynakol.,* 152:127-140, 1932.

HOLMES, E.: *The Metabolism of Living Tissues.* Cambridge University Press, London. 1938.

LEHMAN, C. G.: *Physiological Chemistry.* Vol. II, Blanchard & Lea. 1855.

LINDLAHR, H.: *Philosophy of Natural Therapeutics.* Lindlahr Publishing Co., Chicago. 1922.

LIPMAN, C. B.: "Aspects of Inorganic Metabolism in Plants." *Ann. Rev. Biochem.,* 9:491-508, 1940.

MARTIN, L. F., BALLS, A. K., and McKINNEY, H. H.: "Protein Changes in Mosaic-Diseased Tobacco." *J. Biol. Chem.,* 130:687-701, 1939.

MAST, S. O., and PACE, D. M.: "The Effects of Calcium and Magnesium on Metabolic Processes in Chilomonas paramecium." *J. Cell & Comp. Physiol.,* 14:261- 279, 1939.

——————— "The Effect of Phosphorus on Metabolism in Chilomonas paramecium," *Ibid,* 20:1-9, 1942.

MAZIA, D.: "Enzyme Studies on Chromosomes." *Cold Spring Harbor Symposia on Quantitative Biology,* 9:40-46, 1941.

MAZIA, D., and JAEGER, L.: *Proc. Natl. Acad. Sci. U. S. A.,* 25:456, 1939.

MEYER, K. H.: *Natural and Synthetic High Polymers.* Interscience Publishers, Inc., New York. 1942.

MIRSKY, A. E.: "Chromosomes and Nucleoproteins." *Advances in Enzymology,* Vol. 3, Interscience Publishers, Inc., New York. 1943.

NUTTALL: *Blood Immunity and Blood Relationship.* Cambridge University Press, London. 1904.

PIERCE, W. P.: "The Effect of Phosphorus on Chromosome and Nuclear Volume in a Violet Species." *Bull. Torrey Bot. Club,* 64:345-354, 1937.

REED, H. S.: "Relation of an Essential Micro-Element to Seed Production of Peas." *Growth,* 6:391-398, 1942.

REHM, S.: "Die Wirkung von Elektrolyten auf Die Aufnahme Sauer und Basischer Farbstoffe Durch Die Pfanzenzelle." *Planta,* 28: 359-383, 1938.

RUSSOF, L. L., and GADDUM, L. W.: "The Trace Element Content of the New-Born Rat (As Determined Spectrographically)." *J. Nutrition,* 15:169-176, 1938.

SAEGER, A. C.: "The Concentration of Copper in Nutrient Solutions for Spirodela polyrrhiza." *Am. J. Bot.,* 24:640--643, 1937.

SCHULTZ, J.: "Physiological Aspects of Genetics." *Ann. Rev. Physiol.,* 5:34--62, 1943.

SHIVE, J. W., and ROBBINS, W. R.: "Mineral Nutrition of Plants." *Ann. Rev. Biochem.,* 8:503-520, 1939.

SHOHL, A. T.: *Mineral Metabolism.* Reinhold Publishing Corp., New York. 1939.

SOROKIN, H., and SOMMER, A. L.: "Effects of Calcium Deficiency Upon the Roots of Pisum sativum." *Am. J. Bot.,* 27: 308-318, 1940.

STEINBERG, R. A.: "Role of Molybdenum in the Utilization of Ammonium and Nitrate Nitrogen by Aspergillus niger." *J. Agr. Res.,* 55:891--902, 1937.

STERN, K. G.: "The Relationship Between Prosthetic Group and Protein Carrier in Certain Enzymes and Biological Pigments." *Cold Spring Harbor Symposia on Quantitative Biology,* 6:286-300, 1938.

TANGL, H.: "The Effect of Arsenic on the Growth of Bone." *Mag. orv. Arch.,* 40:439-443, 1939.

TENNANT, R., and LIEBOW, A. A.: "The Effect of Potassium on the Growth of Normal and Malignant Cells in Tissue Culture." *Yale J. Biol. & Med.,* 14:553-560, 1942.

TURCK, F. B.: *The Action of the Living Cell.* The Macmillan Co., New York. 1933.

UNDERWOOD, E. J.: "The Significance of the 'Trace Elements' in Nutrition." *Nutrition Abstracts & Reviews,* 9:515-534, 1940.

WADLEIGH, C. H., and SHIVE, J. W.: *Soil Science,* 47:33, 1939.

WEISMANN, A.: *Essays Upon Heredity and Biological Problems.* Clarendon Press, Oxford. 1892.

WINNEK, P. S., and SMITH, A.H.: "Studies on the Role of Bromine in Nutrition."

J. Biol. Chem., 121:345-352, 1937.

WRINCH, D.: "The Native Protein Theory of the Structure of Cytoplasm." *Cold Spring Harbor Symposia on Quantitative Biology,* 9:218-234, 1941.

———————— also refer to: "The Structure of Proteins," I. Langmuir. General Electric Company Research Laboratory, Schenectady, N. Y., Bulletin No. 966, 1939.

WYCKOFF, R. W. G.: "Some Biophysical Problems of Viruses." *Science,* 101:129-136, 1945.

CHAPTER 2

MORPHOGENS AS REGULATORS OF CELL VITALITY

I. Experimental Basis of their Influences in Cultures

Review in Elementary Principles

WE HAVE POSTULATED a basic biological "determinant" to be termed protomorphogen, meaning the primary organizer of form. A comprehensive philosophy of molecular organization is presented in which we have attempted to establish the need for a primary organizer.

The experimental evidence we have reviewed indicates that a primary organizer of specific proteins does exist. The protomorphogen is shown in these experiments to be characterized by the presence of certain trace mineral groups in a most complex molecular arrangement. Its organizing characteristic may be attributed to this arrangement which, perhaps, spaces "points of tremendous affinity" or "receptors" for organic assemblies. A layman might say that protomorphogen is the "skeleton" for the specific biological protein molecule.

Dynamic State of Living Matter.—We propose that protomorphogen is the primary or basic life organizer because it is responsible for the specificity of the protein molecule. No one would call a specific protein a living entity. Nevertheless, specific protein molecules are the primary building blocks in all animal life. The identity of any organism would not remain intact for long were the specific nature of its proteins to change. This fact alone suffices to emphasize the importance of a protein organizer in maintaining the integrity of living structure. It is necessary to note here that the specific living proteins are constantly being replaced with new molecules; this phenomenon is known as the dynamic state of living matter. Schoenheimer (1942) generalizes that "if the starting materials are available, all chemical reactions which the animal is capable of performing are carried out continually." Rauen (1942) has reviewed the constantly changing metabolism of specific proteins and noted evidence that they are not stable compounds.

We venture the definition that the dynamic state is a state of matter requiring a constant input of energy for its maintenance. This, we believe, is the characteristic difference between living and nonliving protein molecules. We consider the dynamic state an important characteristic of nuclear chromatin. However, Brues, Tracy and Cohn (1942) have reported that the turnover of cytoplasmic nucleic acids is higher than that of nuclear nucleic acids. They reach these conclusions from experiments with isotopic phosphorus. While the dynamic state is probably enhanced in the cytoplasm (we shall indicate this later in our discussion of the cytoplasmic activity of nuclear constituents) it is important to note that these investigators reported a greatly enhanced nucleic acid turnover in the nuclei of leukemic cells, indicating that the dynamic state is enhanced during active mitosis.

It might not be incorrect to suggest that the rate of dynamic interchange is an index of the vital activity of the cell. The more rapidly the protein molecules are replaced the "younger" the average state of the constituent proteins in a tissue.

Miscellaneous Problems Connected with Morphogens.—In postulating the organizer of protein specificity as the primary morphogen, we must explore the possibility of the existence of still simpler and more basic organizers. Substances, simpler in chemical organization than specific proteins, are held intact in their identity by the affinities of the component molecules. The non-specific proteins, albumins, peptides, globulins, proteoses, and peptones are relatively stable when compared to specific proteins. There seems to be no indication of either the necessity for, or existence of, a simpler organizer than protomorphogen.

In the first chapter reference was made to morphogens other than protomorphogen. We introduced the term "cytomorphogen." This term, from its derivation, would signify "organizer of cell form." This is precisely what we wish to imply. By our definition a cytomorphogen is "an extremely complicated assemblage of molecules exhibiting some characteristics of a virus... a determinant tor the morphology of the individual cell."

By comparison, let us remember that protomorphogen is the determinant for the specific protein molecule. The gene is the determinant for the separate characteristics of an organism. But the cytomorphogen organizes the whole morphology of a single cell, certainly no more, probably no less. The cytomorphogen bears the same relationship to a single cell as the chromosomes bear to a whole organism of cells. The cytomorphogen must necessarily contain protomorphogens, since the form of a cell is a function of its specific proteins whose integrity is maintained by a protomorphogen.

Whenever biological products are discussed in terms dealing with their function, rather than their structure, it is impossible to be precise in their definition. Nevertheless, practically all new biological products have been known according to their function when they have been first discovered, and by their chemistry at a much later date. This applies to the morphogens, and as a consequence an examination of the chemical and physical properties of protomorphogens and cytomorphogens brings to light many inconsistencies. We feel that it is only possible to deal with these determinants in terms of function in our present hypothetical treatment of the subject.

Chemical Properties of Morphogens.—A study of the experiments dealing with these simpler morphogens have shown many different and varied chemical and physical properties of substances exhibiting the same biological effects. We feel that cytomorphogen, during the course of cell life, gradually disintegrates into its components (as a consequence of the dynamic nature of protoplasm) and the only organizing portion left is the protomorphogen. We shall review experiments covering this problem later in the chapter but at this point it will suffice to mention that the evidence indicates that morphogens accumulate in the media outside the cell wall and exert a certain influence on the cell which may be species specific. But the degree of specificity seems to vary from one species to another and in no case is as marked as when the nucleus of the cell is broken down and the contained morphogens released directly into the media, as in the experiments of Turck (1933) reported in Chapter 1. In a sense the degree of specificity seems to vary in inverse proportion to the diffusibility of the substance and therefore roughly to the size of the molecule.

This problem will be treated in more detail later in the chapter when experiments are reviewed, but it is necessary to state it here in order that we may report some of the chemical and physical properties of cytomorphogens. It is well to bear in mind that there is a great possibility that from a chemical standpoint there may not be a clear cut difference between protomorphogens and cytomorphogens. Rather it seems likely that a cytomorphogen may go through a great many steps of disintegration, exhibiting different chemical properties at each step, before the basic protomorphogen components are separated. We believe this to be the actual case but to outline the details of such steps is a problem of great complexity which cannot be attempted until the biological effects of the morphogens are more completely classified.

We have noted in the first chapter that protomorphogens are relatively thermostable and owe their organizing characteristics to a mineral pattern. Turck (1933) prepared a form of protomorphogen by firing living tissue at temperatures in the neighborhood of 300 degrees C. He found that this ash lost its biological effect at higher temperatures, becoming inactive if fired at 700 degrees C.

Turck (1933) has extracted protomorphogen from tissue with saline solution. This is of interest in view of the recently discovered fact that nucleic acids may be extracted from tissues with saline solution (Mirsky and Pollister, 1942). It is our conclusion, for which experimental data will be presented later, that nucleic acids are intimately combined with protomorphogens in the cell. In fact, we consider that the living protomorphogen is a nucleoprotein with a basic mineral framework and virus characteristics.

Robertson (1923) has commented upon the probable diffusibility of allelocatalyst (which we suggest is identical with breakdown products of cytomorphogen) because of its autocatalytic effect; and yet, before its secretion from the cell during mitosis, it must be relatively non-diffusible since its effect is not demonstrable until a few divisions have taken place.

Mast and Pace (1946) have demonstrated that the allelocatalyst can pass through a cellophane membrane and calculate its molecular size at less than 6 mμ in diameter. They also report (1938) that its thermolability is roughly proportional to its concentration. Apparently it is destroyed by oxidation, and its rate of disintegration is proportional to the temperature, withstanding 100 degrees C. for 1 to 5 hours.

Robertson also obtained a biological reaction from allelocatalyst (protomorphogen) which had been boiled several times to sterilize it. This confirms Turck's observation of the relative thermostability of this group of substances.

Also of interest in this respect are Robertson's experiments in which he determines that the protomorphogen from yeast is acetone soluble, since after extracting the yeast with this solvent the biological activity of protomorphogens can no longer be demonstrated.

Protomorphogens have generally been found to be soluble in lipoid solvents as well as saline and water. Baker and Carrel (1925) find the growth inhibitor in tissues soluble in alcohol-ether; Wemer (1945) finds the growth inhibitor in cancer tissues soluble in acetone and petroleum ether.

We shall present evidence further in our discussion (see Chapters 4 and 5) that the morphogens (although of nucleoprotein structure) are intimately associated with the lipids and phospholipides. In many cases the phospholipides in particular have been thought to be the causative factors behind reactions which we consider to be strictly a result of morphogen activity.

So-called solvents for protomorphogen, therefore, may simply be a solvent for the particular lipid that it is associated with. This may account for the varying degrees of solubilities reported by various investigators.

Later (Chapter 3) we shall present data which leads to the hypothesis that there is considerable polymerization and depolymerization of protomorphogen as it undergoes its various biological cycles. This change in molecular size may account for the varying degrees of thermostability and permeability reported by various investigators.

In conclusion, we may briefly summarize the chemical and physical characteristics of the protomorphogens as follows:

1. Protomorphogens, as found in living organisms, are virus-like nucleoprotein molecules;

2. The structural characteristics and morphogenic influences of protomorphogen

depend upon the protein moiety of the nucleoprotein molecule;

3. The mineral content of the protein moiety is the stabilizing influence on its structure; the organic constituents may be destroyed by charring, without destroying the basic mineral pattern and its receptor linkages;

4. Being nucleoprotein in nature (in its functional state), the protomorphogens can be handled by chemical techniques acceptable for nucleoprotein, therefore they are soluble in saline solution;

5. Since the protomorphogens are usually associated with lipoids, they may be extracted by solvents effective in extracting the associated lipoid material such as acetone, ether, petroleum ether, and ether-alcohol;

6. The molecular size of protomorphogens varies, depending upon what part of their biological cycle is considered;

7. The thermostability of protomorphogen varies, depending upon the concentration and the nature of the protomorphogen molecule; different degrees of degradation of the protomorphogen molecule have different biological effects, and the thermostability would depend to a considerable degree on the nature of the degradation and the biological effect used to test its presence; generally speaking, protomorphogens are thermostable up to 700 degrees C. if the mineral framework and its protein synthesizing influence is considered, while they are destroyed by boiling at 100 degrees C. in a few hours if their growth influences are considered;

8. Oxidation destroys some of the growth influencing potency of protomorphogen;

9. We shall present evidence later which indicates that protomorphogens polymerize to form chain-like molecules; indeed, this characteristic is one of their most significant in biological processes (Chapter 5);

10. We shall also present evidence later (Chapter 5) indicating that protomorphogens have affinity for fibrin which leads us to suspect that in certain unattached states they are very easily adsorbed on various adsorbents.

Cytomorphogen varies from protomorphogen in the above properties in that it is relatively more complex, thermolabile, specific for species and tissues, usually rigidly associated with the chromatin (never being normally found outside of the nucleus), and of larger molecular size (being therefore less diffusible).

Introduction to Growth Influences of the Morphogens

In our studies leading to this theory of morphogens, we were not concerned with morphogenic effects or causes. We were studying growth and senescence. Our study of growth resulted in a search for and classification of all known factors influencing mitosis in unicellular life. We found that various isolated investigators had reported substances affecting mitosis which had characteristics that set them apart from enzymes, hormones or vitamins. These reports were so consistent in their description of certain characteristics that we gave this varied group of growth substances special study.

Our investigations did not carry us far before we were surprised to learn that the particular growth substances we were studying were also morphogenic in effect. This discovery broadened greatly the aspects of these factors and we therefore carefully studied all available accounts of various unconnected experiments dealing with morphogenic influences. These studies have led to an integration of the experimental work that has been done with these factors, finally resulting in the development of this hypothesis of morphogens with its comprehensive ramifications.

It is fitting to present here the experimental material which is concerned with the influence of the morphogens upon fission. We shall confine ourselves to unicellular life to avoid the confusion of metazoan differentiation. Considerable material has been gathered from workers in the field of cell culture. In particular we recognize the outstanding contribution of T. Brailsford Robertson whose allelocatalyst theory seems to offer the most acceptable explanation of 1;he autocatalytic phenomena in growth.

Factors Influencing Division Rate in Cultures of Protozoa

Relation of Media Volume.—An important axiom in the consideration of the factors influencing unicellular culture growth is the importance of the ratio of the number of cells to the volume of the media. One of the first to report this in modern scientific literature was Wildiers (1901), who demonstrated that the rate of mitosis was dependent upon the ratio of the volume of media to the number of cells. In its natural environment, unicellular life follows the same law that governs its reactions in cultures. A limited volume of media is a necessity in nature as it is in cultures. Cultures cannot be started in running water, likewise colonies of unicellular life in nature are found in stagnant or quiet water rather than in running brooks.

Robertson (1921) has also reported that the reproductive capacity of a cell is determined by the density of cells in a restricted culture media rather than by its age in days.

Countless experiments have provided us with growth curves to indicate that the reproductive rate in a culture varies with time and, after reaching a maximum, rapidly drops off until mitosis ceases entirely.

Upon first consideration of these observations we would logically suppose that as the density of population of a culture increases the available foodstuff per unit cell decreases. We might assume that this decrease of available foodstuffs per cell is the cause of the gradual paralysis of reproductive capacity.

This assumption, however, does not explain the phenomenon reported by Robertson that single cells usually do not survive when transferred into culture media exceeding 1 cc. in volume. Robertson was careful in his experiments to note that old media has a lower osmotic pressure than fresh media and therefore he made his transfers into media neither hypertonic nor hypotonic for the isolated cells. Wildiers also reported the same experimental observations.

About the time of Robertson's report, Peters (1921) commented upon the lethal effect of excessive culture media volume into which a single cell may be transferred. This phenomenon has been experimentally demonstrated many times and is common knowledge among those who work with cultures of single celled life.

If the effect of variation of media volume upon the multiplication rate of cells were solely a product of relative availability of foodstuffs, it would be difficult to explain the detrimental effect of excessive volume of fresh media on single cells.

We must look elsewhere for the factors responsible for the effects of variation of media volume. Investigation discloses the fact that other experimental evidence dovetails nicely into this phenomenon and it is now possible to integrate it all into a complete picture. Robertson (1923) has shown experimentally that the gradual decrease in the rate of multiplication of infusoria is neither due to the exhaustion of foodstuffs nor to the accumulation of toxic products in the media.

When infusoria are removed from an old hay infusion culture by killing them with a temperature above their thermal death point (50 degrees C.) and filtering the media, he found that the media is quite capable of supporting the multiplication of transplanted infusoria. Indeed his experiments show that the transfer of infusoria to an old culture media results in a greater rate of multiplication than if they were transferred to a fresh media. However, if the transferred infusoria have previously inhabited an old culture with a maximum population they will not attain maximum reproductive rate when isolated into an old culture media; they succeed, however, if isolated into a fresh culture media or into an old culture media diluted with distilled water containing two volumes per hundred of m/15 phosphate mixture at pH 7.7.

He concludes from these observations that the limitation of reproduction in infusorian cultures is not due to elimination of foodstuffs from the media, accumulation of toxic products or to inherent inability to reproduce further on the part of the individual.

Allelocatalyst Theory.—There is only one tenable hypothesis in view of these experiments. This is Robertson's conclusion that reproducing cells secrete a substance (allelocatalyst) into the media which exerts a very powerful influence over the vitality and rate of reproduction of the cells inhabiting the culture, but this influence is a product of the relationship of the internal condition of the cell to the concentration of the cell-secretion in the media. When the internal condition of the cells reaches a balance with a high concentration of allelocatalyst in the media, reproduction ceases. However, at lower concentrations in the media, the internal condition of the cell is still in the constructive phase of its life cycle and the allelocatalyst exerts a catalytic action upon synthesis of new protoplasm. This coordinates the experimental results we have just reviewed with the importance of the ratio of the volume of media to the number of inhabitants. Changing the volume of media in a culture will of necessity change the concentration of the cell-secretion and consequently alter its influence on the inhabiting cells. The phenomena of the reciprocal relationship between the internal dynamics of the cell and the concentration of a cell-secretion (allelocatalyst) in the media is known as Robertson's allelocatalyst theory. It is based upon the evidence we have briefly sketched which outlines the following observations: (1) a cell will not grow if the volume of the media is excessive; (2) media in which inhabitants have ceased mitosis will support the growth of fresh cells; (3) old cells grow poorly when transferred to an old media which may, however, support the growth of fresh cells; (4) these same old cells will nevertheless grow when transferred to fresh media.

Robertson's theory at the present time appears to have been developed beyond the stage of a tentative hypothesis. It has been thoroughly tested in the laboratory and we shall now review more of the evidence of its verity.

The rate of growth of cells in a culture is not constant. When a transfer is made, after an interval of no cell division known as "lag period," the reproductive rate varies with time. Successive fissions take place in shorter and shorter intervals up to the point of a maximum attainable population . after which the intervals increase.

Robertson (1923) believes that this acceleration of the rate of fission is a result of a growth catalyst of endocellular origin secreted into the media by the inhabitants of the culture. He terms this substance allelocatalyst. He demonstrated the effects of this endocellular catalyst by tabulating the number of individuals inhabiting a culture started with a single isolated infusoria (Enchelys). After the first cell division had occurred, the rate of net total divisions was three divisions for the first twenty-four hours and seven divisions during the next twenty-four hours. This resulted in a total of eight inhabitants at twenty-four hours and 1024 individuals in forty-eight hours. If the original rate of division had not been autocatalyzed there would have been only six or seven divisions in forty-eight hours or a total of approximately 200 culture inhabitants. These experiments always result in the production of the typical "S" type growth curve which is exactly analogous to the curve of an autocatalyzed chemical reaction. Indeed, all growth responses in animals are shown by Robertson to produce curves of the "S" type similar to the curve of extent of transformation in time of the typical monomolecular autocatalyzed chemical reactions. This relationship is shown in figure 1.

Two methods are used to experimentally prove that this accelerated rate of fission is caused by an endocellular substance (allelocatalyst) secreted into the media. In one experiment (Enchelys), after the

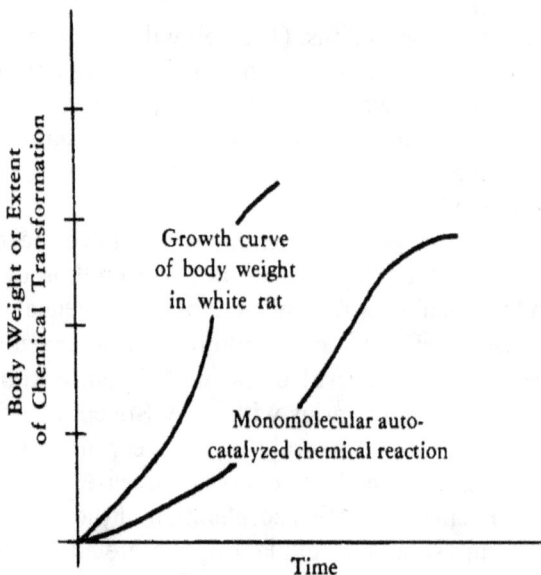

FIGURE 1

Comparison of curves of growth against time and extent of transformation against time in an autocatalyzed monomolecular chemical reaction. (Robertson, T. B., *Chemical Basis of Growth and Senescence*. J. B. Lippincott Co., Philadelphia. 1923.)

cell division had occurred, one of the two daughter cells was isolated again into a fresh medium. The rate of cell division in the culture arising from the isolated cell was much less than that of the undisturbed culture. This effect is noticed even if the isolated infusoria is transferred to a medium comparatively richer in foodstuffs. This demonstrates the existence of allelocatalyst in the original medium, resulting from the first cell division.

Stimulating Influence of Old Media.—The second method of experimental proof is the growth stimulating effect of old media extract when added to fresh media. Robertson has concentrated the allelocatalyst from old media, added it to a new culture and observed a resulting increase in the rate of division. Distilled water with pH adjusted by addition of 0.1 normal sodium carbonate was inhabited by multiplying infusoria for 48 hours. The infusoria were immobilized at 50 degrees C. and filtered off through double filter paper. The pH of the filtrate was adjusted and its volume reduced to one-half by evaporation on a water bath. A precipitate of coagulated protein was filtered off and the resulting filtrate was stored in flasks and boiled several times at

41

daily intervals to sterilize it.

In one experiment allelocatalyst concentrate was prepared by this method from media inhabited by infusoria for a month. This allelocatalyst concentrate was added to buffered distilled water and the resulting rate of reproduction was compared with a control culture of buffered distilled water and another culture of buffered distilled water plus hay infusion.

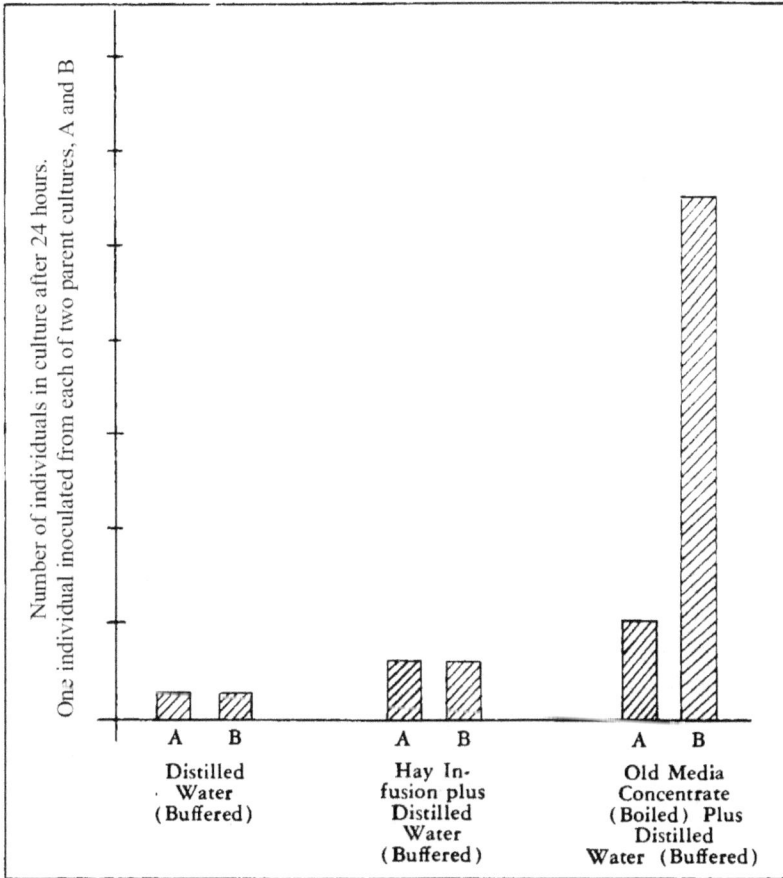

FIGURE 2

Robertson's experiment showing the influence of allelocatalyst concentrated from old culture in stimulating growth rate of new culture inoculated with single infusoria. (Robertson, T. B., *Chemical Basis of Growth and Senescence.* J. B. Lippincott Co., Philadelphia. 1923.)

Each culture was inoculated with a single infusoria from two different parent cultures. After 24 hours the number of infusoria in each culture was determined. The cultures of hay infusion plus buffered distilled water contained twice as many infusoria as the controls. But the medium containing allelocatalyst plus buffered distilled water contained five times as many in one culture and 43 times as many in the other culture. The results are shown in figure 2. The stimulating effect of a concentrate from old media is remarkable. The wide difference between the growth of the two inoculants A and B in buffered distilled water containing allelocatalyst concentrate can be explained by the reciprocal relationship of the culture age of the inoculated cell and the concentration of allelocatalyst in the media. This will be dealt with in detail further on in the chapter.

Dimitrowa (1932) added small amounts of culture media which would no longer support reproduction of its inhabitants, to new cultures of Paramecia caudatum. The reproductive rate of the new culture was accelerated by the addition of this old media. Hall and Loefer (1938) added filtrates of old cultures to bacteria-free cultures of Colpidium campylum and observed the resulting acceleration of reproductive activity. Kidder (1939) has repeated their preliminary experiments on Colpidium campylum and reported the accelerating effect of old medium upon the reproductive rate.

However, it is apparent that a high relative concentration of allelocatalyst in a medium is the factor responsible for the cessation of reproduction in an old culture. Robertson (1923) has demonstrated this fact dearly by diluting old medium into which old cells are transferred, thereby greatly increasing the amount of maximum attainable growth it will support. Maximum attainable growth is reached close to the point where division rate begins to reduce. The reciprocal nature of this phenomenon is clearly demonstrated by the experiments reviewed in this chapter in which it is shown that cells inhabiting a culture media eventually cease mitotic activity, but the media can support maximum growth of new cells, and the old cells will begin to divide as soon as transferred to a fresh media.

Mast and Pace (1938) have repeated Robertson's experiments using cultures of Chilomonas paramecium. They employed the most painstaking care in their technique in an effort to control all variables. The organisms employed had been cultivated in their laboratory under sterile conditions for over five years. The culture media consisted of the purest obtainable inorganic salts in triple distilled water. Pyrex glassware was employed throughout and all cultures were kept sterile.

These investigators concluded that the rate of reproduction of Chilomonas was a function of the reciprocal relationship of the age of the cell and the concentration of a cell-secretion in the media. This cell-secretion inhibited growth and caused cell dissolution in high media concentrations and stimulated growth in low concentrations.

The importance of media concentration was demonstrated by diluting the media of a "dead" culture, thereby causing the remaining chilomonads to become active and increase to the maximum rate of division. The importance of the intra-cellular concentration was strikingly demonstrated by the observation that the rate of reproduction of transplants depends upon the age of the culture from which they were taken.

Arbitrarily assuming that each individual produces an amount of allelocatalyst equal to 0.1 its own volume in four hours, they calculate that the lowest effective concentration in the media would be about one part in 8×10^8 and the optimum one part in 1×10^7.

It is of interest to note their observation that the thermolability of allelocatalyst is in inverse proportion to its concentration. Allelocatalyst from a culture during its maximum reproductive rate is destroyed by heating to 100 degrees C. for 60 minutes, but from a culture which has ceased division, the temperature of 100 degrees C. must be maintained for 90 minutes.

More recently Mast and Pace (1946) have conducted an exhaustive investigation of the allelocatalytic phenomena using Chilomonas paramecium. Their inoculants were obtained from pure cultures with a 12 year pedigree maintained in a sterile chemically pure acetate-ammonium solution.

In their first experiment sacks of chemically cleaned Cellophane membranes were fastened to glass tubes and suspended in Erlenmeyer flasks. The acetate-ammonium solution was placed in the sacks and also in the flask. Measured concentrations of chilomonads were inoculated in the sack only in one case, and in the sack and flask in the other. In the former the chilomonads lived twice as long as in the latter. This indicates that the allelocatalyst produced by the chilomonads in the sack diffused out into the surrounding un-inoculated flask solution in the first case, increasing the time necessary for a lethal concentration in the sack. In the second case, the presence of chilomonads in the flask prevented the allelocatalyst from diffusing out of the sack and therefore the chilomonads in the sack died sooner from a lethal concentration of allelocatalyst. They calculate that the allelocatalyst molecules produced by Chilomonas are less than 6 mμ in diameter.

Further experiments indicate that the growth substance produced by Chilomonas is destroyed by oxidation and its rate of disintegration is proportional to temperature. It is very low at 0 degrees C. and disintegrates in one to five hours at 100 degrees C.

Reciprocal Ratio of Intra- and Extracellular Autocatalyst.—A careful consideratio9 of Robertson's theory of reciprocal relationship between the cell and concentration of allelocatalyst in the media is necessary to establish a few points in culture technique which explain various conflicting experiments. A consideration of lag period phenomena discussed further in this chapter serves to show that young cells have a short lag period and old cells a long lag period when isolated into a fresh media. The lag period is defined as that period of time during which mitotic activity remains at a standstill after a cell is isolated into a new media.

By means of experiments on the infusoria Enchelys, Robertson (1923) has demonstrated that an individual isolated from a one day old culture will produce forty individuals in 24 hours; an individual from a two day old culture will produce five individuals in 24 hours; and an individual from a four day old culture will produce only two individuals in 24 hours. Nevertheless, once the first division has occurred after the lag period, the reproductive rate may be equal for all cultures.

It is apparent that when a cell is isolated into a fresh media there is some change within the cell that brings its internal dynamics back to a common point to recommence reproductive activity. We propose to argue that this point is merely a matter of a low enough internal concentration of allelocatalyst to• permit cell division. We may assume therefore that when a cell is isolated into a fresh media, after the lag period, it reaches a "common denominator" of cell reactivity with the medium, which we may use as a basis for comparison of reciprocal relationship during other periods of culture life. Penfold (1914) has confirmed these experiments on the relationship of lag period to the "age" of the cell through experiments on the culture of Bacillus coli in peptone water. He reports that bacteria, isolated from a parent culture during a period of maximal reproductive activity, exhibit no demonstrable lag period. Burrows and Jorstad (1926) have also found that the lag period is shortest for actively growing tissues and established the principle that the older the transplant, the longer the latent period

Pace (1944) has experimentally reviewed these problems with his strain of pedigreed Chilomonas paramecium maintained in a chemically pure acetate-ammonium solution. He confirms early observations that there is a growth substance produced by Chilomonas which accelerates mitosis in low concentrations and inhibits in higher concentrations, being constantly diffused into the media by the cell. At higher concentrations in the media the substance accumulates in the protoplasm inhibiting mitosis. They conclude that the latent or lag period is the time required for diffusion of the allelocatalyst out of the cell until the concentration within the cell becomes low enough to permit cell division.

The "common denominator" of the internal dynamics of the cell is that point where the first division takes place after transfer to a new media. This "common denominator" condition is approximately the same as the condition within the cell during its maximum rate of reproduction since at this period a transfer exhibits no lag period. It is apparently the optimum internal concentration of allelocatalyst compatible with dynamic cell division.

There are conditions, however, under which the cell has difficulty in achieving the "common denominator" balance not because of an excess of allelocatalyst within the protoplasm that cannot be lost to the media, but because not enough allelocatalyst can be synthesized in the protoplasm to achieve a satisfactory balance. Mast and Pace (1938) have shown that when paramecia are transferred too frequently, the rate of reproduction increases and later decreases. They interpret this phenomenon as being due to a prolonged diffusion and consequent decrease in allelocatalyst in the protoplasm.

We cannot interpret all experiments alike, however, for the "common denominator" protoplasm balance is not consistent when cells are transferred into media containing allelocatalyst. When transferred to a media containing allelocatalyst the internal dynamics of the cell changes to that point where it exhibits mitotic tendencies in the "conditioned" media, which is a different point than it would reach in a fresh media containing no allelocatalyst. Old cells isolated into previously inhabited media have a lower reproductive rate and reach the maximum attainable population sooner than young cells isolated into such a "conditioned" media. Robertson's (1923) experiments comparing the isolation of such "old" individuals into previously inhabited and fresh media, strikingly illustrate this phenomena. When old cells are isolated into old media the maximal population is reached sooner and may be only 10 per cent of that obtained in the same time when they are isolated into fresh media.

In considering these experiments we must remember that maximum population in a culture is reached when a critical balance is obtained between the internal condition of the cell and the concentration of allelocatalyst in the media.

This critical ratio must be distinguished from the optimum ratio mentioned a few paragraphs previous. The optimum ratio is one of high internal concentration and low media concentration inducive to mitosis. The critical ratio is one of higher media concentration compared to internal concentration and this ratio is the point where mitosis either ceases entirely or is balanced by an equal death rate of cells so that the population reaches its peak and becomes static for a period before declining.

Hall and Loefer (1940), working with Colpidium campylum, have reported some conflicting experiments on this subject. In experiment A, when C. campylum were isolated into a fresh media, they reached the maximal population sooner than when isolated into a similar medium to which was added some previously inhabited culture filtrate, although the total population was twice as much and the reproductive rate was greater in the latter. In experiment B, when a greater amount of old medium was added to the fresh medium, however, the maximum population was obtained sooner than in fresh medium alone and was four times as great. Although both controls reached their maximum at about the same time, the control of experiment A exhibited a slower initial velocity of growth and a smaller total population than the control of experiment B, although the investigators point out that the differences are within the limits of experimental error. This might indicate a difference in the internal condition of the inoculum which could be the cause of this discrepancy in the time necessary to reach maximum population. The difference may have been considerable since, when isolated into a fresh media, older cells require more latent rime in which to reach a "common denominator" condition. This lag period would have to be taken into consideration in considering the time necessary to reach maximum population.

Hall and Loefer point out that the maximum attainable population densities were roughly proportional to the amount of old filtrate added to the fresh media. If the termination of mitosis in a culture is determined simply by the relationship between the concentration of allelocatalyst in the media, and the number of cells inhabiting the media, how are we to explain the fact that when the same number of inoculum are added to the same volume of media, different maximum populations may be obtained in different cultures? Reference to hypothetical average curves of culture population in figure 3 will aid in an interpretation of. this phenomenon.

Point A on curve of Culture I and point C on curve of Culture II represent the points at which maximal population is obtained in each culture. This would be the point at which the internal condition of the inhabiting cells reached a "balance" with the concentration of allelocatalyst in the media. It would seem that the cells of Culture II should have reached the same proportionate relationship with the media concentration of allelocatalyst at B, as Culture I does at A, since there are the same number of cells in the same volume of media for each culture at these points. Such, obviously, is not the case.

We shall present additional evidence later in this chapter which suggests that the growth promoting function of small amounts of allelocatalyst is based upon its effect of stimulating the synthesis of new protoplasm from the food stuffs in the media. At point B on the growth curve, culture II is at its maximum reproductive activity, the external source of allelocatalyst stimulating protoplasm synthesis. Because of this catalysis, many more cell divisions may rake place with the internal condition of the cell in a constructive phase before the "balance" with

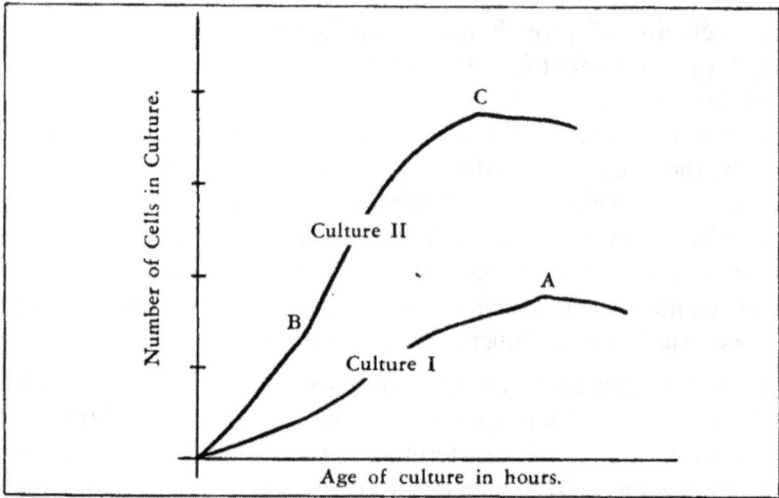

FIGURE 3

Theoretical average growth curves of protozoa cultures. (Derived from Hall, R. P., and Loefer, J. B.: *Proc. Soc. Exp. Biol. & Med.,* 43:128-133, 1940.) Both Culture I and Culture II were inoculated with the same number of individuals from the same parent culture. Both cultures have identical volume. Culture I consists entirely of fresh media, but Culture II contains about 50 per cent of sterilized media previously inhabited by an actively growing culture.

external concentration of allelocatalyst, point C, is reached. Many more cell divisions and a greater utilization of food is made possible than in the corresponding period up to point A, or culture I, which lacks this catalytic stimulus in the media. Our conclusion is that the internal condition of the cells in culture II are identical at point C with those of culture I at point A, but the media in culture II has given up a great deal more foodstuffs due to the catalytic augmentation of mitosis, while in culture I, a greater percentage of foodstuffs are still available but have not been utilized as efficiently in the absence of catalytic stimulation during earlier stages of culture growth.

These experiments serve to emphasize the contention that Robertson's allelocatalyst theory cannot be interpreted simply on the relationship of the number of inhabitants to the allelocatalyst concentration in the media. Rather, his theory is based upon the relationship between the intracellular and extracellular allelocatalyst concentrations. The vitality of

49

the cell is controlled by this dynamic balance and if external stimuli are added during a favorable period of this balance larger populations may be obtained in identical media of the same volume. The greater metabolic activity resulting from such stimuli, however, will effect a more rapid increase in the extracellular allelocatalyst concentration so that the peak of maximum population may be reached sooner.

Hall and Loefer have reported another conflicting observation from this same series of experiments. When *un-inoculated* aged culture medium was added to a fresh culture an accelerating effect was observed. This would seem to indicate that the accelerating effect of small amounts of old culture media may not, as Robertson advances, be due to accumulated cell-secretion. However, they do not eliminate the possibility that the effects of old previously inhabited media and old un-inoculated media may not be due to the same cause, although the growth curves are very similar. With this comment in mind we suggest two explanations which demand further investigation. One is that the old un-inoculated media was not absolutely sterile and bacterial growth resulted in the production of allelocatalyst. Robertson reports that if bacteria are allowed to grow in a media for 24 hours before infusoria are isolated into it, the rate of multiplication is enhanced. He attributes this effect to the enhancement of available foodstuffs. Robertson (1923) also notes that the stimulating effect of allelocatalyst is not specific, a fact which we will discuss more thoroughly later. In such an experiment therefore, the very greatest care must be used to ascertain that no bacterial contamination occurs during the period the un-inoculated medium is allowed to age.

Another possibility is that irradiation was responsible for the growth accelerating effects of aged un-inoculated media. Unpublished reports[1] indicate that synthetic sea water media is rendered equal to natural sea water for plankton by solar irradiation.

The possibility that irradiation may have affected the un-inoculated media is also indicated by the conclusions of Guha and Chakrovorty[2] that adenine acquires the properties of vitamin B1 by irradiation. It is quite

1 Personal communication from Dr. Bernard Chiego, Newark, N. J.

2 Reported in: *The Vitamin and Their Clinical Applications,* Stepp, W., Kuhnau, J. and Schroeder, H. Translation published by The Vitamin Products Co., Milwaukee, Wis. 1938.

possible that the casein-peptone media employed by Hall and Loefer contained the purine derivative adenine which acquired B1 effect from irradiation during the three day ageing period. Vitamin B1 exerts a growth promoting effect on cultures.

The acquisition of growth promoting properties by aged un-inoculated media may supplement Robertson's theory, but it does not deny the reciprocal growth inhibiting effects of old culture medium when aged cells, capable of growth in a fresh medium, are transferred into it. Robertson's theory is based upon this relationship.

The satisfactory evidence in favor of Robertson's allelocatalyst theory makes it mandatory that conflicting evidence be examined very carefully. With this theory in mind, there are many apparently insignificant aspects in a study of cultures, such as: the age of the transferred cell, the old media transferred by the cell if it is not washed, the bacterization of the media, the specificity or non-specificity of the allelocatalyst and lastly, the thermostability of the allelocatalyst, all of which make early conclusions from such experiments a hazardous venture.

Conversely, the influence of accumulated allelocatalyst on a culture must be taken into consideration when studying the effects of other factors on cell growth. Mast and Pace (1939) have recognized this danger in their study of the influence of calcium and magnesium on the growth of Chilomonas paramecium. They noted that this organism secretes growth promoting substances which were toxic in sufficient concentration (Allelocatalyst). They comment that this influence must not be overlooked when drawing conclusions that the death of a culture is due to a deficiency of calcium.

Conjugation and Endomixis.—It is expedient to make here the distinction of the age of a cell in relationship to the age of an individual culture, and the age of a cell in relationship to the period of time between successive conjugations or endomixis. Woodruff (1917) has shown that conjugation or endomixis occurs at normal periodic intervals during the life of pedigreed races of Paramecia aurelia. Calkins (1919) has shown that conjugation is necessary to keep a race of Uroleptus mobilis alive, for in its absence the race will die after 269-349 generations. He has also shown that during the periods between conjugations, the reproductive capacity of a race is gradually lowered, although it has been adequately demonstrated by Jennings (1913) that conjugation diminished the reproductive rate of the conjugants immediately after conjugation.

We shall review experiments on conjugation further on in this discussion and therefore do not wish to discuss this problem in detail at this point. The problems of conjugation and endomixis are quite apart from the phenomenon of allelocatalysis of growth and their ramifications should not be introduced here. It will suffice to say that when making culture determinations with protozoa, it is wise to select the inoculum for all cultures from the same culture, in order to insure that inoculum of the same age (in number of cell divisions from the previous conjugation) are being tested. The results will then be free from the influence of conjugation age on the control and experimental cultures.

Review of Allelocatalyst Theory.—Let us review the salient features of Robertson's allelocatalyst theory. (1) At the time of each division, the cell secretes a substance, termed allelocatalyst, which catalyzes the synthesis of new protoplasm. (2) A small amount of this substance in the media of a culture catalyzes growth and increases the rate of reproduction. (3) If this substance is excessively diffused, the culture will not grow. (4) The allelocatalyst in the protoplasm of the cell has a reciprocal relationship with the allelocatalyst in the media and this determines the rate of reproduction. (5) When the cell has undergone a certain number of divisions and its media has therefore acquired a relatively high concentration of allelocatalyst, a critical ratio between protoplasm and media allelocatalyst is reached which prevents any further mitosis. Later we shall aduce that this critical ratio approaches the condition where the media concentration impairs further excretion of allelocatalyst from within the cell, this excretion being necessary to the continued vitality of the latter. (6) This is the only limiting influence on the population of a culture, provided adequate foodstuffs are available. (7) Neither the cells nor the media in a self limited culture are rendered incapable of participating in mitotic phenomena if they are separated. (8) There is a reciprocal relationship between the concentration of allelocatalyst in the cell and that in the media which controls this phenomena.

Further Notes on the Allelocatalyst Theory.—A complete and candid review of the allelocatalyst theory has recently been presented.[1] This theory of autocatalytic control over growth as presented by T. Brailsford Robertson[2] has been variously rejected, acknowledged and accepted by investigators. In the interest of completeness we present here a brief outline of the opinions expressed by various investigators-as discussed in the recent reviews.

1. The following investigators have suggested that waste products of cell metabolism accumulate in the media and depress the division rate.

WOODRUFF, L. L.: *J. Exp. Zool.,* 10:557-581, 1911; *J. Exp. Zool.,* 14:575-582, 1913.

MYERS, E. C.: *J. Exp. Zool.,* 49:1-43, 1927.

CALKINS, G. N.: *Biology of the Protozoa.* Lea & Febiger, Philadelphia, 1926.

GREENLEAF, W. E.: *J. Exp. Zool.,* 46:143-167, 1926.

PETERSEN, W. A.: *Physiol. Zool.,* 2:221-254, 1929.

DI TOMO, M.: *Boll. di Zool.,* 3:137-140, 1932.

BEERS, C. D.: *Arch. f. Protistenk,* 80:36-64, 1933.

LWOFF, A., and ROUKHEIMAN, N.: *Compt. rend. Akad. Sci.* (Paris), 183:156-158, 1929.

2. The following investigators suggest that the depression of division rate with age is due solely to sub-optimum food concentration and the toxic effects (if any) of waste products are not demonstrable until concentrations greatly in excess of those reported in earlier experiments are reached.

PHELPS, A.: *J. Exp. Zool.,* 70:109-130, 1935.

TAYLOR, C. V., and STRICKLAND, A. G. R.: *Arch. f. Protistenk.,* 90:396-409, 1938; *Physiol. Zool.,* 12:219-230, 1939.

KIDDER, G. W., and STUART, C. A.: *Physiol. Zool.,* 12:329-340, 1939.

JOHNSON, W. H., and HARDIN, G.: *Physiol. Zool.,* 11:333-346, 1938.

3. The investigators listed below have studied the phenomena of the relationship of the number of cells to unit media volume.

A. The following have suggested that overcrowding hinders cell division:

ALLEE, W. C.: *Biol. Rev.,* 9:1-48, 1934; *Animal Aggregations.* University of Chicago Press, Chicago, 1931; *The Social Life of Animals.* W. W. Norton, New York, 1938.

JAHN, T. L.: *Biol. Bull.,* 57:81-!06, 1929.

1 Hall, R. P.: "Populations of Plant-Like Flagellates." *Biol. Symp.,* 4:21-39, 1941. Johnson, W. H.: "Populations of Ciliates." *Ibid,* 4:40-59, 1941.

2 *Chemical Basis of Growth and Senescence,* T. B. Robertson. J. B. Lippincott Co., Philadelphia. 1923.

Sweet, H. E.: *Physiol. Zool.,* 12:173-200, 1939.
Mast, S. O., and Pace, D. M.: *Physiol. Zool.,* 11:359-382, 1938.
Ludwig, W., and Boost, C.: *Arch. f. Protistenk.,* 92:453-484, 1939.

B. The following have suggested that isolated transfers do not grow as well as "grouped" cells:

Johnson, W. H.: *Physiol. Zool.,* 6:22-54, 1933.
Chejfec, M.: *Acta Biol. Experimentalis,* 4:73-118, 1929.
Barker, H. A., and Taylor, C. V.: *Physiol. Zool.,* 4:620-634, 1931.
McPherson, M., Smith, G. A., and Banta, A. M.: *Anat. Rec.,* 54 (suppl.):23, 1932.
Sweet, H. E.: *Physiol. Zool.,* 12:173-200, 1939.
Mast, S. O., and Pace, D. M.: *Physiol. Zool.,* 11:359-382, 1938.
Peterson, W. A.: *Physiol. Zool.,* 2:221, 1929.

4. The following investigators have observed the allelocatalytic effects of cell-secretion (stimulation of initial division or inhibition of division in more concentrated amounts).

Mast, S. O., and Pace, D. M.: *Physiol. Zool.,* 11:359-382, 1938.
Pace, D. M.: *Physiol. Zool.,* 17:278-289, 1944.
Mast, S. O., and Pace, D. M.: *Physiol. Zool.,* 19:224-235, 1946.
Dimitrowa, A.: *Zool. Anz.,* 100:127-132, 1932.
Hall, R. P., and Loefer, J. B.: *Anat. Rec.,* 72 (suppl.):50, 1939.
Kidder, G. W.: *Science,* 90:405-406, 1939.
Mast, S. O., and Pace, D. M.: *Anat. Rec.,* 75 (suppl.):77, 1939.
Reich, K.: *Physiol. Zool.,* 11: 347-358, 1938.
Hall, R. P., and Loefer, J. B.: *Proc. Soc. Exp. Biol. & Med.,* 43:128-133, 1940.
Ludwig, W., and Boost, C.: *Arch. f. Protistenk.,* 92:453-484, 1939.
Robertson, T. B.: *Biochem. J.,* 15:595-611, 1921; *Chemical Basis of Growth and Senescence.* J. B. Lippincott Co., Philadelphia. 1923.

5. The following investigators have attempted to duplicate Robertson's allelocatalyst experiments and have either not been able to duplicate them, or they have felt that their investigations did not substantiate Robertson's allelocatalyst theory:

Cutler, D. W., and Crump, L. M.: *Biochem. J.,* 17:174-186, 1923.
Greenleaf, W. E.: *J. Exp. Zool.,* 46:143-167, 1926.
Calkins, G. N.: *Biology of the Protozoa.* Lea & Febiger, Philadelphia. 1926.
Myers, E. C.: *J. Exp. Zool.,* 49:1-43, 1927.
Grumwald, E.: *Acta Biol. Experimentalis,* 3:81-100, 1928
Darby, H. H.: *J. Exp. Biol.,* 7:308-316, 1930.
Di Tomo, M.: *Boll. di Zool.,* 3:137-140, 1932.
Beers, C. D.: *Arch. f. Protistenk,* 80:36-64, 1933.
Yocum, H.B.: *Biol. Bull.,* 54:410-417, 1928.
Petersen, W. A.: *Physiol. Zool.,* 2:221-254, 1929.

We refer the reader to the excellent recent reviews on this phenomenon[1] from which the bibliography in this resume has been obtained. The fact that a cell secretion accumulates in the media

---

[1] Biological Symposia, Vol. IV, Jacques Cattell Press, Lancaster, Pa. 1941.

exerting toxic effects and that a "grouping" of transfers accelerates the division rate seems to be generally accepted. More recent investigations have shown that what was previously interpreted as the toxic effects of a cell-secretion is actually either a change in the relative media-cell volume, or a result of a sub-optimum food supply. It is apparent that more time than was previously thought is necessary for the toxic effects to occur.

We believe that these discrepancies are the inevitable result of a type of investigation in which the food supply, pH, relative conjugation "age" of the cells, cell media relationship and isotonic conditioning cannot be controlled with the exactitude necessary for conclusive results. Much progress in technique is necessary before this controversial problem of allelocatalytic phenomena can be solved satisfactorily.

Meanwhile, we have perhaps presumptuously accorded Robertson's theory a token acceptance *in toto* and have attempted to utilize it as a working hypothesis in our study of other biological phenomena. e believe future experimental evidence will not prove our confidence misplaced.

Factors Influencing Division Rate in Tissue Cultures

Thus far we have discussed the effects of the allelocatalyst (protomorphogen) only insofar as they have been experimentally demonstrated with cultures of single celled infusoria. In order to gain a perspective of the broad basis upon which the theory of auto-catalysis by protomorphogens is founded, we shall now review the experimental evidence gleaned from work with tissue cultures, a relatively recent technique.

A comprehensive study of the methods of tissue culture brings to one's attention the similarity between the biological principles governing the culture of organs *in vitro* and those influencing the culture of infusoria. An examination of experimental data strongly suggests that the allelocatalyst theory applies to all cell life.

Toxic Concentrations of Allelocatalyst in Media.—One of the most important indications that cultures of infusoria accumulate a toxic allelocatalyst concentration in the media is the observation that infusoria cultures may be indefinitely maintained by periodic changes of the medium. Similarly, the successful culture of tissue *in vitro* requires that the media be changed periodically to remove the accumulating inhibitory substance. Alexis Carrel (1924) states, "The first technique by which cells could be kept indefinitely in a condition of constant activity consisted in removing the tissue fragment frequently from its medium, washing it in Ringer solution, and transferring it to a fresh medium." In this technique he transferred the tissue to a new medium about every forty-eight hours. It should be noted that in this time there had not been any marked disintegration of nutritive substances in the medium. The change, therefore, was not necessary to furnish a fresh supply of foodstuffs, but rather to prevent the. accumulation of an inhibitory concentration of allelocatalytic substance in the medium.

More recent work in tissue culture technique emphasizes the importance of frequent renewal- of media to eliminate toxic substances. Parker (1936) has succeeded in culturing the breast muscle of chicken embryo for a year. He particularly notes that the medium was changed at least twice weekly.

The growth inhibiting factor that accumulates in the media of tissue cultures has been investigated by various students of this technique. Carrel and Ebeling (1922) have identified a growth stimulating and growth inhibiting factor in the serum of young animals. Upon heating the serum the relative potency of the inhibitor is increased, due to the thermal destruction of the stimulator. The inhibitor substance is relatively thermostable and its concentration increases with age.[1] Later, Carrel[2] reports that the growth inhibiting substance in the serum increases as an exponential function of time, indicating its autocatalytic nature.

Werner (1945) has demonstrated that the growth inhibiting factor in rat tumor tissue is soluble in acetone. We have reviewed Robertson's comments that the allelocatalyst is soluble in acetone.

Simms and Stillman (1937) have identified a growth inhibitor substance from tissue fluids which withstands temperatures of 58 degrees C. It is destroyed at 100 degrees C. and by the action of trypsin. Its relative ability to permeate membranes varies.

---

[1] Carrel, A. and A. H. Ebeling: "Antagonistic Growth Principles of Serum and Their Relation to Old Age." *J. Exp. Med.*, 38:419-425, 1923.

[2] Carrel, A.: "Tissue Culture and Cell Physiology." *Physiol. Rev.*, 4:1-17, 1924.

In a series of investigations concerning the limit of growth in tissue cultures, Mayer (1935) demonstrated that the maximum growth period is not obtained unless the marginal clot is periodically removed. He suggests that the efficacy of this method is a result of the removal of decomposition products, enabling a supply of food to reach the cells. In the course of his experiments, Mayer noted that in spite of the removal of the clot the cultures ceased growing in a period of two or three weeks. Only a transfer to a fresh media was effective in renewing growth. It is likely that the inhibitory substance accumulates in the clot and most of it is removed with the clot; however it probably diffuses from the clot into the medium and after a prolonged period further removal of the clot will not effect renewed growth because enough inhibitor has diffused into the medium to prevent it.

Mayer reviews the observations of Olivo (1932) who concludes that there is a very definite toxic substance present in an old clot. He demonstrated that when an old clot is mixed with a fresh clot the inhibition of growth was more pronounced. If, as we suggest, the inhibitory substance is allelocatalyst, then, although a fresh clot might not contain inhibitory concentrations, its addition to an old clot simply adds more allelocatalyst to that already contained therein, resulting in greater inhibition.

We must, however, have more evidence of an allelocatalytic growth controlling substance in tissue cultures than the accumulation in the media of a relatively thermostable growth inhibitor. Robertson's allelocatalyst theory predicates the growth activating effects of more dilute amounts of this substance. Burrows (1913) noted that transplants *in vitro* will not divide in a media free from a special stimulating substance unless crowded into a small stagnant drop. After mitosis has begun it can be stopped by separating the cells, diluting the media or washing the cells with a stream of serum or salt solution without disturbing the oxygen supply (Burrows, 1923).

Relation to Media Volume.—Fischer (1923) (1925) also has noted that isolated tissue cells do not begin division in a large volume of medium. He made hundreds of attempts to obtain the division and proliferation of a connective tissue cell isolated into a fresh medium but without success. In his experiments the cells assumed a spindle shape, filled up with vacuoles and fat granules, and degenerated within a few hours of the transfer. His observations on connective tissue cells in vitro are identical with those of Robertson on infusoria. The latter also reported the extreme difficulty of cultivating a single transfer except into a small volume of stagnant medium.

Stimulating Influence of Diluted Allelocatalyst.—Many investigators have found the growth inhibiting substance to be a growth stimulating substance when diluted. Burrows has substantiated many of Robertson's contentions through his work on culture of cells in vitro. He (Burrows, 1916-17) has demonstrated the lyric effects of a high concentration of a cell-secretion that accumulates in the media with age. He terms this substance "archusia." His experimental analysis of the digestion of cells in the center of a fragment cultured *in vitro* indicates that it is not due to autolysis from the absence of oxygen, but rather is the result of an excess concentration of this cell-secretion.

Burrows and Jorstad (1926) have concluded, as a result of extensive experimental investigations, that the cell-secretion (archusia) exerts no effect in extremely dilute amounts; in medium concentrations the cells engorge with proteins and in greater concentrations divide and grow.[1] In still higher concentrations, cell division is prevented; the concentration gradually increases to the point where it causes autolysis and dissolution. Burrows and Jorstad have investigated the ramifications of the archusia hypothesis especially in its relationship to cancer and a study of their works is to be recommended to all those who are interested in this subject.

Brues, Subbarow, Jackson and Aub (1940) have investigated various methods of extracting growth inhibitors from tissue. Of particular significance is their observation that the growth inhibitors extracted from liver with saline become less efficacious when diluted and exert no inhibitory effect in dilutions of 1:4.

---

1 In a recent communication, however, Burrows has advised us that he does not believe archusia should be classed as an autocatalytic substance. Nevertheless, it seems to us that a substance secreted by the cells which in turn stimulates their mitotic rate, provided the media concentration does not become too high, can safely he classed as such. The similarity of the phenomena accompanying Burrows' archusia and the allelocatalyst of Robertson strongly indicates that we are dealing with the same basic substance, although in Burrows' case it is a product of differentiated metazoan cells and in Robertson's a product of protozoan individuals.

Turck (1933) has strikingly demonstrated the growth promoting effects of the growth inhibitor when diluted. He terms his thermostable product of tissue autolysis and the biologically active tissue ash described in Chapter I of his book, "cytost." He has shown this substance to exert both growth inhibiting and growth stimulating effects *in vitro*. The extract for his experiments was prepared by autoclaving 10 gms. of autolyzed tissue with 10 cc. of water. Transfers were made from stock cultures into hanging drop cultures, one series as a control, and the other inoculated with cytost by touching the plasma drop with a platinum needle previously immersed in the autolyzed solution. In a series of 308 successive transplants of human and chick tissues the inoculated cultures showed definite evidence of greater mitosis after 48 hours incubation when compared with controls.

To test the effects of more concentrated amounts of cytost, transplants were made from stock cultures into two drops of homologous cytost solution prepared as above. After 48 hours incubation all these transplants were dead, while the controls remained healthy.

He has demonstrated the same concentration phenomena with tissue ashed at 300 degrees C. He extracted the black ash with distilled water and filtered it to remove insoluble material. When a platinum needle was immersed in this solution and then dipped in hanging drop cultures, mitosis was stimulated as in the previous experiment; but when the media of the cultures consisted of two drops of homologous plasma and one drop of the ash tissue extract all the cultures were dead at the end of 48 hours incubation. The controls remained healthy. Heterologous ash solutions had comparably little effect on cultures.

These experiments provide evidence that the species specificity exhibited by Turck's ashed tissue, described elsewhere, is apparent in the growth effects *in vitro,* as well as the shock producing effects *in vivo*. The relative biological effects of various concentrations, the remarkable thermostability, and species specificity of Turck's cytost indicate that it is a cell product analogous to Robertson's allelocatalyst.

Universal Nature of Allelocatalyst Phenomenon.—Thus, experiments with the culture of tissue *in vitro* have established several facts which suggest that the allelocatalyst phenomenon is universal among animal cells. From a brief review of experiments we see that: (1) the media must be periodically replaced in order to eliminate the concentration of a growth inhibiting cell-secretion; (2) when diluted, this inhibitor stimulates, and is necessary to the growth of transfers; (3) this substance is relatively thermostable and (4) it is somewhat specific, depending upon the method of preparation.

The studies of Simms and Stillman (1937) on the lag period preceding the initial cell division of a transfer have suggested further that the cells must undergo diffusion of the protoplasm content of an inhibitor before growth begins. Digestion of the cells with trypsin followed by washing shortened the lag period considerably. This suggests the removal of cytoplasmic allelocatalyst and hastening of the processes by which the internal conditions of the cell reach the "common denominator" point with its medium. The reader will recall that the "common denominator" hypothesis was introduced in our discussion of the division rate of infusoria to illustrate the reciprocal relationship between the cell and medium allelocatalyst concentration.

Carrel (1913) noted that extracts of adult tissues stimulate cell division in tissue cultures. However, Carrel and Ebeling discovered a distinct difference in the growth stimulating effects of adult tissue extracts and embryonic extracts.[1] While embryo extracts stimulate proliferation and extension of the growth period, homologous adult extracts at first stimulated division of fibroblasts *in vitro,* but the fibroblasts soon cease proliferation and die. They conclude that embryo extract is necessary for the culture of tissue *in vitro.* Doljanski and Hoffman (1943), in a masterful reanalysis of this problem, were able to duplicate Carrel's and Ebeling's experiments and indefinitely cultivate adult chicken heart muscle, substituting homologous adult tissue extract for embryo extract. An important part of their technique was the addition of fresh extract after washing the tissues every four days and a transfer every 16 to 20 days to a fresh medium.

With a keen discernment of the significance of their experiments, these workers point out that Carrel's work was carried out with hanging drop

[1] Carrel, A. and A. H. Eberling: "Action on Fibroblasts of Extracts of Homologous and Heterologous Tissues." *J. Exp. Med.,* 23:499-511, 1923.

cultures, while theirs was conducted in Carrel flasks. They recognize that difference of the environment can make considerable difference in the effects of the same growth factor. In a Carrel flask, the increased volume of media will not allow the allelocatalytic agent to concentrate as rapidly as in a hanging drop culture.

Whatever the nature of the growth stimulating factors in adult tissue, it is obvious that when extracts of homologous adult tissue are employed to stimulate growth *in vitro,* care must be taken to insure adequate volume of media, otherwise the allelocatalyst, which must contaminate any crude extract of adult tissue, will complicate the results by exerting its inhibitory effect.

Miszurski (1939-40) tested the effects of embryo extract on growth of tibia from a 7 day old embryo chick *in vitro.* The nature of his observations indicates that he was concerned with growth factors of an allelocatalytic nature. The extract was prepared from homologous embryos 7, 13 and 19 days old by grinding and mashing, followed by centrifuging. He reports that when the extract was diluted to 50 per cent concentration, it was toxic to cultures. However, when diluted to 12 per cent concentration it stimulated mitosis. These experiments demonstrated that juice prepared from 13 or 19 day old embryos had to be diluted more than that from seven day old embryos before it became the equal as a growth stimulator.

The conclusion from these experiments is that there is a substance that accumulates in the cell with age, in measurable amounts even in the embryo stage, which stimulates growth when present in small amounts in the media, and exerts a toxic effect when present in the media in greater concentrations. Note should be made of Miszurski's observation that concentrations slightly higher than the optimum for growth, promoted differentiation in the form of ossification of the tibia *in vitro.*

In discussing these experiments we have attempted to review the work in which substances consequent to cell metabolism exert an influence upon the growth of the cell. Kusano (1937-38) has presented a summary of the effects of tissue extracts on the growth of cells *in vitro.* Any brief review of the experimental evidence suffices to indicate the plurality of growth stimulating substances present in embryo and tissue extracts. For the purposes of our hypothesis we are at present only interested in those which exhibit the allelocatalytic phenomena of stimulating growth when small amounts are in the media, and inhibiting growth when present in higher concentrations.

The environment necessary for cell multiplication in a culture of infusoria is different from that necessary for the culture of tissue in vitro in one particularly pertinent circumstance. Both culture techniques demand a periodic change of media to eliminate toxic substances which accumulate, as shown by Robertson (1923) for infusoria and Carrel (1924) for tissue culture. However, Carrel has shown that tissue culture technique demands the addition of thermolabile embryo juice which is not necessary for successful growth of infusoria. More recently, Doljanski and Hoffman (1943) have shown that it is not strictly true that embryo juice must be present for successful cultivation of tissue cells *in vitro,* but an extract from adult tissues (not of an allelocatalytic nature) exerted a similar necessary stimulus to growth.

A detailed review of the various factors in tissue extracts that stimulate growth *in vitro* is not pertinent at this point of our discussion but will follow later.

Thus it. becomes apparent that Robertson's allelocatalyst theory applies universally to animal cells. All animal cells secrete allelocatalyst into their surrounding medium and are in a constant reciprocal relationship with the intercellular concentration of this substance. This relationship is the basic influence over the division rate and the vitality of the cell; it is primarily responsible for the degenerative processes of senescence.

The Allelocatalytic Growth Substance is a Morphogen Group

Turck (1933) has supplied the key to our hypothetical classification of all growth substances whose presence in the media stimulates the division rate when dilute and inhibits it when concentrated. His experiments, reviewed earlier in this work, have demonstrated that tissue ash, carrying species specificity as demonstrated by serological tests. also stimulates and inhibits growth of homologous tissue *in vitro* depending upon its concentration. The allelocatalytic class of growth factors differs from the embryo juice factors in several fundamental ways. The allelocatalytic groups we choose to term morphogens, inasmuch as Turck's work (Chapter 1) has shown these factors to be identified with those that we term protomorphogens or cytomorphogens.

The morphogens are distinguished in their growth effects by the phenomenon of stimulating growth when present in dilute amounts and inhibiting growth when present in more concentrated amounts in the media. The experiments reported by Fischer, Carrel, Mayer, Parker, Burrows, Turck and Robertson, which we have reviewed herein have indicated that the growth substances under observation have exhibited this phenomena. This characteristic of being indispensable to growth in dilute amounts and inhibiting in concentrated amounts is unique, we believe, with the morphogens. Embryo juice growth promoters do not exhibit this effect, and in this manner are significantly different than the morphogens.

The morphogens are more thermostable than other growth factors found in tissue. Turck has shown them to be active even after ashing at 300 degrees C. Robertson also reported that the allelocatalytic substance retained its potency after boiling at 100 degrees C.

Embryo juice growth promoters. on the other hand, have been universally reported to suffer destruction from heat. In an extensive group of controlled experiments, Lasnitzki (1937) has shown that the growth promoting action of embryo extract is considerably decreased after heating at 70 degrees C. for 30 minutes. In 1913, Carrel came to the conclusion that embryonic extracts begin to lose their effectiveness when heated to 56 degrees C. Fischer (1941), in a recent publication, has adequately reviewed the properties of the embryo juice factor. He mentions that the embryo juice growth promoter is adversely affected by temperature above 56 degrees C. The thermolability of the embryo juice growth promoter is in marked contrast to the relative thermostability of the morphogens.

Protomorphogen, therefore, is as much a growth factor as it is a determinant of morphology. We suggest that any experimentally demonstrated growth factor that stimulates or inhibits mitosis depending upon its concentration, depends for its effect on its protomorphogen content. (We might digress to mention that a determinant that establishes the specificity of a living protein would by this function alone be necessary for growth, since protein synthesis could not occur in its absence.)

Again we point out that our terminology is based upon physiological action rather than chemical structure. The protomorphogen class ranges from complex nucleoprotein structures to the extremely stable mineral ash demonstrated by Turck. As we have mentioned, it is inevitable that any class of substances with such a range of chemical structure will exhibit varying degrees of specificity, thermostability, solubility and molecular size. Because this class of morphogens is identified by biological action, dissimilar chemistry does not necessarily preclude assignment of a growth factor to this group. In fact, the chemistry probably would be necessarily different for every type of living protein.

Experiments with growth factors should carefully eliminate all variables so as to definitely establish to what class the factor belongs, and whether another class of growth factors is complicating the result. For instance, Doljanski, as reviewed earlier, was able to maintain growth in tissue cultures without embryo factor, but with adult tissue extracts, because his media volume was greater than that of Carrel who reported negative results. Miszurski, also reviewed earlier, clearly demonstrated the significance of accumulating morphogens in embryo substance which exerted a varying influence on tissue cultures due to the difference in concentration at various ages of both the donor and the recipient embryo tissue.

Any set of experiments on the growth of tissue *in vitro,* in order to be conclusive, should be conducted with the same volume of media per cell; the tissue or embryonic extracts should be made from cells of exactly the same age and grown under exactly the same conditions; the transplants should be of exactly the same age from the time of the previous transplants and come from cultures having exactly the same relative volume of media to the number of cells. This warning is also mentioned in our review of the morphogenic factors influencing infusorial cultures. It is no less important at this point.

If the allelocatalytic growth factors are protomorphogens, then by reason of the determinant nature of the latter, we would suspect these growth factors to exhibit a high degree of species or at least tissue specificity in the experiments which have been reviewed. Such is not necessarily the case, the various investigators reporting varying degrees of specificity.

The two extremes in specificity are reported by Turck and by Robertson. Turck (1933) showed that the ash of tissues heated 300 degrees C. stimulated growth of homologous tissues *in vitro* in dilute amounts and inhibited growth in concentrated amounts. The ash had little effect on heterologous tissues. Autolyzed extract of paramecia influenced the growth of paramecia in the same manner, but heterologous autolyzed extract had no effect.

Robertson (1923) on the other hand has demonstrated that the stimulating effect of the morphogens is not necessarily specific for species. For instance, heat stable, acetone soluble, autolyzed yeast extract was effective in stimulating the growth of infusoria. Also, Robertson demonstrated that a boiled and filtered media, which had previously been inhabited by infusoria, stimulated the multiplication of heterologous infusoria inoculated into it. He concluded from these and other observations that the morphogens secreted by growing cells stimulate growth of cells of all species. In no case, however, have we noted any evidence that completely heterologous morphogens can *prevent or inhibit* the growth of another species *in vitro, whatever their concentration.*

There is an important and basic difference in the manner in which these two investigators obtained their morphogen solutions. Turck prepared morphogens from the cells by autolysis or by ashing. Robertson on the other hand studied the morphogens present in the media; these had been secreted by the inhabiting cells. It is logical to postulate that morphogens, which are secreted into the media, have been broken down by cell metabolism further than those which are obtained from the autolyzed cell substance. In order for them to be secreted they would of necessity be reduced in molecular size to permeate the cell wall (with an accelerated loss during cell division). Apparently the effect of autolysis on intracellular morphogens is different from the normal metabolic influence. This is to be anticipated.

(It might, in fact, be supposed that ashing *would* break down the determinant character and strict specificity, less than the enzyme reactions involved in either normal metabolism or autolysis. The ashing process may destroy the organic components alone *leaving* the basic mineral linkage patterns intact; enzymatic action, however, can be counted upon to break into the important linkages in the molecular chain.)

Inasmuch as Robertson's morphogen preparation was a product of normal cell metabolism, we must conclude that normally protomorphogen in the media is specific in its inhibitory effects of concentration, but non-specific in its stimulating influence when diluted.

In review, let us state the conclusions reached thus far in our study of growth substances:

1. All cells secrete, as a product of their metabolism, a relatively thermostable substance.

2. Dilute amounts of this thermostable substance must be present in the media for initial cell division to occur. Addition of this substance to the media stimulates the rate of multiplication of a transplant after the completion of the lag period.

3. The accumulation of this substance in the media inhibits mitosis in proportion to its concentration, excessive amounts resulting in lysis of the inhabitants.

4. We have applied the term morphogen to all substances exhibiting this phenomena.

5. There is a relationship between the amount of morphogens in the protoplasm and the amount in the media. All cell transplants must undergo a lag period during which the intracellular morphogens are reduced to a "common denominator," at which point cell division may commence. Extracellular morphogens have no influence on this lag period, but determine the time necessary for extracellular morphogens to reach a critical inhibitory balance with intracellular morphogens; thus they determine the total number of cell multiplications before mitosis ceases in the culture.

6. The primary controlling factor is the concentration of morphogens in the cell. These, being continually produced must be constantly excreted or the vitality of the cell will be impaired. Increased media morphogen concentration apparently impairs this excretion, resulting in an increased concentration in the protoplasm.

7. Morphogens, secreted as a result of cell metabolism, only inhibit the growth of their own homologous species but can exert stimulating effects on any species.

8. Morphogens, obtained from the protoplasm by autolysis or ashing, are likely to be cytomorphogens and therefore may only exert stimulating or inhibitory influence on homologous species.

9. The morphogens also are the determinants of the protein molecular structures of the cell and may exert their growth influence because of this action.

BIBLIOGRAPHY

BAKER, L. E. and CARREL, A.: "Lipoids as the Growth-Inhibiting Factor in Serum." *J. Exp.. Med.*, 42:143-154, 1925.

BRUES, A., M., SUBBAROW, Y., JACKSON, E. B., and AUB, J. C.: "Growth Inhibitor by substances in Liver." *J. Exp. Med.*, 71:423-438, 1940.

BRUES, A. M., TRACY, M. M., and COHN, W. E.: "The Relation Between Nucleic Acid and Growth." *Science*, 95:558-560, 1942.

BURROWS, M. T.: *Trans. Cong. Am. Phys. & Surg.*, ix:77; *XVII Internatl. Cong. Med., Gen. Path. & Path. Anat.*, 217-237, 1913.

———— "Some Factors Regulating Growth." *Anat. Rec.*, 11-12:335-339, 1916-17.

————*Proc. Soc. Exp. Biol. & Med.*, 21:94, 1923.

BURROWS, M. T., and JORSTAD, L. H.: "On the Source of Vitamin B in Nature." *Am. J. Physiol.*, 77:24-37, 1926.

———— "On the Source of Vitamin A in Nature." *Ibid*, 77:38-50, 1926.

CALKINS, G. N.: "Uroleptus Mobilis Engelm. II. Renewal of Vitality Through Conjugation." *J. Exp. Zool*, 29:121-156, 1919.

CARREL, A.: "Artificial Activation of the Growth in Vitro of Connective Tissue." *J. Exp. Med.*, 17:14-19, 1913.

———— "Tissue Culture and Cell Physiology." *Physiol. Rev.*, 4:1-17, 1924.

CARREL, A., and EBELING, A. H.: "Heat and Growth-Inhibiting Action of Serum." *J. Exp. Med.*, 35:647-656, 1922.

DIMITROWA, A.: *Zool Anz.*, 100:127, 1932.

DOLJANSKI, L., and HOFFMAN, R. S.: "The Growth Activating Effect of Extract of Adult Tissue on Fibroblast Colonies in Vitro: III. The Cultivation for Prolonged Periods." *Growth*, 7:67-72, 1943.

FISCHER, A.: "Contributions to the Biology of Tissue Cells. II. The Relation of Cell Crowding to Tissue Growth in Vitro." *J. Exp. Med.*, 38:667-672, 1923.

———— "A Functional Study of Cell Division in Cultures of Fibroblasts." *J. Cancer Res.*, 9:50, 1925.

———— "Nature of the Growth-Promoting Substances in the Embryonic Tissue Juice." *Acta Physiol. Scand.*, 3:54-70, 1941.

HALL, R. P., and LOEFER, J.B.: "Effect of the Addition of Old Culture Medium on the Growth of Colpidium campylum." *Anat. Rec.*, 72:50, 1938.

———— "Effects of Culture Filtrates and Old Medium on Growth of the Ciliate, Colpidium campylum." *Proc. Soc. Exp. Biol. & Med.*, 43:128-133, 1940.

JENNINGS, H. S.: "The Effect of Conjugation in Paramecium." *J. Exp. Zool.*, 14-15: 279-391, 1913.

KIDDER, G. W.: "The Effects of Biologically Conditioned Medium on the Growth Rate and Population Yield of Certain Ciliated Protozoas." *Science*, 90:405, 1939.

KUSANO Y.: "Influence of Cell Constituents of Kidney and Other Organs on Growth of Kidney Tissue in Vitro." *Jap. J. Exp. Med.*, 15-16:209-234, 1937-38.

LASNITZKI, I.: "Action of Heated Embryo Extract Upon Growth of Fibroblast Cultures." *Skandinav. Arch. f. Physiol.*, 76:303-312, 1937.

MAST, S. O., and PACE, D. M.: "The Effect of Substances Produced by Chilomonas paramecium on Rate of Reproduction." *Physiol. Zool.*, 11: 359-382, 1938.

———— "The Effects of Calcium and Magnesium on Metabolic Processes in Chilomonas paramecium." *J. Cell. & Comp. Physiol.*, 14:261-279, 1939.

——— "The Nature of the Growth-Substance Produced by Chilomonas paramecium." *Physiol. Zool.,* 19:223-235, 1946.

MAYER, E.: "Experiments on the Limit of Growth in Tissue Cultures." *Skandinav. Arch. f. Physiol.,* 72:249-258, 1935.

MIRSKY, A. E., and POLLISTER, A. W.: "Nucleoproteins of Cell Nuclei." *Proc. Natl. Acad. Sci. U.S. A.,* 28:344-352, 1942.

MISZURSKI, B.: "Researches on the Influence of Embryo Extract of Different Ages on Growth and Differentiation of Cartilage and Bone in Vitro." *Arch. D'Anat. Microscop.,* 35:223, 1939-40.

OLIVO, O. M.: *Monit. zool. ital.,* 42:suppl. 147, 1932.

PACE, D. M.: "The Relation Between Concentration of Growth-Promoting Substance and Its Effect on Growth in Chilomonas paramecium." *Physiol. Zool.,* 17: 278-289, i944.

PARKER, R. C.: "The Cultivation of Tissues for Prolonged Periods in Single Flasks." *J. Exp. Med.,* 64:121-130, 1936.

PENFOLD, W. J.: "On the Nature of Bacterial Lag."*J. Hyg.,* 14:215-241, 1914.

PETERS, R. A.: "The Substances Needed for the Growth of a Pure Culture of Colpidium colpoda." *J. Physiol.,* 55:1-32, 1921.

RAUEN, H. M.: "Dynamics of Proteins." *Umschau,* 44:547-8, 1940.

ROBERTSON, T. B.: "LXXII. Experimental Studies on Cellular Multiplication. I. The Multiplication of Isolated Infusoria." *Biochem. J.,* 15:595-611, 1921.

——— *Chemical Basis of Growth and Senescence.* J. B. Lippincott Co., Philadelphia. 1923.

SCHOENHEIMER, R.: *The Dynamic State of Body Constituents.* Harvard Press, Cambridge, Mass. 1942.

SIMMS, H. S. and STILLMAN, N. P.: "Substances Affecting Adult Tissue in Vitro. I. The Stimulating Action of Trypsin on Fresh Adult Tissue." *J. Gen. Physiol.,* 20:603-619, 1937.

——— "II. A Growth Inhibitor in Adult Tissue." *Ibid,* 20:621-629, 1937.

——— "Production of Fat Granules and of Degeneration in Cultures of Adult Tissue by Agents From Blood Plasma." *Arch. Path.,* 23:316--331, 1937.

TURCK, F. B.: *The Action of the Living Cell.* The Macmillan Co., New York. 1933.

WERNER, H.: "Growth-Promoting and Growth-Inhibiting Factors in Rat Tumor Tissue." *Proc. Soc. Exp. Biol. & Med.,* 59:128-129, 1945.

WILDIERS, E.: *La Cellule* (Louvain), 18:313, 1901.

WOODRUFF, L. L.: "Rhythms and Endomixis in Various Races of Paramecium aurelia." *Biol. Bull.,* 33:51-56, 1917.

CHAPTER 3

MORPHOGENS AS REGULATORS OF CELL VITALITY

II. BIODYNAMIC INFLUENCES ON CELL METABOLISM

The Nucleus as the Seat of Control of Fission

NOW THAT WE HAVE INTEGRATED the experimental evidence into coordinated principles of the effects of the morphogens on cell division, we propose to study the biodynamics of cell metabolism in an attempt to explain the manner in which these effects are accomplished.

Cell division is preceded by changes in the nucleus. The nuclear changes, involving a duplication of nuclear material, are the first phenomena observed in the complicated course of events leading to complete fission. The duplication of the nucleus, previous to complete mitosis, has been demonstrated in a spectacular manner by Jacques Loeb (1906). He demonstrated that hypertonic sea water prevents the cell wall rupture which allows complete fission to occur. It does not for a limited time, however, interfere with the normal metabolism of the nucleus. By placing an egg in hypertonic sea water he was able to observe as many as forty complete nuclear divisions without one complete fission. When the egg was transferred back to normal sea water there were forty immediate complete fissions. Each nuclear division gave rise to a complete cell. The number of nuclear divisions depended upon the time the egg was exposed to the hypertonic solution.

This nuclear influence has also been noted by Mast and Pace (1939) in their studies of the influence of minerals on Chilomonas paramecium. Monsters were formed when Chilomonas were grown in a culture medium containing a sub-optimum magnesium concentration. If the differentiation were not allowed to carry too far, the organisms would revert back to normal morphology when transferred to a medium containing an optimum concentration of magnesium. The monster differentiation occurred in the cytoplasm as a result of the deficient media, but the organizing abilities of the nucleus were not influenced and succeeded in bringing the cell morphology back to normal.

The existence of certain changes in the nucleus which initiate cell division is indicated by the sudden increase in nuclear size shortly before division. Robertson (1923) has reviewed this phenomena and included a growth curve which demonstrates that the cytoplasm undergoes steady increase in mass after cell division but the nuclear increase is slower until shortly before subsequent division, at which time there is a sudden acceleration, completely absent in the cytoplasm. (See Fig. 4)

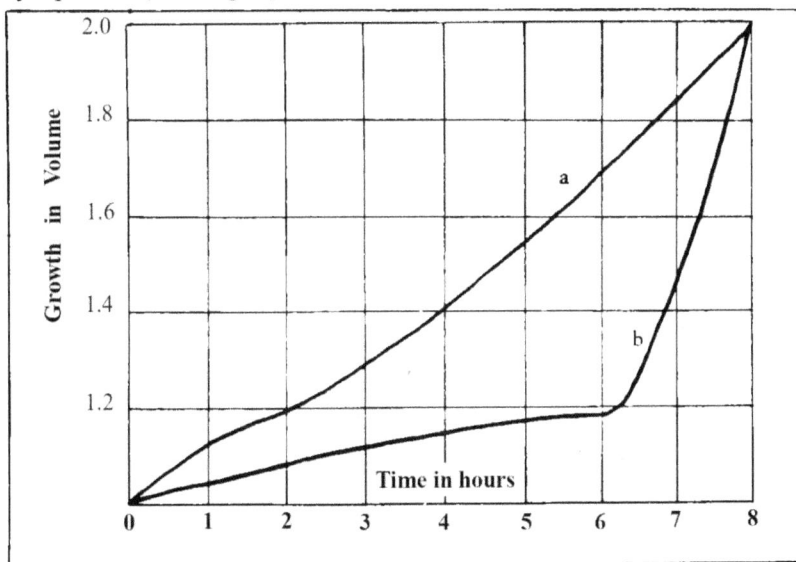

FIGURE 4

Comparison of the growth of cytoplasm (a) and nucleus (b). The abscissa represent the time in hours since the last nuclear division, and ordinates the growth in volume. (Robertson, T. B., *Chemical Basis of Growth and Senescence.* J. B. Lippincott Co., Philadelphia. 1923.)

There has been controversy over the problem of the location of the mechanism responsible for the perpetuation of the cell. The great predominance of evidence seems to favor the nucleus. Claude (1943) mentions the axiom that any substance which possesses the property of self-duplication also contains nucleic acid. He mentions that small cytoplasmic particles have been shown to contain ribonucleic acid but cautions that there is a possibility that these arise from nuclear metabolism.

The conclusions of Mirsky and Pollister (1943) leave little doubt but that nucleic acids and nucleoproteins are exclusively of nuclear origin. Quantitative determinations of nucleoprotein from different tissues indicate that more nucleoprotein is obtained from cells with relatively larger nuclear volume; this is exactly reverse of what would exist if nucleoprotein were produced in the cytoplasm. By removing the cytoplasm of cells and subjecting the residue to a highspeed mixer, the nuclei are caused to form threads which are considered to be solid masses of chromatin. Analysis has shown these threads to be almost 100 per cent nucleoprotein. These facts lead these investigators to conclude that nucleoprotein is the substance of the chromosome.

There is an interchange of nuclear and cytoplasmic constituents shortly before mitosis. At this time nuclear chromatin particles are "shed" into the cytoplasm. This phenomenon has been adequately reviewed by Jennings (1940). The phenomenon of the "shedding" of chromatin macronuclear material at each fission is also reviewed by Diller (1941).

The fact that chromatin or nucleoproteins are secreted into the cytoplasm in minute amounts at each division not only explains the puzzling occurrence of nucleic acids in the cytoplasm, but also suggests that photomorphogen determinants are split from the genes during this process and are utilized by the cytoplasm in the organization of its structure. The influence of chromatin in this respect is further suggested by the observation of Loeb that the final mass of protoplasm in a cell is directly proportional to the mass of chromatin in a given species.

The Nucleus as the Seat of Cell Vitality

Not only is the nucleus of. vital importance as that pan of the cell which is responsible for fission and reproduction, but also it appears to be the seat of the complicated chemistry that provides the cell with energy to maintain its metabolic processes and structural integrity.

Electrical Potentials in the Cell.—Dr. George W. Crile of the Cleveland Clinic Foundation has provided us with a series of brilliant experiments which establish a foundation upon which to build a coordinated hypothesis of cell vitality. His researches, critically received by his contemporaries as unwarranted speculation, have been shown to be extremely prophetic in the light of later developments.[1]

Working in Crile's laboratory, Telkes (1931) demonstrated by means of micromanipulator needles, that there is an electrical potential between the surrounding media and cytoplasm of an amoeba in the order of 15 millivolts. This potential is apparently in proportion to one existing between the nucleus and the cytoplasm.

Autosynthetic Cells.—Crile and his co-workers speculated upon the function of this potential difference in the organization of the cell. They observed that lipoid and protein derivatives from brain tissues have a potential difference of approximately 60 millivolts when measured against an electrolyte solution. By combining the lipoid and protein extracts of brain in the presence of a suitable electrolyte solution, Crile, Telkes and Rowland (1931) were able to obtain an immediate organization of life-like individuals which they termed autosynthetic cells.

The lipoid solution was obtained by ether extraction of calf's brain which had been minced and rapidly dried at a temperature determined by evaporation. The sterile protein solution was obtained from the ether extracted residue of calf's brain by mixing it with a solution of electrolytes in the same concentration as found in calf's brain, allowing it to stand, and then boiling for ½ hour followed by filtering. The pH is adjusted to 7.8 by addition of 0.2 N HCl and the boiling and filtering repeated several times, care being taken to prevent dilution.

An alternate method of preparation of the sterile protein solution was offered. Mix the residue with 10 per cent NaCl solution, filter acidulate with acetic acid and saturate with ammonium sulphate. The mixture is refrigerated and centrifuged. This process is repeated several times. The filtrate is finally dissolved in electrolyte solution so as to contain about 0.5 per cent protein.

When the lipoid and protein solutions were mixed, 0.2 to 0.5 cc. lipoid with 10 cc. of sterile protein, immediate organization of autosynthetic cells was observed under the microscope. Crile and his co-workers (1931) observed that many investigators have succeeded

1 Seifriz, W.: "The Physical Properties of Protoplasm." *Ann. Rev. Physiol.,* 7: 35-60, 1945.

in producing "synthetic" cells exhibiting various phenomena peculiar to living organisms, but none have been able to maintain this reactivity for prolonged periods.

The autosynthetic cells formed by Crile's method grew slowly and multiplied both by budding and direct fission. The cells were nucleated and took vital stains. The cells were kept alive in cultures for several months by the addition of sterile protein solution from time to time, and by constant subdivision of the cultures. During this period they continued to exhibit the phenomena of life, such as reproduction, growth, movement and respiration. The oxygen consumption was measured as high as 14 cubic mm. per 2 cc. of cell mixture and was shown to be dependent upon the availability of protein in the substrate for metabolic. purposes. The respiratory quotient was measured at 0.7 to 0.98 and was considerably increased by the addition of glucose to the media. The cells were dependent upon oxygen supply to maintain their metabolism since when separated from oxygen they gradually disintegrated into amorphous masses with yellow droplets. Free movement, similar to Brownian movement, could be detected in the cytoplasm under microscopic examination.

The form of the cells depended upon the integrity of the lipoid solution. Intense radium radiation of the lipoid prevented the cells from forming, although radiation of the protein or electrolyte had no demonstrable effect. Lipoid solution that had been stored or to which cyanide had been added did not form cells, but rather large "fatty" droplets reminiscent of fatty degeneration. Lipoid obtained from dogs that had died of distemper, rabbits which had died from exhaustion of insomnia, or from cancerous tissue, would not produce organized cells but bizarre structures with fatty droplets similar to those formed from irradiated lipoid. Although the protein moiety could be severely treated, the effectiveness of the lipoid was destroyed by heating the dried substance to 50 degrees C. It was very necessary to obtain the lipoid from healthy animals or imperfect cells were formed.

It becomes increasingly obvious that the brain lipoid supplied some substance that was responsible for the morphological organization of the autosynthetic cells. In their natural environment morphogen groups are often associated with a lecithin or fatty molecule. The fact that they seem to be associated thus in these experiments conforms to the observation of other experimenters. It is quite likely that the greater success of Crile and his co-workers in the production and cultivation of "synthetic" cells is due to the fact that their technique introduces protomorphogen groups with organizing characteristics as a part of the substrate.

These workers seemed to realize the organizing importance of the mineral groups (protomorphogens) for they noticed that when potassium was eliminated from the electrolyte mixture, the formation of the cells was delayed. They comment that the potassium and phosphorus compounds adsorbed on the lipoid were important for their morphogenic influences and speculate that no cells would be organized if these minerals could be eliminated from the lipoid as well as the electrolyte. We suggest that the potassium and other minerals present in the electrolyte are not a part of the original organizing influences, but are necessary as substrate material for the protomorphogens in the lipoid without which the formation of the cells would be impaired.

Although protein and electrolyte solutions could be effectively prepared from other organs, only brain lipoid could be successfully employed. We might wonder why it is not possible to obtain morphogen groups from all organs, and why they are so peculiarly effective when obtained from brain. As a possible suggestion we postulate that the metabolism of nervous tissue, brain in particular, requires the presence of a polymerizing factor in order to meet the demands of the constant production of association fibres and memory paths. We shall introduce evidence later that protomorphogens have the property of inducing the formation of just such thread molecules as we are concerned with here. (Further experiments show brain cephalin to be one of the best sources of active protomorphogens. Chapter 5.)

The influence of external factors on the morphology of the autosynthetic cells suggests that the morphogens concerned do not exert the comprehensive organizing characteristics of cytomorphogens but are limited to the more basic functions of protomorphogens. (If the primary purpose of brain morphogens were simply to facilitate protein polymerization, this is what we would expect.) Although nucleated cells were formed within the range pH 4.0 to pH 8.5, ciliated structures were formed only at pH 7.2 to pH 7.8. The addition of adrenal protein solution caused the production of amoeboid cells with active pseudopodia. Our investigation of the literature has failed to produce any evidence that such pH changes or substrate additions influence the form of normal cells, whose morphological integrity is maintained by cytomorphogen.

Electrical Potentials and Cell Vitality.—These experiments from the Cleveland Clinic Foundation have thrown considerable light on the problems of cell senescence. Crile, Rowland and Telkes (1928) demonstrated the significance of the electrical potential between the subcutaneous fascia and injured tissue in living animals. When dying, the electrical potential dropped to a minimum, increasing a few minutes after death and was zero within a few hours.

Telkes (1931), in her studies on the electrical potential of the Giant Amoeba, observed that when the electrical potential between the protoplasm and media was reduced to zero the cell died. She observed a "striking parallelism" between this potential difference and the viability of the amoeba. Reports from the Cleveland Clinic state that when the potential difference was decreased to zero by the introduction of an equal current of opposite polarity, movement ceased and the cell assumed a spherical form. Granular material began to move through rents in the cell membrane and the cell disintegrated. If the normal potential difference were restored by introduction of current of proper polarity before the tear in the membrane occurred, the cell assumed its healthy form and became active. These carefully conducted experiments can lead only to the conclusion that the structural integrity and electrical potential of the cell are in most intimate interdependence and one cannot exist without the other.

We should note that this phenomena is not observed in plant cells in the same manner. Osterhout and Harris (1927-28) have shown that in some types of plant cells it is possible to kill part of the protoplasm with chloroform and measure the potential difference between the living and dead moieties. Blinks (1929) has reduced the potential difference of Halicystis to zero without observing the loss of form and death of the organism. It is apparent that there are certain basic differences in the biodynamics of plant and animal cells which must be kept in mind.

Particularly significant is the fact that plants are distinguished by a clearly defined structural cell wall, whereas the animal cell substitutes the mere surface potential of its protoplasm. Information agreeing with the amoeba experiments and to an extent introducing an integrating factor with the plant measurements is presented by the experiments of Damon and Osterhout (1930). They were able to observe potential difference changes in Valonia as a result of variation of the NaCl concentration in the medium, and these changes were reversible to an extent; however, if allowed to proceed too far the potential difference changes were irreversible and accompanied by permanent cell damage.

The observation of Lundegardh (1940) on the transfer of inorganic ions across the root cell membranes in plants may offer some light on this problem. He states that, while the fundamental respiration in the plant cells is a product of combustion processes not directly involved in ion absorption, there is a respiration dependent upon the degree of surface potential and ion absorption, or vice versa. He terms this anion respiration. This is an oxidation-reduction equilibrium in which the surface potential of the cell is balanced with the ion transport across this surface.

Raising the surface potential by application of an e. m. f. induces an increased respiration and decreased ion absorption. These experiments can be correlated with those on animal cells to the extent that the cell surface potential seems directly proportional to a fundamental respiration of the cell.

We have noted that Dr. Maria Telkes (1931) determined that the potential difference between the cytoplasm of the Giant Amoeba and its media averages 15 mv. Crile, Telkes and Rowland (1931) (1932) made similar determinations on autosynthetic cells. The potential difference between the cytoplasm and the media was about 40 mv., the negative charge being on the cytoplasm. They also measured the potential directly between the nucleus and the cytoplasm and determined it to be about 20 mv., also with the negative charge on the cytoplasm. If the potential difference is checked when the culture is a few days old, it is found to have decreased about 10 per cent and by 30 per cent after three or four weeks. Some process of ageing, therefore, is responsible for the potential difference decrease. That a toxic factor in the media may be responsible is strongly hinted by the observation that washing freshly prepared autosynthetic cells increased the potential difference by about 20 per cent, from ¬45 mv. to ¬55 mv., and washing amoeba with distilled water increased the potential difference by 100 per cent from 20 mv. to 40 mv. as well as increasing their motility. We wish to call the reader's attention to the oft reported observation that stagnant tissue cultures, or infusorial cultures, may sometimes be reactivated by replacing the media and washing the cells in distilled or tap water. This procedure, we believe, removes the morphogens from the media and external surface of the cell where they are becoming concentrated and inhibiting the cell's vital metabolism.

Telkes (1931) was able to cause a decrease in the potential difference of the Giant Amoeba by the addition of concentrated salt solutions to the media. The same phenomena was noted when salt solutions were added to autosynthetic cells. Blinks (1929) also demonstrated the effects of various electrolytes in reducing or eliminating the potential difference. Upon casual speculation the reader might suggest that the toxic influence of intense concentrations of allelocatalyst (morphogens of mineral nature) is due to its effect of raising the electrolyte concentration of the media. Although the toxicity of high electrolyte concentrations cannot be denied, this phenomena cannot be the basis of allelocatalyst toxicity since Robertson (1923) demonstrated that media that has become toxic for its inhabitants will nevertheless support transplants. We must look elsewhere for a suggestion concerning the mechanism of the allelocatalyst.

Permeability and Electrical Potentials.—Lillie (1923) reviews that a loss of permeability of the cell wall is followed by a disappearance of the potential between the cell and surrounding media. The electrical potential existing across the nuclear membrane has also been suggested by Hardy (1913-14). He mentions that the colloidal particles inside the nucleus are negative and those in the cytoplasm are positive. Holmes (1938) has supplied us with a pertinent commentary on the potential difference between the inside and outside of the crab nerve fiber. During an impulse the nerve membrane is depolarized and the resulting diffusion of ions drops the potential to zero. This potential was calculated on the tenfold difference in potassium content of the nerve and surrounding fluids to be about 60 mv. Direct measurement shows it to be in the neighborhood of 40 mv. By experimentally raising the potassium content of the surrounding fluid the potential difference may be caused to disappear and even be reversed. Thus, this potential is proven to be dependent upon the integrity of the nerve membrane and its ability to become selectively polarized to potassium.

Beumer (1934) has experimentally demonstrated that the electrical potential in living tissues is a result of the integrity of the semi-permeable membrane. He advances a well founded hypothesis which, among other things, establishes the principle that the potential across such a membrane varies with the permeability of the membrane. He considers the chemical interaction between the phase boundaries of the solution to be basic in his explanation of the potential.

Donnan's theory of membrane equilibria is briefly outlined by Harrow (1943). Donnan bases his theory on the impermeability of the membrane to one of the ions present. He postulated that an equilibrium potential would exist across the membrane that could be expressed in mathematical terms; Loeb confirmed this theory by experiment. Hober (1924) considers the electrical potential across a membrane to be a function of the electric charges of the particles or pores of the membrane. Many workers have advanced tenable hypotheses to explain this phenomenon. We are not prepared to discuss here the intricacies of the physicochemical problems involved. It will suffice to advance the obvious conclusion that the electrical potential is co-existent with the membrane and, from the experiments of Crile on the Giant Amoeba, the membrane depends upon the electrical potential. The two phenomena are inseparable and both are essential to the living cell.

As a cell ages, the surface boundary of protoplasm undergoes a distinct and measurable change. The work of Kopac has been reviewed by Chambers (1940) who described the method of determining the relative surface energy of the cell membrane by determining the size of an oil drop necessary to coalesce with the cell.

By means of this ingenious method it has been experimentally shown that the surface potential of an egg that has been allowed to stand in sea water for several hours is lowered to as little as 1 per cent of normal. This indicates the gradual disintegration of the vitelline membrane with time.

Kopac (1938) has developed a method of measuring the interfacial tension by means of allowing an injected oil drop to reach a given size and gradually reducing its size with a pipette. The amount of protein adsorbed on the surface of the oil drop can be estimated by the critical diameter of the oil drop at which the crinkling effect appears. The adsorbed material gathers on the surface of the oil drop, and if the diameter is reduced below the critical point at which there is a monomolecular layer of the adsorbed substance, its concentration is increased and the crinkling appears.

By means of this method, Kopac demonstrated that no proteins were adsorbed on the oil drop from a healthy cell. The proteins in a healthy cell therefore, are bound and cannot freely accumulate on its surface. At the time of cytolysis the membrane potential is zero, and the spontaneous deformation of the drop as a result of protein adsorption is observed. Other investigators have subjected the problem of cytolysis and ageing to a similar scrutiny and come to the same conclusions.

Ruzicka (1922) states that as age advances the pH approaches the isoelectric point. Hitchcock (1926) states that the permeability of a membrane is greatest near the isoelectric point of the proteins, and varies with the pH. Thus, generally speaking, the permeability increases with age.

We believe, that the pH variation is a result of this membrane change. Some students of geriatrics may object to this statement. Senescence is generally considered to be accompanied by gradual decreases in the permeability of various cells, particularly those of the vascular wall. We are using the term "permeability" loosely to describe that tendency of a cell boundary to maintain a potential difference. This function depends upon a selective permeability to certain ions, and it is probably this selective permeability which is increased while permeability to other substances, not concerned with maintaining the potential difference, may be increased or decreased independently.

Review, Electrical Potentials and Vitality.—Reviewing the evidence herein presented, we believe we are correct in postulating that the vitality of a cell is in direct proportion to the electrical potential existing between it and its surrounding medium, and probably to that co-existent between the nucleus and the cytoplasm. The pH of the cytoplasm is directly proportional to the electrical potential across the cell membrane.

The electrical potential is a function of the integrity of the surface boundary of the cell and likewise the nuclear-cytoplasmic potential is a function of the integrity of the nuclear membrane. As the cell is allowed to age, the surface boundary and probably also the nuclear boundary gradually disintegrates and loses its integrity. The result is a gradual lowering of the electrical potential and cytoplasmic pH. Eventually the isoelectric point of the proteins is reached or approximated; the surface potential becomes nearly zero, and the cell undergoes lysis. As we have shown earlier, this is the inevitable fate of a culture of single cells when the culture media is not periodically diluted or replaced, irrespective of the continued availability of foodstuffs.

Our problem now is to organize the experimental evidence in such a way as to present an hypothesis to indicate both the cause of the gradual disintegration of the cell membrane and the mode by which the eventual lysis occurs.

Effect of Protomorphogens on Electrical Potentials.—Burrows and Jorstad (1926) have demonstrated that there arc changes in the cell membrane in consequence of the secretion of a toxic product by the cell itself. They have shown that when cells are caused to exist in sterile culture flasks, morphogens[1] accumulate which exert an effect on the cell. When the morphogens have reached a moderate concentration in the media, they cause the cell to secrete a soap-like substance which they term "ergusia." This substance decreases the surface tension of the cell membrane and apparently is responsible for the change in the membrane, resulting in loss of potential difference across it.

Because of the accumulation of fatty droplets in ageing cells filled

1 Burrows and Jorstad call this substance "archusia" but inasmuch as we have identified it as morphogens, we shall employ this term.

with waste materials, we suggest that this ergusia consists of concentrated morphogens whose electrolytic mineral properties are protected by a lecithin or fatty envelope. Simms and Stillman (1937) have experimentally induced the formation of fat granules in the cytoplasm of cells cultured in vitro by means of the addition to the medium of a factor extracted from plasma.

A brief review of the methods and chemistry of morphogens will recall that fat solvents, such as acetone and alcohol-ether, are used to extract them from cells. Robertson was able to deprive autolyzed yeast cells of allelocatalytic effect by repeated washings in acetone. Carrel found that washing cell extracts with alcohol-ether reduced the growth inhibiting effects of the extracts to a remarkable extent.

There seems to be no possibility of preventing the disastrous effects of morphogens in reducing the integrity of the cell surface boundary. Baker has been able to rejuvenate inactive cultures by the addition of embryo juice and serum. He was able to cause extreme mitotic activity in old cultures by this method. Baker (1938) concludes from this work that the inactivity of old cultures is not due to accumulation of toxic substances, but in one instance, where exceedingly active mitosis occurred, the culture cells suddenly "dissolved." In spite of the driving effect of such growth promoters as embryonic juice extracts, the accumulated morphogens succeeded in lowering the cell potential to the extent that lysis occurred.

Robertson (1923) also demonstrated that once cells had reached the limit of morphogen concentration, addition of excess foodstuffs did not forestall the inevitable degeneration of the cells.

We have seen that the gradual disintegration of the cell membrane with age is due to the accumulation of morphogens in the surrounding media. This then, is probably the universal mode of action by which the accumulating morphogens inhibit mitosis and eventually cause destruction of the cell.

Biochemical Systems in the Cell

Reversible Nature of Dynamic Cell Enzyme Systems.—Elementary cell biodynamics teaches that the protein moiety, at least, of the protoplasm is in a dynamic state of constant catabolism and anabolism. Rauen

(1942) has advanced this conception by reviewing the constant changes occurring within a certain framework of living proteins. He is concerned with the need for such a dynamic activity to enable the protein to engage in specific immunization reactions. Schoenheimer (1942) generalizes that "all chemical reactions which the body is capable of performing are carried out continually." His conception of the dynamic state of protein metabolism, in particular, is adequately founded upon numerous experiments with nitrogen isotopes.

The cell enzymes catalyze these reactions, which are constantly in a state of equilibrium. As the cell ages, its metabolic processes slow up, and such indications of vitality as mitotic activity are impaired and finally cease. It has often been observed and commented upon that aged tissues do not repair as rapidly as youthful structures.

This gradual diminishing of the rate of cell metabolism can be explained on the basis of a change in the equilibrium of the dynamic reactions of the proteins of the protoplasm. As the cell ages, the constructive or anabolic phase of this reaction assumes less importance than the destructive or catabolic phase.

The critical factor in the determination of the balance between the constructive and destructive phase of protein metabolism may be the pH of the cell fluids exerting its influence over the catalyzing enzymes. Haldane (1930) states that if an enzyme accelerates any reaction, it accelerates the reverse action to about the same extent. He mentions that hydrogen ions may influence a reaction either by affecting the rate of change or by destroying the enzyme.

The reversibility of enzyme action is extensively discussed in reviews and textbooks. Mathews (1939) discusses in detail the frequency of reversible enzyme reactions which respond to the mass-action law. Waksman and Davison (1926) mention that enzymes accelerate both hydrolytic and synthetic processes and theoretically all enzyme reactions are reversible, depending upon the limitations of their environment. Starling's (1936) work on physiology reviews reversible enzyme phenomena and mentions the constructive action of pepsin and trypsin at pH values different from the optimum for hydrolytic action.

Hydrogen ion potential has been reported in the work of Mullins (1942) to effect the binding of potassium by protein molecules. This binding is probably catalyzed by enzyme activity and is of special interest m our work because it is a phenomena analogous to the organization of protein structure by mineral protomorphogens.

The gradual reduction of cytoplasm pH values with age, as outlined herein, is quite conceivably the influence which causes the dynamic equilibrium of protein metabolism to emphasize the destructive phase. This change in pH as a consequence of membrane instability could either swing the reversible enzyme activity more to the catabolic side of the equilibrium or inhibit the anabolic enzymes in their balance with the catabolic, depending upon what type of enzyme equilibria is responsible for the dynamic metabolism of cell proteins.

There is a fertile field of experimental investigation in the uncharted phases of enzyme activity responsible for this phenomenon and the influence of environmental factors, such as pH, on this equilibrium.

There is an unquestioned swing to lower pH values in the protoplasm of cells undergoing dissolution and autolysis. Turck (1933) reviews this phenomenon, calling attention to the fact that autolysis may be caused to cease if the tissues are made sufficiently alkaline. He mentions that no evidence of autolysis in mammalian tissues, either *in vitro* or *in vivo,* can be detected if the pH values are held close to that of the animal's normal blood. In a discussion of the autolysis of tissue proteins, Lloyd and Shore (1938) generalize that all the changes occurring in a living cell are reversible while those in a dead cell occur only in one direction. They mention that the reduced pH in a dead cell supplies a more favorable environment for the action of the autolytic enzymes.

Although we hold that this lowering pH is a gradually occurring phenomenon, once it has reached a critical stage it is enhanced to a prodigious degree by the production of ionizable carboxyl groups as a result of the irreversible dissociation of protein macromolecules to lower molecular weights.[1]

1 *The Cell and Protoplasm,* Chambers, R.: Publ. Am. Assoc. Adv. Sci., No. 14. 1940.

It is of interest to note Chambers' comment (1940) that during autolysis there is a vigorous shift to lower pH values in the cytoplasm, while that of the nucleus tends to persist. We must conclude that there are two electrical potentials and consequently two pH values of importance in the cell. The first is the electrical potential across the surface boundary of the cytoplasm between the cytoplasm and the media. The second is the electrical potential across the nuclear membrane between the nucleus and the cytoplasm. It is apparent, therefore, that the effects of morphogens accumulating in the media are primarily expended upon the surface boundary of the cell and exert their primary influence on this phase of the cell potential. The secondary effect, as a consequence of the changes in the cytoplasm itself, is expended upon the nuclear wall. Therefore, it is seen that the nuclear pH would tend to persist after the cytoplasmic pH begins to change. This is the experimentally observed fact.

Review; Adverse Influence of Protomorphogen Accumulations.
—We may now expand our working hypothesis concerning the changes induced in the cell as a result of accumulating morphogens in the surrounding media. The morphogens, a product of cell activity, accumulate in the surrounding media and protoplasm. They progressively destroy the surface boundary of the cytoplasm resulting in a lowered potential difference between the cell and its surrounding media. As a consequence of this action the pH of the cytoplasm, and secondarily, the pH of the nucleus, is lowered.

The lowered pH in the cytoplasm and nucleus interferes with the equilibrium between the constructive and destructive phases of enzyme activity on the cell's proteins. The dynamic metabolism of cell proteins becomes therefore, less competent to maintain repair and energy producing reactions. Gradual ageing and eventual dissolution of the cell results.

It is obvious that there is a group of very complicated energy mechanisms in the cell that is suppressed by this chain of events. From a review of the literature it may be possible to speculate on some of these energy chains that seem to be more particularly linked with morphogen metabolism. Not one of the least important of these is the phosphatase-phospho-creatine system.

Phosphatase - Phosphagen System. — Huggins (1943) has commented that the phosphatases are most important in the energy production of the cell. He mentions that phosphatases are abundant in rapidly dividing tissue. Willmer (1942) has observed that the chromosomes of cells undergoing mitosis *in vitro* give a strongly positive reaction for phosphatase.

The presence of the enzyme phosphatase in the chromatin material is of singular importance to our study. The suggestion that it is concerned with the energy mechanism and associated strongly with rapidly dividing cells prods us to search for a possible substrate which is concerned with energy metabolism.

Such a substrate is phosphagen (also called phosphocreatine) which was isolated by Fiske and Subbarow (1929). It was shown that during muscle contraction this phosphocreatine was hydrolyzed, losing its phosphate group. This reaction is a part of the complicated chemistry of muscle contraction involving the conversion of glycogen.

Myers and Mangun (1936) reported their observations on the muscle of guinea pig and dog in which they noticed that potassium holds a constant ratio to the creatine or phosphocreatine. They noticed that there was excess K over the formula requirement but its ratio remained the same in all species and in all tissues tested. This work was an outgrowth of Myer's comment in 1922 that glycogen, creatine, phosphoric acid and potassium are closely associated in active muscle.

Myers and Mangun (1940) conclude that, because of the phosphocreatine power of neutralizing organic acids formed during muscle contraction, phosphagen exists in muscle in the form of the dipotassium salt of creatine-phosphoric acid.

In his studies of cell injury and inflammation, Menkin (1940) (1943) has shown that the injured cell releases sugar and potassium ions concomitant with glycolysis. Danowski (1941) has demonstrated that potassium is lost from cells at the point where glycolysis ceases and the phosphate esters begin to break down.

These observations strongly suggest that potassium creatine hexosephosphate is present in all cells and the energy reaction of the cell is concerned, among other things, with the synthesis and destruction of this substance by the phosphatases present in the chromatin. It is very likely that these reactions are in a dynamic state of equilibrium in the nucleus of the cell.

Bradfield (1946) has recently commented on the activity of alkaline phosphatase and the production of fibrous proteins. He recognizes that this synthesis involves more than reversal of proteolytic activity and includes some participation of nucleic acids, but the exact nature of phosphatase influence can only be determined by further experimentation. This activity of phosphatase is significant since the morphogens come under the classification of fibrous proteins. (See discussion in Chapter 5.)

It is noted that the key to these reactions, the enzyme phosphatase, is abundantly found in the chromosomes, especially of rapidly dividing cells. These phosphagen reactions are quite likely linked with the self-duplication of the chromosomes, but further experimental investigations are needed to clarify the interrelationships. Since the chromosomes are probably the primary source of morphogens, it is fitting that we speculate further on the reactions concerned with their self-duplication in an effort to establish a working hypothesis explaining the synthesis and disposition of new morphogen molecules.

Radioactivity.—It is difficult to overestimate the importance of potassium in the biochemistry of the cell. The organization of autosynthetic cells (Crile, Telkes and Rowland, 1931, 1932) was considerably delayed when the potassium content of the electrolyte was reduced. This has led us to suggest that this ion is an indispensible constituent of the morphogen molecule. Turck (1933) found potassium a consistent constituent of incinerated tissue ash. We have reviewed evidence that it is linked with the phosphagen energy cycle.

The radioactivity of the potassium molecule has been commented upon as an important participant in biological activities.[1] Reports have even been published that potassium may be substituted for in various cells by other radioactive elements such as rubidium, or even by rays themselves transmitted to the cells. They seem to awaken the life of the cell catalytically. In this respect it is interesting to note that Turck found a surprising occurrence of rubidium in incinerated tissue ash.

1 *Experimental Pharmacology as a Basis for Therapeutics,* H. H. Meyer and R. Gottlieb. 2nd Edition, J. B. Lippincott Co., Philadelphia. 1926.

A possible avenue of future research would be the investigation of the relationship of these radioactive elements to the Brownian movement commonly observed in protoplasm. This radioactive energy might possibly be the source of the electrical potential of colloidal particles which prevents their flocculation. Sedimentation rates, believed to be due to variation in the colloidal state[1] vary with the dynamics of protein metabolism,[2] and indicate the loss of integrity of colloid systems in disease and senescence. The whole field of investigation of the interrelationships between radioactive elements and colloidal systems as indicated by variations in Brownian movement and sedimentation rate promises some clarification of the problems of potassium metabolism. We mention this phenomena in passing since it seems to be one possible nuclear energy cycle that has received little attention.

Nucleoproteins and Chromosome Metabolism.—Of greater interest, in our present study, are those reactions which are more intimately connected with the metabolism of the chromatin material, of which the determinant morphogens are a part.

In an excellent review of the problems of chromosomes and nucleoproteins, Mirsky (1943) has mentioned the importance of nucleoproteins in the reproduction of the chromosomes. He comments that the discovery that plant viruses consist of self-duplicating nucleoproteins suggests the possibility that chromosome nucleoproteins are also self-duplicating.

We have emphasized that the nucleus is the seat of the vital activities of the cell. We reported the experiments of Loeb in which, by altering the ionic concentration of the surrounding media, the nucleus of a cell divided into as many as forty parts without cell division. Various experimenters have successfully separated living cell fragments from their nuclei, and these denucleated fragments were able to engage in many of the activities of living tissue, but they could not reproduce and eventually died. It is interesting at this point to recall the comment of Wilson (1900) who stated forty-seven years ago "A fragment of a cell deprived of its nucleus may live for a

1 Ropes, M. W., Rossmeisl, E., and Bauer, W.: "The Relationship Between the Erythrocyte Sedimentation Rate and the Plasma Proteins." *J. Clin. Invest.,* 18:791-799, 1939.

2 Lichtman, S. S.: "The Influence of Intravenous Glucose Injections on Abnormal Erythrocyte Sedimentation Speed in Relation to Activity of Infection." *Ann. Internal Med.,* 13:1297-1305, 1940.

Kobryner, A., and Glajchgewicht, Z. U.: *Sang,* 13:979-985, 1939.

considerable time and manifest the power of coordinated movement without perceptible impairment. Such a mass of protoplasm is, however, devoid of the power of assimilation, growth and repair, and sooner or later dies. In other words, those functions that involve destructive metabolism may continue for a time in the absence of the nucleus; those that involve constructive metabolism cease with its removal."

Muller (1922) predicted that bacteriophage and the so-called bacterial virus were substances intimately related to genes and was the first to recognize this analogy which Mirsky states is constantly before the minds of investigators.

For the purpose of our hypothesis we assume that the nucleoprotein moiety of the chromosome is a self-reproducing virus, as indeed much of the evidence suggests. We might go further and state that since nucleoprotein is the only substance which has been demonstrated to be self-duplicating, the synthesis of new chromatin nucleoprotein is the first step in the chain of events leading to cell division.

Although the experimental work is far from complete, enough is known about the biochemistry of the chromatin nucleoprotein to enable us to off er a possible solution to some puzzling biological problems and to speculate on the position of the morphogen molecule in this system.

Wrinch (1935) considers the genetic identity of the chromosome to he a function of its characteristic protein moiety. Mazia and Jaeger (1939) have demonstrated that the structural integrity of the chromosome does not depend upon nucleic acid, but rather upon its protein moiety. Enzymatic destruction of the nucleic acid with nuclease left a clearly visible and stainable chromosome network. Treatment with trypsin, however, results in the complete disintegration of the chromosome structure, indicating that it is maintained by the protein component. Mazia (1941) considers this protein component to he a histone complex, since pepsin digestion did not destroy the chromosome form, hut only decreased its volume. Histones are susceptible to trypsin digestion but are simply degenerated to "histopeptones" by pepsin. Mazia considers that pepsin attacks a "matrix" protein causing shrinking, but not a destruction of chromosome integrity.

Stedman and Stedman[1] have presented conclusions on a characteristic protein constituent of the chromosomes which they term "chromosomin." Their arguments have been disputed by Callan. They regard "chromosomin" as the chemical basis of inheritance and supply supportive evidence in the form of enzymatic analyses for their contention that "chromosomin" rather than nucleic acid or histone is the characteristic chromosome substance responsible for the determinant influences of the chromosome.

Mirsky and Pollister (1946) have extracted a substance from the nucleus termed "chromosin" which consists of desoxyribonucleic acid, histone and a non-histone protein. It apparently carries the genetically active chromosome constituents since it is active in transforming pneumococcus types. These investigators caution that it should not be considered a definite chemical compound and it is not yet established experimentally whether protein is a necessary constituent of the transforming agent.

Lately the stand that protein rather than nucleic acid is responsible for the genetic basis of the chromosome has been threatened as a result of the discovery[2] that a highly polymerized protein free desoxyribonucleic acid can cause inheritable transformations in the characteristics of certain types of pneumococci. Such nucleic acid is antigenically inert when brought into contact with immune sera. It seems, however, that nucleic acid is concerned with metabolic transformations of the chromatin (see later discussion in this chapter) and perhaps this phenomena will resolve itself into an expression of nucleic acid influence over the polymerization of chromosome morphogens. (This and similar experiments are discussed in Chapter 4.)

Darlington and La Cour (1945) have studied the influence of nucleic acid over chromosome structure. They conclude that the stability of a chromosome is dependent upon its nucleic acid charge, preventing the depolymerization induced by X-rays or promoting a prompt reunion of the broken ends of the chromosome.

1 Callan, H. G., Stedman, E., and Stedman, Mrs. E.: "Distribution of Nucleic Acid in the Cell." *Nature*, 152:503-504, 1943. Stedman, E.: "The Chemistry of Cell Nuclei." *Biochem. J.*, 39:lviii-lvix, 1945.

2 Avery, O. T., McLeod, C. M., and McCarty, M.: "Studies on the Chemical Nature of the Substance Inducing Transformation of Pneumococcal Types." *J. Exp. Med.*, 79:137-158, 1944.

Important advances in the knowledge of nucleoprotein chemistry are being constantly reported. Many excellent reviews of the subject exist, probably the most pertinent to our studies being that of Mirsky (1943). Much general information on the nature of the bonds between nucleic acid and various protein components is adequately reviewed therein.

The identification, of the protein moiety of nucleoprotein with the morphogen groups is strongly suggested by the morphogenic influence of the former, and the growth influences of both *in vitro*.

We shall shortly review evidence establishing nucleoprotein as a growth stimulant *in vitro*. It is of interest to note, however, that Burrows (1925) reports that he was easily able to extract archusia (morphogens) with salt solution. Mirsky and Pollister (1942) have recently reported a method for the extraction of nucleoprotein that consists of treatment with saline solution. After such treatment, microscopic examination shows the nuclei devoid of chromatin. This evidence seems to link morphogens with chromatin nucleoprotein.

Also significant in establishing the morphogen groups with the protein moiety of chromosome nucleoprotein is the observation that the structure of the chromosome is destroyed by trypsin but not by nuclease. Simms and Stillman (1937) mention that the inhibitory product of cell metabolism which accumulates in the media is destroyed by trypsin. This inhibitory substance consists primarily of morphogens. In spite of their comparative stability, the morphogen groups seem especially susceptible to trypsin digestion.

We postulate therefore, that the essential nucleoprotein of the chromosome contains a protein moiety that carries the morphogen groups and is concerned with its determinant activity, and a nucleic acid moiety which, as we shall now discuss, is responsible for its self-duplicating characteristic.

Wrinch (1936) has suggested a similar hypothesis of the molecular structure of the chromosomes. Recognizing the dual nature of the chromosome, basic protein and nucleic acid, she comments that the protein is the disruptive element while the nucleic acid is the synthesizing or constructive element of the pair.

Lately it has been demonstrated that actively dividing cells have a high concentration of ribonucleic acid while resting cells contain a greater amount of desoxyribonucleic acids.[1] This suggests that the chemical reactions of the nucleic acid components of chromatin may be concerned with mitosis.

Thanks to Feulgen, who established a valuable reaction to desoxyribonucleic acid, it has been demonstrated that all nuclei contain this substance. This reaction has been observed in all cells investigated, but no evidence of desoxyribonucleic acid has been found in the cytoplasm (Mirsky, 1943).

On the other hand, ribonucleic is considered to be the nucleic acid of the cytoplasm. This suggestion was first made by Feulgen and Rossenbeck (1924). Later Feulgen, Behrens and Mahdihassan (1937) separated desoxyribonucleic acid from the nuclei of rye embryo cells and Behrens isolated ribonucleic acid from the remaining cytoplasm (Behrens, 1938).

More recently Davidson and Waymouth (1944) have presented more conclusive experimental evidence that desoxyribonucleic acid is located in the nucleus and ribonucleic acid mainly in the cytoplasm. They further report that the concentration is particularly high in rapidly growing tissues.

Because of the intense concentration of ribonucleic acid at the nuclear membrane of the sea-urchin egg, Caspersson and Schultz (1940) concluded that it is synthesized at this point. These investigators (Caspersson and Schultz, 1939) also noted the intense nucleic acid absorption spectrum in rapidly dividing cells as compared with resting cells. Caspersson and Thorell (1941) found high concentrations of nucleic acid in the cytoplasm of the most actively dividing cells in the chick embryo. Caspersson speculated upon the link between nucleic acid concentrations in the cytoplasm and the synthesis of new protein. With his colleagues (Caspersson, Landstrom-Hyden and Aquilonius, 1941) he was able to demonstrate that the cytoplasm of glands concerned with the synthesis of protein is high in ribonucleic acid; specifically they report the increase in ribonucleic in the cytoplasm of pancreas stimulated into intense activity. The increase was located particularly in the region of the nuclear membrane. Thorell (1944) comments that the content of ribonucleic acid in the cytoplasm is an index of the intensity of

---

1 Jorpes, E.: "Analysis of Pancreatic Nucleic Acids." *Acta Med. Scand.,* 68:503-573, 1928. Caspersson, T., and Schultz, J.: "Pentose Nucleotides in the Cytoplasm of Growing Tissues." *Nature,* 143:602-603, 1939.

metabolism and the protein synthesizing capacity of the system. Some cytoplasmic nucleic acid is found in basophilic granules. Van Herwerden (1913) caused cytoplasmic granules to disappear after treatment of the cell with nuclease. Brachet (1940) demonstrated, by use of a nuclease specific to ribonucleic acid, that the basophilic granules in cytoplasm contain this substance. Menke (1938) has concluded that the chloroplasts contain ribonucleic acid. Claude (1941) has isolated cytoplasmic granules by centrifugation which contain nucleoprotein and lipid.

Fischer (1939) has corroborated these observations by demonstrating that the non-specific growth promoting properties of beef embryo nucleoproteins are primarily due to the ribonucleic acid fraction rather than the desoxyribose fraction.

In a recent review of cytoplasmic nucleoproteins Davidson (1945) remarks that the phospholipin-ribonucleoprotein particles of the cytoplasm lend themselves admirably as organs of protein synthesis. Their similarity with genes and viruses is touched upon. An extremely significant statement from the standpoint of the morphogen hypothesis is his comment: "It appears probable, therefore, that the phospholipin-ribonucleoprotein macromolecules constitute some of the fundamental units out of which living matter is built."

Review of Nucleoprotein Metabolism.—To recapitulate, we are faced with the following experimentally demonstrated phenomena: (1) desoxyribonucleic acid is exclusively found in the nucleus, probably in the chromatin; (2) ribonucleic acid is found both at the nuclear wall where it is synthesized, and in certain basophilic cytoplasmic granules; (3) it is the ribonucleic acid fraction of the nucleoprotein which is a growth promoter when added to tissue *in vitro,* and (4) an increase in the cytoplasmic content of ribonucleic acid is associated with protein synthesis and cell division.

We venture to speculate as follows: (1) the ribonucleic acid synthesized at the nuclear wall is convened into desoxyribonucleic acid in the process of chromatin synthesis, and becomes a pan of chromatin nucleoprotein; (2) as a consequence of chromatin metabolism it is again broken down into the ribose form in the nucleoprotein complement of the chromatin granules secreted into the cytoplasm, whose purpose is to serve as determinants for cell structure; (3) as the morphogen moiety of the chromatin granule is utilized as a determinant, the ribonucleic acid becomes available again for more chromatin synthesis at the nuclear wall, (4) an alternative suggestion would embrace the hypothesis that chromatin metabolism preceding cell division results in the nuclease conversion of desoxyribonucleic acid into ribonucleic acid and its transfer into the cytoplasm. As a consequence of this reaction the morphogen moiety of the chromatin finds its way into the cytoplasm where its determinant activity is expressed. The inclusion of ribonucleic acid in the media of a cell would facilitate its appearance in the cytoplasm and thus act as a growth promoting substance. This hypothesis would integrate the observation of greatly increased amounts of ribonucleic acid appearing in the cytoplasm at the time of fission with the probability that chromatin synthesis is the key activity associated with mitosis.

The nucleic acid cycle described above is indicated in part by the observations of Davidson and Waymouth (1944) that the ratio of desoxyribonucleic acid (nucleus) to ribonucleic acid (cytoplasm) varied widely in different organs but was *possibly higher in the adult than the embryo.* Evidently the constructive processes are not proceeding as rapidly in the adult (nuclear conversion of ribo- into desoxyribonucleic acid) and therefore ribonucleic acid is now at a higher level in the cytoplasm.

The evidence suggests that the chemistry of nucleic acid synthesis is connected with the phosphatase destruction of phosphagen. The knowledge of the high concentration of phosphatase in chromatin, in addition to the energy coupling reaction of phosphagen, should stimulate investigation of this possibility. The release of hexose, as a result of phosphatase influence on phosphagen, could possibly supply the carbohydrate moiety for ribonucleic acid synthesis. That morphogen metabolism is intimately linked with glycolysis is indicated by reports that the glycogen content of embryos decreases 35 per cent during invagination, the period of most intense morphogen action (Heatley, 1935).

Inasmuch as we postulate that the nucleic acid constituent is concerned with the energy and synthesizing reactions of the chromatin (the protein being the factor concerned with structure) the interesting observations of Mazia and Jaeger (1939) that nuclease releases nucleic acid into the cytoplasm while leaving the protein framework intact is extremely significant. It is possible that physiological nuclease activity is responsible for the appearance of ribonucleic acid in the cytoplasm of rapidly dividing cells. This nuclease activity may be a necessary precursor of the extrusion of chromatin granules into the cytoplasm during cell division, instead of a concomitant phenomenon.

The conception of nucleic acid metabolism offered above has been suggested in part by others. Caspersson (1938) has observed that the quantity of nucleic acid in a dividing spermatozoa increases in prophase and decreases in telophase. He suggests that nucleic acid is therefore concerned in some manner with the synthesis of the genes. He reports elsewhere (1938) that desoxyribonucleic acid is found exclusively in the chromosomes and present there in large amounts before cell division. This strongly indicates the correlation between desoxyribonucleic acid metabolism and synthesis of new chromosome material.

Brachet (1933) has suggested that desoxyribonucleic acid is formed at the expense of ribonucleic acid. He observed the increase in desoxy- and decrease in ribonucleic acids after fertilization of the sea-urchin egg. His contentions have been supported by ultra-violet absorption measurements made on the sea-urchin egg by Caspersson and Schultz (1940). The suggestion of the determinant effect of cytoplasmic granules receives support from the observations of Caspersson and Brandt (1941). Ultra-violet absorption measurements of volutin granules in yeast show that when the cell begins to grow the resting granules swell, multiply and disappear with a concomitant increase in the absorption of ultra-violet light in the hyaloplasm. This indicates a loss of nucleic acid from the granule into the hyaloplasm. Further substantiation of this suggestion is contained in the comments of Jennings and of Diller, reviewed earlier, concerning the "shedding" of macromolecular fragments into the cytoplasm from the nucleus.

Further knowledge on this subject is possible as a result of the conceptions advanced by Schultz, Caspersson and Aquilonius (1940), concerned with heterochromatin. Heterochromatin is the framework of the chromosome which remains visible during the period of the mitotic cycle when the chromosome loses its nucleic acids and becomes lost to sight. It is not considered to contain any genes.

The functions of the nucleolus seem to be connected with the heterochromatin. It is thought to exert an influence over the production of nucleic acids and to be inert only in the respect that it does not carry genes to or influence morphology. More work on this conception has been added by Darlington (1942).

It is quite possible that the heterochromatin in the nucleolus is concerned with the synthesis of nucleic acid and morphogen groups for the construction of a new genetically potent chromosome.

We have discussed the synthesis of ribose nucleic acid at the nuclear wall, and its possible conversion into the desoxyribonucleic acid of the chromatin. The biochemistry of the protein component of chromatin nucleoproteins is less well charted.

Fischer (1941) in his review of the growth promoting substance in embryo juice states that neither the nucleic acid nor the protein moiety of nucleoproteins is an active growth promoter by itself. When combined, however, nucleoprotein in embryo juice is a nonspecific growth promoting factor. We have mentioned previously that nucleoprotein is a powerful growth promoter and in general is non-specific in its activity. A recent review discusses the growth stimulating effects of chromatin *in vitro*.[1]

Werner (1944) has performed similar experiments. Saline extracts of heterologous tissues were tested for growth promoting effects *in vitro*. (The reader will recall the fact that nucleoproteins are extracted with saline solution.[2]) No growth stimulation was observed unless further extracted with acetone or petroleum ether (morphogens are removed by these solvents). Their presence contaminated the nucleoproteins and prevented their use as a growth stimulant in heterologous species.

The reader may question our suggestion that the heat-stable protein component of chromatin nucleoprotein contains morphogen groups because its growth promoting effects are non-specific. It would seem that the morphogens would limit the activity of the compound to homologous species. In the first place nucleoprotien is

1 "Mitotic Stimulation of Wound Healing." *J. A. M. A.*, 128:290-291, 1945.

2 Mirsky, A. E., and Pollister, A. W.: "Nucleoproteins of Cell Nuclei." *Proc. Natl. Acad. Sci. U. S. A.*, 28:344-352, 1942.
Burrows, M. T.: "Studies to Determine the Biological Significance of the Vitamins." *Proc. Soc. Exp. Biol. & Med.*, 22:241-245, 1925.

a most complex molecule as shown by the fact that It must be hydrolyzed to respond with Feulgen's reaction.[1] And, in the second place, it has been shown that ultracentrifuged extracts of the macromolecular fraction of chick embryo, which is chromatin nucleoprotein, have greater growth promoting properties after heating.[2] It is possible that the relatively less heat-stable morphogen groups are broken down by this process into simpler forms that do not exert such strict specificity, which would restrict the biological effects of nucleoprotein.

The chemical structure of chromatin is so complex that we must be guided as much by biological observations and consequent deductions as by the insignificant knowledge we may have of the chemistry involved. It is probable that in the above cases the inclusion of the specific morphogen in the nucleoprotein molecule does not prevent it from exerting its growth promoting effects on a variety of species. The reader must not confuse the direct growth promoting and inhibiting effects of morphogens with the growth promoting effects inherent in nucleoproteins. When the two functions are combined, as they would be in chromatin nucleoprotein, many possibilities can be envisioned. There may be a competition between the growth inhibiting effects of the morphogens and stimulating effects of the nucleoproteins if the test is made on a related species. Many experiments reviewed earlier have shown that the inclusion of morphogens in an embryo extract may even result in an inhibition of growth. This effect would depend upon the complexity, concentration and relative specificity of the morphogens involved, and thus its appearance in experimental observations will vary widely with the various techniques employed.

To summarize our contentions: (1) the nucleoprotein of chromatin is the key to cell vitality and cell division, it being the only self-duplicating molecule in living organisms; (2) this nucleoprotein consists of nucleic acids and protein; (3) the nucleic acids are concerned with the synthesis of new molecules and with the energy mechanisms of the nucleus; (4) the protein moiety contains the structural basis of the chromatin and the morphogens;

1 Van Camp, G.: "Role of an Enzyme, Endosomase, in Cell Division." *Bull. Soc. Chem. Biol.,* 17:169-179, 1935.

2 Tennant, R., Liebow, A. A., and Stern, K. G.: "Effect of Macromolecular Material from Chick Embryos on Growth Rate of Mouse Heart Fibroblast Cultures." *Proc. Soc. Exp. Biol. & Med.,* 46:18-21, 1941.

(5) new nucleoprotein is synthesized in the nuclei as the initial stage of cell reproduction.

Morphogen Cycles

We have briefly postulated on the biochemistry of the nucleic acid component, but have little to off er in the form of an hypothesis for the protein part. We must be concerned, however, with the mineral substrate that must be present in order for the mineral "skeleton" of the morphogens to be created. Duplication of chromatin nucleoprotein infers synthesis of new morphogens, and at this point we revert to our hypothesis of morphogen metabolism. It is obvious that minerals for morphogen synthesis must come from the media, which is the source of foodstuffs for the cell, but in order to come in contact with the nuclear wall they must pass through the cytoplasm. The physical chemistry of membrane potential prohibits any suggestion that inorganic ions could pass through the cytoplasm without disastrous effects on the integrity of the cytoplasmic surface boundary, on the nuclear membrane and on the colloidal nature of the cytoplasm itself.

The conception has been advanced that the colloidal state is dependent upon the adsorption of inorganic ions on protein molecules. When the electrolytes are removed from a colloid it is denatured and polymerized even to the extent of coagulation.[1] We have advanced the hypothesis that no biological protein can exist without its structure being organized by a "skeleton" of mineral morphogens.

The structural proteins of the cytoplasm are necessarily a great deal more simple than those in the nucleus. In the absence of any suggestion of a better source, we anticipate that the nucleoprotein morphogenic material is constructed at the nuclear wall from the simpler specific proteins present in the cytoplasm. These proteins contain morphogens of a simpler nature, already partially constructed, for the further synthesis into cytomorphogens in the chromatin. This, then, is the source of simpler morphogen molecules containing mineral elements for the synthesis into chromatin in the nucleus.

We have mentioned the dynamic state of body proteins and its universal occurrence. The proteins of the cytoplasm are no exception

---

1 *A Handbook of Colloid-Chemistry,* Ostwald, W. P. Blakiston's Son & Co., Philadelphia. 1915.

and their constant replacement is likely the source of "unbound" protomorphogens, protein derivatives and mineral elements for the primary reactions of chromatin synthesis at the nuclear membrane. We doubt that these protomorphogens are lost into the media without first undergoing metabolic reactions in the nucleus. By utilizing them for chromatin synthesis, the nucleus reorganizes their structure and corrects slight changes that may have occurred so that when they are again present in the media, they may organize cytoplasmic proteins without introducing changes in structure incompatible with the cell's integrity. This problem is discussed in detail in our review of the organizing influence of the morphogens.

Our problem then, is simply transferred to the question of where the cytoplasmic proteins obtain the mineral elements for the construction of the protomorphogen determinants necessary for their synthesis. We believe the answer to this problem is that elementary protomorphogens must be present in the media in order for cytoplasmic proteins to be formed. *The cytoplasmic proteins can only be formed where all the nutrient substrate is present, namely, at the surface boundary of the cell.* New mineral elements and protein components from the media must also be utilized for cytoplasmic synthesis at the cell wall.

Later in our discussion (Chapter 4) we shall see that in certain types of cells which engulf food particles from the media, the synthesis of new proteins may occur in the food vacuole inside the cytoplasm, the morphogens being supplied by the mitochondria associated with such vacuoles. This represents a special evolutionary case, however, for only a limited class of cells ingest food in this manner.

Morphogens and Nuclear Synthesis.—The influence of media protomorphogens over chromatin synthesis is strikingly illustrated in the experiments reported by Iwanitzskaia (1939-40). He demonstrated that when chicken heart fibroblasts were cultivated in heterologous (dog and cat) plasma, the chromosomes were in small fragments, distorted and weakly colored. From these observations we may deduce that the heterologous protomorphogens in the media were not universal enough to act as a competent substrate for the synthesis of cytoplasmic proteins which supply morphogen factors for chromatin synthesis. Hence, the synthesis of genetic material in the chromatin was hindered. That there was a deficiency of genetic material and not simply of structural molecules is indicated by the observation that the cells with distorted chromosomes exhibited nine times as many abnormal mitoses as normal cells cultivated in homologous plasma.

At this point we should digress for a moment to mention that the more complex differentiation that occurs in the cytoplasm may. be determined by the cytomorphogens which exert their effect directly from the nucleus. This hypothesis will be expanded later in our discussion. However, for the present, we are only concerned with those elementary proteins constantly being formed in the cytoplasm and the elementary protomorphogens necessary for their formation. There is a distinction between media protomorphogens which organize cytoplasmic proteins and chromatin protomorphogens which control cytoplasmic differentiation that should be emphasized.

We suggest that these basic elementary proteins in the cytoplasm are such that their specificity is not very well defined. The morphogens, to assist in their construction, need not be so complex as to exhibit distinct species specific properties.

The reader will now no doubt suspect that the growth stimulating properties that we have attributed to morphogens in the surrounding media are due to their availability for protein synthesis at the surface boundary of the cytoplasm. They are, indeed, a distinct necessity and probably no cytoplasmic protein synthesis can occur if they are not present in the media.

It will be well to briefly recall here the experimental evidence which indicates the necessary presence of these factors in the media for cytoplasmic protein synthesis.

We have reviewed the experimental evidence which suggests the growth promoting properties of cell metabolic products•when present in dilute quantities in the surrounding media. Several workers have succeeded in demonstrating this phenomenon — Montrose T. Burrows, Fenton B. Turck and T. Brailsford Robertson being those whose work we have most extensively discussed. We have also reported experimental evidence which indicates that transplanted cells will not grow if the volume of the media is too large, thus excessively diluting the protomorphogen.

We are now ready to present the hypothesis that these phenomena are a consequence of the activity of protomorphogen fragments in the surrounding media, and their necessity for the construction of new cytoplasmic protein at the cell wall. We have mentioned that mitosis is initiated by chromatin synthesis and not cytoplasmic activity, but the synthesis of some cytoplasmic proteins must precede chromatin synthesis since the mineral components of the protomorphogens are supplied to the nucleus by the cytoplasmic proteins.

Excretion of Morphogens by the Cell.—We are now presented with a seeming paradox. For one thing we have presented the idea that morphogens are synthesized only in the chromatin. Then we offer the paradoxical idea that chromatin synthesis cannot succeed if the cytoplasmic proteins do not supply them with morphogens— but cytoplasmic proteins cannot be synthesized without the presence of morphogen particles in the media.

The only answer to this paradox is that the cytomorphogens of the chromatin must disintegrate and "leak" into the media. Mast and Pace (1938) have shown that the elimination of the allelocatalyst into the media is not necessarily dependent upon mitosis. In order to test this suggestion it will be necessary to supply experimental evidence of the occurrence of unbound morphogens in the cytoplasm which find their way into the media. It will also be necessary to suggest an explanation for this phenomenon.

Kidder and Claff (1938) have observed in ciliates the discarding of a portion of macronuclear material during division. Since this occurs most extensively in cells shortly before conjugation they conclude that this is a phenomenon of elimination of waste substances. The mechanism of this discarding of nuclear material varied in different species observed.[1]

We have mentioned that the morphogens are extracted from tissues with fat solvents. We have tentatively brought forth the suggestion that free morphogens in the cytoplasm are prevented from exerting lethal effects by a lecithin or fatty envelope. There is more evidence that these waste morphogens accumulate in the

1 Kidder, G. W., and Diller, W. F.: "Observations on the Binary Fission of Four Species of Common Free-Living Ciliates, With Special Reference to the Macronuclear Chromatin." *Biol. Bull.* 67:201-219, 1934. Calkins, G. N.: "Factors Controlling Longevity in Protozoan Protoplasm." *Biol. Bull.* 67:.410-431, 1934.

cytoplasm, forming visible fatty vacuoles as they become more concentrated. The formation of fat vacuoles in the cytoplasm is a consistent occurrence in ageing cells.

Probably the most significant fact in this respect is the discovery that the growth inhibitor in serum can be removed to a considerable extent by extraction with ether-alcohol.[1] Werner (1944) (1945) has reported that the growth inhibitor in both normal adult and tumor tissue must be removed with fat solvents before growth stimulating factors can be demonstrated. Robertson was able to extract morphogens from yeast with acetone. Fischer (1923) has shown that as cells grow old in a restricted media the protoplasm gradually fills up with vacuoles and fat granules, after which the cell disintegrates. MacNider (1942) reviews that as age advances there is an accumulation of stainable lipoid in liver and kidney tissues, and that this is correlated with the ability of ether and chloroform to exert local toxic effects on these organs. We interpret this to indicate that the ether and chloroform are able to release the stainable chromatin material (morphogens) from the fatty protection and thus allow it to exert toxic effects.

The reader will recall that Crile and his co-workers (1931) found that the lipoid extract contained the organizing influence for the production of autosynthetic cells. We are emphasizing that the evidence links cytoplasmic morphogens with lipoid molecules. It seems probable that the protomorphogens discarded as nuclear metabolic products appear linked with fat in the cytoplasm.

The shedding of nuclear material into the cytoplasm is also indicated by the reorganization of the chromatin which has been observed in various infusoria. There are various means by which cells have been observed to accomplish this, one of which consists of a deeply stainable band passing along a half-moon-like mass of chromatin from one end to the other, becoming more deeply stained as it passes, and finally being shed into the cytoplasm where it disintegrates and, we believe, accumulates under the protection of a fatty envelope.[2]

1 Baker, L. E., and Carrel, A.: "Lipoids as the Growth-Inhibiting Factor in Serum." *J. Exp. Med.*, 42:143-154, 1925.

2 Problems of Ageing, Cowdry, E. V., Williams & Wilkins Co., Baltimore, d. 1939. Summers, F. M.: "The Division and Reorganization of the Macronuclei of Aspidisca lynceus Muller, Diophrys appendiculata Stein, and Stylonychia pustulata Ehrbg." *Arch Protistenk.*, 85:173-208, 1935.

Horning and Miller (1930) discuss the phenomenon of chromidiosis[1] and its significance in the cells of higher metazoa. Although it is associated with conjugation and reproductive phenomena in protozoa, it is linked with different processes in metazoan cells.

They present evidence that in metazoan cells chromidia pass through the nuclear membrane by exosmosis and this activity varies directly with the metabolism and growth rate of the cells, particularly in neoplasms. They comment that the chromidia take part in the liberation of growth promoting substances found in the cultures of malignant tissues.

This process (chromidiosis) is thought to upset the tension forces normally existing between the cytosome and nuclear plasm, thus affecting the stability of the nuclear membrane. This concept may be integrated with Wrinch's hypothesis (1936) that the variable permeability of the nuclear membrane (consisting of proteins and lipoids) is concerned with mitosis.

Mazia and Jaeger (1939) have demonstrated that nuclease releases nucleic acid from the chromatin and it is then found in the cytoplasm. The protein constituent of the chromosome, however, remain intact; this evidence may or may not be significant in the study of the release of morphogens from the chromatin into the cytoplasm. Upon cursory examination, it would seem that the fact that the protein constituent remains intact would preclude any such assumption, since the morphogens are associated with the protein in the chromosome. Nevertheless all such data must be examined and integrated into any complete hypothesis; it is apparent from these reports that the nucleoprotein constituent may react with nuclease to release ribonucleic acid into the cytoplasm as a part of the energy mechanism preceding the elimination of morphogen during the mitotic cycle.

Tittler (1935) has described the reorganization band which passes through each macronucleus of a protozoan, Urostyla grandis, at each division. He considers this reorganization a "purification" process resulting in a phase reversal of chromatin and nuclear colloids. We postulate that this process may be universal in one form or another and that it represents the excretion of chromatin materials (morphogens) into the cytoplasm.

1 Chromidiosis: An outpouring of nuclear substance and chromatin into the cell protoplasm.

This presents some experimental evidence concerning the elimination of morphogens as waste material into the cytoplasm. Although it accumulates there with age under the protection of a lecithin or fatty envelope, the fact that it also accumulates in the media in a disintegrated form strongly suggests that it further leaks through the cytoplasmic surface boundary.

The question may be logically asked as to what purpose does the excretion of degenerated morphogens of chromatin serve? How is it worn out? Our suggestion for future research revolves around the fact that there is a dynamic metabolic equilibrium in the chromatin that is intimately connected, in some manner, with the essential energy reactions of the cell, especially the dynamic replacement of protein molecules. We have discussed this in more detail previously. This dynamic metabolism produces waste chromatin which disintegrates into unbound protomorphogens that are excreted into the media in the manner we have reviewed.

We have comprehensively reviewed the experimental evidence which indicates that the accumulation of morphogens in the media inhibits growth. In addition, we have presented a tentative hypothesis concerning the manner in which they cause the cell to age and eventually undergo lysis.

Adverse Influences of Excreted Morphogens.—We have also reported the experiments of workers who have demonstrated the accumulation of morphogens in the cytoplasm as age progresses. Here again we are presented with the paradox that there is an accumulation of morphogens in the media with age, and also an accumulation in the cytoplasm, yet, both phenomena are demonstrated by experimental evidence that we dare not ignore. Coupled with this is the gradual disintegration of the cytoplasmic wall and consequent probable increase in permeability. This eliminates a possible explanation that the cytoplasmic morphogens cannot get into the media due to decreased permeability of the cell membrane. It seems probable that, as the media concentration of morphogens increases, it influences the cell in some manner by hindering further secretion of "waste" morphogens from the cytoplasm. This can be the only mode of lethal influence of high media concentrations because it has been demonstrated that the effects of these high concentrations can only be observed if there is also a high concentration of "waste" morphogens in the cytoplasm. Robertson's allelocatalyst theory is based upon this reciprocal relationship.

It is apparent that the primary lethal effect is due to the cytoplasmic accumulation. The major influence of the media concentration is probably secondary in that it results in this cytoplasmic accumulation of protomorphogens. This is borne out by Robertson's observations that a high protomorphogen concentration in the media is only toxic to transplants in that it shortens the time before the lethal cytoplasmic concentration will occur. Thus, the time before the inhibiting balance will appear is shortened and the maximum attainable population will be thereby reduced. This explains the reciprocal nature of the allelocatalyst influence.

There does not seem to be enough experimental evidence available to suggest the method by which the high media protomorphogen concentrations inhibit normal secretion of "waste" protomorphogens from the cytoplasm. Experiments with cell surface potential indicate that it decreases with age, suggesting the disintegration of the cell "membrane." It is difficult to reconcile these observations with the hypothesis that cell permeability decreases with age. Quite the reverse seems to be generally true. Therefore, the inhibition of protomorphogen secretion is probably not due to changes in the permeability of the surface boundary, rather vice versa, as we have suggested.

This inability of the cell to secrete "waste" protomorphogens when the media contains a high concentration of the latter, may be a result of a change in the protomorphogen molecules themselves. This is suggested by the fact that higher concentrations of protomorphogens require more prolonged exposure to high temperatures for destruction. Mast and Pace (1938) have demonstrated this phenomenon, working with the cell-secretion of Chilomonas paramecia.

We have mentioned the "chain-molecule" properties of protomorphogen. This property will be discussed in greater detail later in our discussion of platelet physiology and blood coagulation. Burrows (1926) has noted that protomorphogens (ergusia) have the property of forming threadlike molecular combinations. We suggest the possibility that increased concentration causes protomorphogens to form chain molecules, i.e., polymerize. These larger size molecules may "clog" the surface boundary of the cell, and thereby prevent any further escape of "waste" protomorphogen from the cytoplasm.

Mirsky (1943) has published a comprehensive review of nucleoproteins and their relation to chromosome material. He emphasizes that the desoxyribonucleic acid component is highly polymerized. He ascribes the physical properties of the chromosomes largely to their tendency to polymerize. He mentions the depolymerase present in many tissues and sera, also commenting on the ability of proteins to depolymerize nucleic acid. We shall discuss the importance of depolymerizing substances in maintaining cell vitality in Chapter 5 of this volume.

In a review of structural proteins Astbury (1945) discusses the chain folding nature of various fibrous proteins. Investigation with the electron microscope implies a structure of "...patterns within patterns, of successive levels of organization proceeding from the basic plans ...up to relatively enormous fibrils, and thence to combinations of fibrils." This phenomenon is ascribed to a "molecular template action." It illustrates the rather loose manner in which we employ the term polymerization in ref erring to protomorphogen determinants.

More recently (Tipson, 1945) it has been emphasized that nucleic acids appear in a polymerized form. The degree of polymerization depends on the method of preparation, various specimens of desoxyribonucleic acid being reported with molecular weights from 1,500 to over 1 million. We have emphasized protomorphogens are intimately associated with nucleoproteins and probably are linked with them in all biological processes. This known polymerizability of nucleic acids, therefore, lends credance to our suggestion that extracellular protomorphogens may polymerize as they become more concentrated.

Mitogenetic Radiation.—Our discussion of cell dynamics would not be adequate if we did not offer a brief review of the phenomenon known as mitogenetic radiation and its possible connections with the problems under consideration.

The Russian investigator, A. Gurwitsch (1924), was struck by the appearance of cell proliferation surrounding an injury to the cornea of a frog's eye. He rejected the explanation that this was due wholly to the diffusion of an "injury hormone," since a second wound prevented the extension of proliferation and caused a distinct "shadow." The classical experiment reviewed in his comprehensive report (Gurwitsch, 1932) demonstrated, using onion root tips, that radiations exciting proliferation can be projected in straight lines through quartz and reflected in the manner of light. After Baron (1926), the use of yeast cultures as detectors of mitogenetic radiation became more popular. This is discussed by Gurwitsch (1932) and a critical examination of the method has been supplied by Richards and Taylor (1932).

It is only fair to comment that many negative results have been reported by those who have attempted to duplicate Gurwitsch's experiments. Schwartz (1928) and Rossman[1] have reponed negative results while Magrou, Magrou and Croucroun[2] have reported both positive and negative conclusions. Bateman (1935) has reviewed other experiments of interest. .

The predominance of evidence, we believe, tends to support the contentions of the Gurwitsch school. Note must be made of the extremely critical technique required for successful demonstration of the existence of mitogenetic radiation.

Gurwitsch (1929) eventually concluded that the production of mitogenetic rays is not a function of dividing cells but of concurrent chemical and enzymatic reactions which may be duplicated *in vitro*. He divides the types of reactions producing radiation as follows: oxidations producing radiations from 2200 A to 2340 A; proteolytic reactions producing radiations from 1940 A to 2130 A and from 2200 A to 2420 A; glycolytic reactions resulting in radiations from 1900 A to 1970 A and from 2170 A to 2180 A (Gurwitsch, 1932). Other reports[3] have associated the production of mitogenetic radiation with glucolysis in hemolyzed blood, particularly with the decomposition of hexosediphosphoric acid. This reaction may be linked with the glucolytic reactions occurring in the cell nucleus concerned with the hydrolysis of creatine-hexose phosphoric acid (phosphagen) which we

1 Rossman, B.: "Umersuchungen ueber die Theorie der mitogenetischen Strahlung." *Roux' Arch. Entwicklungsmech. Org.,* 113:346, 1928;
———— "Mitogenetic Induction with Yeast as Indicator." *Ibid,* 114:583-586, 1929

2 Magrou, J., Magrou, M., and Croucroun, F.: "Effect at a Distance of Bact. tumefaciens on the Development of Sea-Urchin Eggs." *Compt. rend. Acad. Sci.,* Paris, 188:733-735, 1929. ———— *Ibid,* 186:802, 1928.

3 Gurvich, A.: "Chemistry of Mitogenetic Radiation of Blood." *Russ. J. Physiol.,* 16:495-500, 1933.

have discussed previously.

Various modes of action for mitogenetic radiation have been suggested from the experimental work performed with ultra-violet light in the significant frequencies. Some reports consider them necessary for the synthesis of proteins.[1] Others have indicated that mitogenetic radiation is concerned with the action and production of "desaminase."[2] Heyroth (1941) has enlarged upon an excellent review of the chemical effects of ultra-violet frequencies. Much information on the effects of these radiations upon enzyme activity is contained therein.

Giese (1939) in his review of this phenomenon, has indicated that since the nucleoproteins absorb ultra-violet more strongly than the cytoplasm, the lethal effect of these radiations may be exerted through their influence on the nucleoproteins. Loofbourow and others (1941) have concluded that the mitogenetic effect is due to growth promoting factors released by living cells injured by ultraviolet light. These factors are probably released from cell nucleoproteins. Sperti, Loofbourow and Lane (1937) have demonstrated that the release of cell proliferating substances by injured cells is probably a universal biological phenomenon. Loofbourow, Cook, Dwyer and Hart (1939) have shown that it is liberated by cells subjected to mechanical injury as well as ultra-violet wave lengths. Loofbourow and Morgan (1940) have demonstrated that toxic factors are also produced by such treatment and this undoubtedly has a bearing upon the conflicting results reported by investigators attempting to demonstrate the existence of mitogenetic phenomena by ultra-violet irradiation of cultures.

The effects of ultra-violet irradiation vary somewhat with the wave length and intensity. The reports are not consistent; one review mentioned rays from 2900 A to 3300 A as lethal,[3] other investigators reporting stimulating effects at 2370 A,[4] and still others finding no

1 Gurvich, A., and Gurvich, L.: "Polymerization of Peptides Under the Influence of Mitogenetic Radiation." *Arch. Sci. Biol.* (Leningrad), 54:89-94, 1939.

2 Gurvich, A. G., and Gurvich, L. D.: "Excitation of Polymerization Processes by Mitogeneric Radiations. II. Effects on Amino Acids and the Formation of a 'Desaminase' by Irradiation." *Acta Physicochim. U.R.S.S.,* 13:690-696, 1940.

3 Hollaender, A., and Schoeffel, E.: "Mitogenetic Rays." *Quart. Rev. Biol.,* 6:215-222, 1931.

4 Frank, G. M., and Gurwitsch, A.: "Identity of Mitogenetic and Ultra-Violet Rays." *Roux' Arch. Entwicklungsmech. Org.,* 109:451-454, 1927.

effect at this wave length.[1] Glaser and Schott (1936) have suggested that living cells may be refractory to radiation in excess of the optimum for cell division.

Gurwitsch (1929) has suggested that the effects are exerted by means of polymerizations induced by the radiations. Heyroth (1941) reviews much data on the influence of ultra-violet radiation on polymerization phenomena. Of particular interest is one comment that polymerization by radiation of wave lengths longer than 3300 A may be reversed by rays of shorter wave length.

It is not our purpose to enter into a detailed discussion of mitogenic rays or to present an adequate review of the literature. With the scarce and conflicting data available it is even quite difficult to present any suggestion concerning their connection with morphogen phenomena. We might venture the comment that mitogenetic radiation may interfere with the polymerization of accumulating protomorphogens. This possibility should be experimentally investigated. In this respect it is of interest to note the report of Oster (1934) that 20 per cent to 50 per cent more ultra-violet energy is required to kill an old resting yeast cell than a young active one. If the effect of these rays is due to their influence over the polymerization of the nucleoproteins or morphogens, this phenomenon could be explained by the greater polymerization in aged cells predicated by our experimental hypothesis. The polymerization of proteins in living organisms has been demonstrated to be partially under the control of mitogenetic radiation which has been caused to catalyze the polymerization of peptides into proteins even in the absence of the polymerizing enzymes.[2] Careful experiments will indicate the exact nature of this proposed control of polymerization by mitogenetic radiation.

More recently Gurvich and Gurvich (1940) have demonstrated that these radiations reciprocate between those produced by enzymatic reactions in the media and those produced by reactions within the nucleus. Of special interest to us is their comment that they may be quenched by products in the media of a "fatigued" culture. These "fatigue" products very likely consist of polymerized protomorph-

1 Kreuchen, K. H., and Bateman, J. B.: "Physikalische und Biologische Untersuchungen ueber, motigenetische Strahlung." *Protoplasma,* 22: 243-273, 1934.

2 Gurvich, A., and Gurvich, L., "Polymerization of Peptides Under the Influence of , Mitogenetic Radiation." *Arch. Sci. Biol.* (Leningrad), 54:89-94, 1939.

ogens accumulated in the media. In absorbing and thereby "quenching" the mitogenetic rays the energy of the latter exerts a chemical effect, possibly resulting in the depolymerization of these protomorphogens.

Lag Period.—In our previous discussion of the lag period before the initial cell division in a transplant, we discussed the experimental evidence which indicated that the amount of time lag depended upon the concentration of morphogens in the new media, and also on the relative age of the culture from which the transfer is obtained., We mentioned that the transferred cell had to reach a "common denominator" before its first division. This "common denominator" is the ratio at which the "unbound" morphogen content of the cytoplasm reached a value low enough to permit mitosis to occur.

Any influence that reduces lag period, therefore, would tend to depolymerize and eliminate the unbound protomorphogens if our hypothesis is to be substantiated. Such seems to be the case. Washing the cells at time of transfer slightly shortens the lag period. By eliminating the influence of polymerized protomorphogens on the cell surface, this washing would cause the secretion of the protomorphogens from the cytoplasm. Although the evidence is not clear-cut, it may be that mitogenetic radiation reduces the lag period of a transplant: We have suggested that the effect of this radiation may be due to depolymerization of protomorphogens. Most significant perhaps, is the demonstration of Simms and Stillman (1937) that limited treatment of cells with trypsin before transfer reduces the lag period. We have previously introduced evidence that trypsin decomposes protomorphogens. The limited treatment of aged cells with trypsin would have the effect of lowering the protomorphogen concentration in the cytoplasm.

Determinant Morphogen Cycle.—Our discussion would not be complete without a brief mention of the determinant effects of morphogens on the cell structure and suggestions on the mechanism by which they are effected. We intend to cover this more completely in the following chapters, but a brief mention is necessary here in order to avoid confusion.

Our hypothesis states that degenerated protomorphogens are cleared from the chromatin and lost to the cytoplasm at each cell division. (We do not believe that the determinant effects of *cytomorphogen* are exerted over the cell in this manner.) Jennings (1940) has extensively reviewed the problem of the mixing of cytoplasmic material with the chromatin at each cell generation and subsequent extrusion of the material back into the cytoplasm with many minute chromatin particles. He mentions that this mixing of nuclear and cytoplasm constituents varies with different species. For instance in paramecia at fairly regular intervals of 20 to 60 generations, great masses of nuclear material (the macronucleus) are absorbed into the cytoplasm. He states, "The transfer of so great a quantity of nuclear material into the cytoplasm must greatly affect the nature and physiological activity of the cytoplasm; some of its presumable results we shall see later in the genetics of these organisms."

The phenomena of chromidiosis, observed in protozoa and in cells of higher metazoans may also be concerned in the determinant cycle of the morphogens. We have reviewed (page 198) the comments of Horning and Miller (1930) on this problem. Although chromidiosis (extrusion of chromatin into the cytoplasm) in protozoa is associated with conjugation and thus the organization of cell structure, these workers present evidence that in metazoan cells it is more likely associated with the metabolic processes of growth and cell division.

Thorell (1944) comments that the nucleolar masses diminishes during differentiation becoming very small and surrounded by an intense ring of nucleoprotein. This indicates the loss of determinant material during differentiation.

A detailed study and discussion of the organizing and determinant effects in single celled individuals entails considerable review of experiments and hypotheses covering every phase of single celled life, including conjugation and endomixis and is irrelevant at this point of our discussion.

We simply wish to distinguish between the elimination of morphogens into the cytoplasm as protomorphogen fragments consequent to the dynamic nuclear energy metabolism (consistent with the dynamic state of proteins) and the orderly mixing of cytoplasm with chromatin for the purpose of placing cytomorphogens and protomorphogens in a position to exert a determinant effect. The first is the *metabolic cycle* and puts protomorphogens into the media where they are utilized as determinants for the *protein molecule* synthesized at the cell wall. The second is the *determinant cycle* and puts cytomorphogen fragments into the cytoplasm where they are utilized as determinants for *histogenesis.*

We observed before that the growth stimulating effect of morphogens in the media is not specific, but that of the morphogens extracted from autolyzed tissue is specific. In one case we are dealing with protomorphogen fragments and in the other with cytomorphogens and protomorphogens complex enough to serve as determinants for biological structure and therefore exhibiting species specificity.

Outline of the Metabolic Morphogen Cycle.—A brief outline of the hypothesis we have presented in this chapter follows. Reference to figure 5 will facilitate the presentation of this hypothesis.

1. The chromatin of the nucleus is the only molecule in the cell capable of self reproduction. It consists of nucleoprotein which in turn is composed of a desoxyribonucleic acid and various protein moieties containing the cytomorphogen determinants for the cell structure.

2. The reproduction of this chromatin is necessary for cell division, and also occurs constantly together with the dynamic state of living proteins, as a part of the vital energy cycle of the cell.

3. The energy cycle is connected with phosphatase hydrolysis of the dipotassium salt of creatine-hexose phosphoric acid; the synthesis of nucleic acids, and their change from ribo- to desoxyribo- forms; and possibly the radioactivity of potassium in the morphogen molecule and its influence on the stability of the biological colloids.

4. The morphogens which have participated in this constant energy reaction are broken-down or split and are eliminated into the cytoplasm at each division, or at a slower rate in non-dividing cells. (The fact that after division ceases, ageing still proceeds with eventual lysis, proves that morphogens are still being "shed" by the nucleus.)

5. These split morphogens must not be confused with the cytomorphogens and protomorphogens which are secreted by the nucleus into the cytoplasm at intervals to exert their histogenetic determinant effects.

6. The split morphogens in the cytoplasm are prevented from exerting lethal effects by a fatty or lecithin envelope, and are further discharged into the surrounding media.

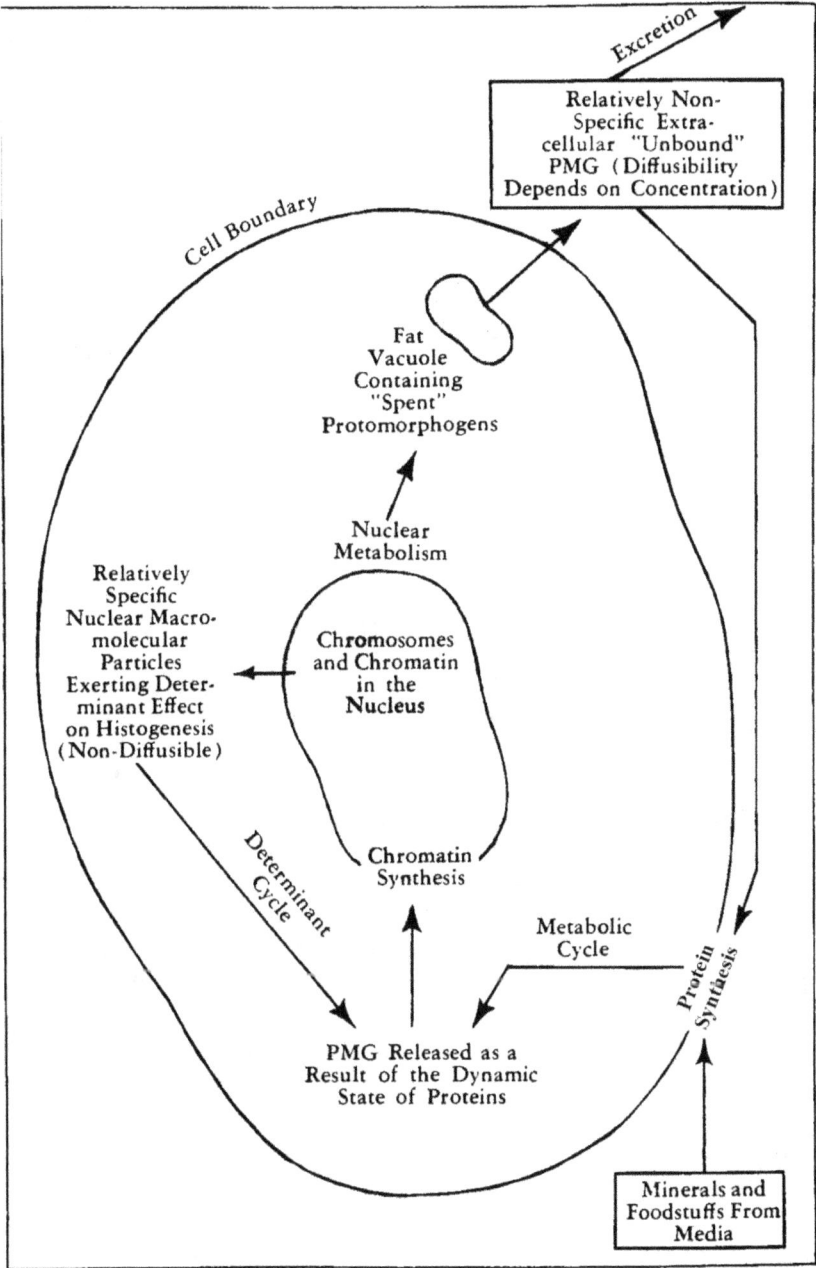

FIGURE 5

Schematic diagram of the dynamics of the mineral portion of the morphogen molecule as suggested by the morphogen hypothesis.

7. The split protomorphogens in the media are available as determinants for cytoplasm protein synthesis at the cell wall; in this manner they are necessary to, and stimulate, growth. (In this effect they are not species specific for they are simplified to the point where their complexity no longer allows it. If the cytomorphogens and protomorphogens present in the cytoplasm for determinant activity are experimentally extracted, they will also stimulate this protein synthesis, but will retain the species specificity as a consequence of their complexity.)

8. The split protomorphogens are now back in the cytoplasm as a part of cytoplasmic protein. (Other mineral elements are also present in the cytoplasmic protein which were supplied by the nutritive media.)

9. These protomorphogens, along with mineral elements and protein from the media, are now utilized at the nuclear wall for the synthesis of new cytomorphogen and chromatin material. The dynamic state of cytoplasmic protein insures a constant supply of protomorphogens at this point for chromatin synthesis

10. Due to the fact that nuclear cytomorphogen synthesis utilizes many minerals associated with cytoplasmic protein but not necessarily as morphogen linkages, the split protomorphogens in the media accumulate if the media is stagnant.

11. As the split protomorphogens in the media become concentrated, they tend to polymerize and thus increase in molecular size. This inhibits the loss of these factors from the cytoplasm, resulting in their polymerization and accumulation there.

12. The increase in cytoplasmic and media concentration of polymerized protomorphogens causes a gradual breakdown of the integrity of the surface boundary of the cell and consequently of the nuclear membrane. (The primary lethal effect of split protomorphogens is exerted by the concentration in the cytoplasm. The toxic effect of the concentration in the media is simply due to its influence in preventing further discharge from the cytoplasm.)

13. This degeneration of the membrane results in a lowering of the electrical potential between the protoplasm and the media, and between the cytoplasm and the nucleus.

14. Concomitant lowering of pH values in the cytoplasm and nucleus inhibits the constructive phase of the protoplasmic enzymes, prevents repair and facilitates a general lowering of cell vitality.

15. As a consequence of lowered vitality and inhibition of protein synthesis, the synthesis of chromatin is impaired and mitosis ceases.

16. Eventually the vitality and cell potential drop to the point where the integrity of the cell can no longer be maintained against the physical forces of the environment; the cell "dies" and undergoes lysis.

17. This cycle may be broken by removal or dilution of the accumulating morphogens from the stagnant media, preventing the development of lethal concentrations.

18. When cells are transferred to new cultures, they exhibit a time lag before commencing mitosis. This is the time necessary for the cytoplasmic protomorphogens to be eliminated into the media in sufficient amounts to restore the catalytic balance of intra- and extracellular protomorphogens necessary for commencement of mitosis. If the cell contains depolymerized protomorphogen at the peak of its diffusibility, this time is negligible.

We make no pretense at holding this hypothesis to be invulnerable. Any attempt to review and organize such tremendous mass of experimental evidence, covering so many diversified fields that it is practically impossible for us to have more than a smattering of knowledge in many of them, is bound to fall short of perfection. We are presenting merely a working hypothesis that experimental research may be enabled to guide its activities into the most lucrative prospects, and not be vulnerable to projects which do not take all matters into account.

BIBLIOGRAPHY

ASTBURY, W. T.: "The Structural Proteins of the Cell." *Biochem. J.*, 39:lvi-vii, 1945.

BAKER, L. E.: "Causes of the Discontinuity of Growth of Fibroblasts Cultivated in Embryo-Juice." *Proc. Soc. Exp. Biol. & Med.*, 39:369-371, 1938.

BARON, M. A.: "Mitogenetic Radiations in Protista." *Roux' Arch. Emwicklungsmech. Org.*, 108:617-633, 1926.

BATEMAN,]. B.: "Mitogenetic Radiation." *Biol. Rev.*, 10:42-71, 1935.

BEHRENS, M.: *Z. physiol. Chem.*, 253:185, 1938.

BEUTNER, R.: *Physical Chemistry of Living Tissues and Life Processes.* Williams & Wilkins Co., Baltimore, Md. 1933.

———— "The Nature of the Vital Battery System." *Arch. f. exp. Zellforschung*, 15:217-235, 1934.

BLINKS, L. R.: "Protoplasmic Potentials in Halicystis." *J. Gen. Physiol.*, 13:223-229, 1929.

BRACHET, J.: *Arch. biol.*, 44:519, 1933.

———— *Compt. rend. soc. biol.*, 83:88, 1940.

BRADFIELD, J. R. G.: "Alkaline Phosphatase in Invertebrate Sites of Protein Secretion." *Nature*, 157:876, 1946.

BRUES, A. M., TRACY, M. M., and COHN, W. E.: "The Relation Between Nucleic Acid and Growth." *Science*, 95:558-560, 1942.

BURROWS, M. T.: "Studies to Determine the Biological Significance of the Vitamins." *Proc. Soc. Exp. Biol. & Med.,* 22:241-245, 1925.

———— "Energy Production and Transformation in Protoplasm as Seen Through a Study of the Mechanism of Migration and Growth of Body Cells." *Am. J. Anat.,* 37: 289-349, 1926.

BURROWS, M. T., and JORSTAD, L. H.: "On the Source of Vitamin A in Nature." *Am. J. Physiol.,* 77:38-50, 1926.

CASPERSSON, T.: "Uber die Rolle der Desoxyribosennucleinsaure bei der Zellteilung." *Chromosoma,* 1:147-156, 1938.

———— "Die Eiweissverteilung in den Strukturen des Zellkerns." *Chromosoma,*1:562-604, 1940.

CASPERSSON, T., and BRANDT, K.: *Protoplasma,* 35:507, 1941.

CASPERSSON, T., and SCHULTZ, J.: *Nature,* 143:602, 1939.

———— *Proc. Natl. Acad. Sci. U.S. A.,* 26:507, 1940.

CASPERSSON, T., and THORELL, B.: *Chromosoma,* 2:132, 1941.

CASPERSSON, T., LANDSTROM-HYDEN, H., and AQUILONIUS, L.: *Chromosoma,* 2: 111,1941.

CHAMBERS, R.: "The Micromanipulation of Living Cells." *The Cell and Protoplasm, Publ. Am. Assoc. Adv. Sci.,* No. 14, p. 20. 1940.

CLAUDE, A.: "Particulate Components of Cytoplasm." *Cold Spring Harbor Symposia on Quantitative Biology,* 9:263-271, 1941.

———— "The Constitution of Protoplasm." *Science,* 97:451-456, 1943.

COWORY, E. V.: *General Cytology.* University of Chicago Press, Chicago, Ill. 1924.

CRILE, G. W., ROWLAND, A. F., and TELKES, M.: "Changes in the Electric Potential of Tissues in Living Animals Under Normal and Pathological Conditions." *Am. J. Physiol.,* 85-86:362-363, 1928.

CRILE, G. W., TELKES, M., and ROWLAND, A. F.: "An Experimental Investigation of the Physical Nature of Death." *Proc. Am. Philosoph. Soc.,* 68:69-81, 1929.

———— "The Nature of Living Cells." *Arch. Surg.,* 23:703-714, 1931.

———— "Autosynthetic Cells." *Protoplasma,* 15:337-360, 1932.

DAMON, E. B., and OSTERHOUT, W. J. V.: "The Concentration Effect With Valonia: Potential Differences With Concentrated and Diluted Sea Water." *J. Gen. Physiol.,* 13:445-457, 1930.

DANOWSKI, T. S.: "The Transfer of Potassium Across the Human Blood Cell Membrane." *J. Biol. Chem.,* 139:693-705, 1941.

DARLINGTON, C. D.: "Chromosome Chemistry and Gene Action." *Nature,* 149:66-69, 1942.

DARLINGTO, C. D.: and LA COUR, L. F.: *J. Genetics,* 46:180-267, 1945.

DAVIDSON, J. N.: "Cytoplasmic Ribonucleoproteins." *Biochem. J.,* 39:59-61, 1945.

DAVIDSON, J. N., and WAYMOUTH, C.: "Tissue Nucleic Acids. I. Ribonucleic Acids and Nucleotides in Embryonic and Adult Tissue." *Biochem. J.,* 38:39-50, 1944.

DILLER, W. F.: "Cytology, Genetics, and Evolution." University of Pennsylvania Bicentennial Conference, University of Pennsylvania Press, Philadelphia, Pa. 1941.

FEULGEN, R., and ROSSENBECK., H.: *Z. physiol. Chem.,* 135:203, 1924.

FEULGEN, R., BEHRENS, M., and MAHDIHASSAN, S.: *Z. physiol. Chem.,* 246:203, 1937.

FICHER, A.: "Contributions to the Biology of Tissue Cells. I. The Relation of Cell Crowding to Tissue Growth in Vitro." *J. Exp. Med.*, 38:667-672, 1923.

———— "Nature of the Growth-Accelerating Substance of Animal Tissue Cells." *Nature*, 144:113, 1939.

———— "Nature of the Growth-Promoting Substances in the Embryonic Tissue Juice. A Review of the Author's Investigations." *Acta Physiol. Scand.*, 3:54-70, 1941.

FISKE, C. H., and SUBBAROW, Y.: "Phosphocreatine." *J. Biol. Chem.*, 81:629-680, 1929.

GIESE, A. C.: "Ultra-Violet Radiation and Cell Division. Nuclear Sensitivity: Effect of Irradiation of Sea Urchin Sperm." *J. Cell. & Comp. Physiol.*, 14:371- 382, 1939

GLASER, O., and SCHOTT, M.: "Mitogenetic Radiation." *The Phenomena of Life*, G. Crile. W. W. Norton & Co., New York. 1936.

GURVICH, A. G., and GURVICH, L. D.: "Excitation Activation and Initiation of Polymerization Processes By Mitogenetic Radiations." *Acta Physicochim. U. R. S. S.*, 10:771, 1939.

———— "Mitogenetic Radiation." *Ibid*, 13:683-689, 1940.

GURWITSCH, A.: "Die Natur des spezifischen Erregers der Zellteilung." *Arch. f. mikr. Anat.*, 100:11, 1924.

———— "Uber den derzeitigen Stand des Problems der mitogenetischen Strahlung." *Protoplasma*, 6:449-493, 1929.

———— Die Mitogenetische Strahlung. Berlin, Springer. 1932.

HALDANE, J.B. S.: *Enzymes*. Longmans, Green & Co., New York. 1930.

HARDY, W. B.: "Note on Differences in Electrical Potential \Within the Living Cell." *J. Physiol.*, 47:108-111, 1913-14.

HARROW, B.: *Textbook of Biochemistry*. W. B. Saunders Co., Philadelphia, Pa. 1943.

HEATLEY, N. G.: "CCCVI. The Distribution of Glycogen in the Regions of the Amphibian Gastrula; With a Method for the Micro-Determination of Glycogen." *Biochem. J.*, 29:2568-2572, 1935.

HEYROTH, F.: *The Chemical Action of Ultraviolet Rays*, C. Ellis and A. A. Wells. Reinhold Publishing Corp., New York. 1941.

HITCHCOCK, D. I.: "The Effect of pH on the Permeability of Collodion Membranes Coated With Proteins." *J. Gen. Physiol.*, 10:179-183, 1926.

HOBER, R.: *Physikalische Chemie der Zelle und der Gewebe*. 5th Edition, Leipzig. 1924.

HOLMES, E.: *The Metabolism of Living Tissues*. Cambridge University Press, Cambridge. 1938.

HORNING, E. S., and MILLER, I. D.: "Chromidial Extrusion and Its Relationship to Atypical Nuclear Phenomena in Tumour Cells." *Austr. J. Exp. Biol. & Med. Sci.*, 7:151-160, 1930.

HUGGINS, C.: "Endocrine Control of Prostatic Cancer." *Science*, 97:541-544, 1943.

IWANITZSKAIA, A. Th.: "Influence des Homo- et des Hétéroplasmas sur la Division des Cellules dans les Cultures de Tissus." *Arch. Anat. Microsc.* (Paris), 35:283-293, 1939-40.

JENNINGS, H. S.: "Chromosomes and Cyroplasm in Protozoa." *The Cell and Protoplasm*, Publ. Am. Assoc. Adv. Sci., No. 14. 1940.

KIDDER, G. W., and CLAFF, C. L.: "Cytological Investigations of Colpoda Cucullus." *Biol. Bull.,* 74:178-197, 1938.

KOPAC, M. J.: "Micro-Estimation of Protein Adsorption at Oil-Protoplasm Interfaces." *Biol. Bull.,* 75: 372, 1938.

LASNITZKI, I.: "Action of Heated Embryo Extract Upon Growth of Fibroblast Cultures." *Skandinav. Arch. f. Physiol.,* 76:303-312, 1937.

LILLIE, R. S.: *Protoplasmic Action and Nervous Action.* University of Chicago Press, Chicago, Ill. 1923.

LLOYD, D. J., and SHORE, A.: *Chemistry of the Proteins.* P. Blakiston's Son & Co., Inc., Philadelphia, Pa. 1938.

LOEB, J.: *The Dynamics of Living Matter.* Columbia University Press, New York. 1906.

LOOFBOUROW, J. R., COOK, E. S., DWYER, C. M., and HART, M. J.: "Production of Intercellular Hormones by Mechanical Injury." *Nature,* 144:553-554, 1939.

LOOFBOUROW, J. R., DWYER, C. M., and CRONIN, A. G.: "Proliferation-Promoting Intercellular Hormones. II. Evidence for Their Production by Living Cells as a Response to Injury." *Biochem. J.,* 35:603-609, 1941.

LOOFBOUROW, J. R., and MORGAN, i\l. N.: "Investigation of the Production of Growth-Promoting and Growth-Inhibiting Factors by Ultra-Violet Irradiated Microorganisms." *J. Bact.,* 39:437-453, 1940.

LUNDEGARDH, H.: "The Absorption and Accumulation of Inorganic Ions." *Lantbruks Hogskol. Ann.,* 8:233-404, 1940.

MACNIDER, W. DEB.: "The Ageing Process and Tissue Resistance." *Sci. Month.,* 54:149-154, 1942.

MAST, S. O., and PACE, D. M.: "The Effect of Substances Produced by Chilomonas paramecium on Rate of Reproduction." *Physiol. Zool.,* 11:359-382, 1938.

——— "The Effects of Calcium and Magnesium on Metabolic Processes in Chilomonas paramecium." *J. Cell. & Comp. Physiol.,* 14:261-279, 1939.

MATHEWS, A. P.: *Physiological Chemistry.* 6th Edition, Williams & Wilkins Co., Baltimore, Md. 1939.

MAZIA, D.: "Enzyme Studies on Chromosomes." *Cold Spring Harbor Symposia on Quantitative Biology,* 9:293, 1941.

MAZIA, D., and JAEGER, L.: "Nuclease Action, Protease Action and Histochemical Tests on Salivary Chromosomes of Drosophila." *Proc. Natl. Acad. Sci. U. S. A.,* 25:456-461, 1939.

MENKE, W.: "Protoplasm of Green Plant Cells. I. Isolation of Chloroplasts From Spinach." *Z. physiol. Chem.,* 257:43-48, 1938.

MENKIN, V.: *Dynamics of Inflammation.* The Macmillan Co., New York. 1940.

——— "Gluconeogeriesis and Cellular Injury. A Further Inquiry Into the Mechanism Involved in Diabetes Enhanced by Inflammation." *Am. J. Physiol.,* 138:396-407, 1942-43.

MIRSKY, A. E.: "Chromosomes and Nucleoproteins." *Advances in Enzymology,* Vol. 3, Interscience Publishers, Inc., New York. 1943.

MIRSKY, A. E., and POLLISTER, A. W.: "Nucleoproteins of Cell Nuclei." *Proc. Natl. Acad. Sci. U. S. A.,* 28:344-352, 1942.

——— "Fibrous Nucleoproteins." *Biol. Symp.,* 10:247, 1943.

——— "Chromosin, a Desoxyribose Nucleoprotein Complex of the Cell Nucleus." *J. Gen. Physiol.,* 30:117-148, 1946.

MULLER, II. J.: *Am. Naturalist,* 56:32, 1922.

MULLINS, L. J.: "Selective Accumulation of Potassium by Myosin." *Fed. Proc.,* 1:61, 1942.

MYERS, V. C.: "Body Tissues and Fluids." *Endocrinology and Metabolism.* Ed. by L. F. Barker, D. Appleton & Co., New York. 1922.

MYERS, V. C., and MANGUN, G.: "The Potassium Content of Muscle and Its Possible Relation to Muscle Creatine." *J. Biol. Chem.,* 114:lxxv, 1936.

———— "Comparative Studies on Creatine, Phosphorus, and Potassium in Various Muscle Tissues." *Ibid,* 132:701-709, 1940.

OSTER, .R. H.: "Results of Irradiating Saccharomyces With Monochromatic Ultraviolet Light. I. Morphological. and Respiratory Changes." *J. Gen. Physiol.,* 18:71-88, 1934.

OSTERHOUT, W. J. V., and HARRIS, E. S.: "Protoplasmic Asymmetry in Nitella as Shown by Bioelectric Measurements." *J. Gen. Physiol.,* 11:391-400, 1927-28.

RAUEN, H. M.: "Dynamics of Proteins." *Umschau,* 44:547, 1940.

RICHARDS, O. W., and TAYLOR, G. W.: "Mitogenetic Rays—A Critique of the Yeast-Detector Method." *Biol. Bull.,* 63:113-128, 1932.

ROBERTSON, T. B.: *Chemical Basis of Growth and Senescence.* J. B. Lippincott Co., Philadelphia. 1923.

RUZICKA, V.: *Arch. Ges. Physiol.,* 194:135-148, 1922.

SCHOENHEIMER, R.: *The Dynamic State of Body Constituents.* Harvard University Press, Cambridge, Mass. 1942.

SCHULTZ, J., CASPERSSON, T., and AQUILONIUS, L.: *Proc. Natl. Acad. Sci. U. S. A.,* 26:515, 1940.

SCHWARTZ, W.: "The Theory of Mitogenetic Rays." *Biol. Zentralbl.,* 48:302-308, 1928.

SIMMS, H. S., and STILLMAN, N. P.: "Substances Affecting Adult Tissue in Vitro. I. The Stimulating Action of Trypsin on Fresh Adult Tissue." *J. Gen. Physiol.,*20:603-619, 1937.

———— "II. A Growth Inhibitor in Adult Tissue." *Ibid,* 20:621-629, 1937.

———— "Production of Fat Granules and of Degeneration in Cultures of Adult Tissue by Agents From Blood Plasma." *Arch. Path.,* 23:316-331, 1937.

SPERTI, G. S., LOOFBOUROW, J. R., and LANE, M. M.: "Effects of Tissue Cultures on Inter-Cellular Hormones From Injured Cells." *Science,* 86:611, 1937.

STARLING, E. H.: *Principles of Human Physiology.* Ed. by C. L. Evans, Lea & Febiger, Philadelphia. 1936.

TELKES, M.: "Bioelectrical Measurements on Amoebae." *Am. J. Physiol.,* 98:475-483, 1931.

THORELL, B.: "Behavior of the Nuclear Apparatus During Growth and Differentiation of the Normal Blood Cells in the Adult Stage." *Acta Med. Scand.,* 117:334-375, 1944.

TIPSON, R. S.: "The Chemistry of the Nucleic Acids." *Advances in Carbohydrate Chemistry,* I: 193-245, 1945.

TUTLER, I. A.: "Division, Encystment and Endomixis in Urostyla grandis With an Account of an Amicronucleate Race." *Cellule,* 44:189-218, 1935.

TURCK, F. B.: *The Action of the Living Cell.* The Macmillan Co., New York. 1933.

VAN HERWERDEN, M.A.: *Arch. Zellforsch.,* 10:431, 1913.

WAKSMAN, S. A., and DAVISON, W. C.: *Enzymes.* Williams & Wilkins Co., Baltimore, Md. 1926.

WERNER, H.: "Concurrence of Growth-Promoting and Growth-Inhibiting Factors in Extracts of Adult Rat Tissues." *Nature,* 154:827, 1944.

——— "Growth-Promoting and Growth-Inhibiting Factors in Rat Tumor Tissue." *Proc. Soc. Exp. Biol. & Med.,* 59:128-129, 1945.

WILLMER, E. N.: "The Localization of Phosphatase in Cells in Tissue Cultures." *J. Exp. Biol.,* 19:11-13, 1942.

WILSON, E. B.: *The Cell in Development and Heredity.* The Macmillan Co., New York. 1900.

WRINCH, D. M.: "The Chromosome Micelle and the Banded Structure of Chromosomes in the Salivary Gland." *Nature,* 136:68, 1935.

——— "On the Molecular Structure of Chromosomes." *Protoplasma,* 25:550-569, 1936.

——— "The Pattern of Proteins." *Nature,* 137:411, 1936.

CHAPTER 4

MORPHOGENS AS DETERMINANTS

Review of Morphogenic Factors

The term "morphogen" signifies a factor that promotes the organization of structure. Thus far we have advanced our hypothesis to include definitions of the primary organizer of protein specificity, protomorphogen and the primary organizer of the form of a single cell, cytomorphogen. We have suggested that the characteristics of these primary organizers depend singularly upon a stable and complex mineral factor that appears to be utilized in the synthesis of protein and affords a "skeleton" of patterned mineral linkages for the protein molecule.

We have reviewed experimental evidence indicating that the morphogens are constantly synthesized and destroyed as a part of cell metabolism. They are constantly "lost" into the surrounding media where their gradual accumulation hinders further discharge from the cytoplasm. As a consequence, the increasing concentration in the cytoplasm progressively inhibits cell metabolism resulting in the ultimate dissolution of the cell.

We have postulated that a cytomorphogen is composed of an organized assemblage of protomorphogens and a gene is composed of an organized assemblage of cytomorphogens. We have suggested that these factors are the individualized genetic units of the chromosome. It is the purpose of this chapter to correlate information on the mode of genetic influence and speculation on the role to be assigned to the morphogens.

Morphogens as the Organizer Substance of the Chromosomes

Mineral Distribution in Dividing Cells.—The first experimental evidence we found which indicated that morphogens are the organizer factors in the chromosome, is the remarkable pattern of mineral distribution in the developing embryo. The reader will recall our conclusions in Chapter 1, particularly those made after reviewing

121

Turck's experiments, which postulate that patterned mineral groups are very important constituents of the basic biological determinants. Horning and Scott (1932) have demonstrated that the differentiation of the chick embryo is accompanied by an organized differentiation of protoplasmic minerals while there is no demonstrable differentiation of mineral salts in adult tissues. Scott (1937), working with microincinerated preparations, has demonstrated that the cytoplasmic mineral pattern of rapidly developing cells undergoes changes, seemingly of a predetermined nature. He has observed that during some phases of mitosis the mineral content is confined to the cell wall and the nucleus. This discovery is significant in respect to our hypothesis, since we have postulated that there is an intense morphogen metabolism at these points concerned with the synthesis of cytoplasmic proteins and nuclear chromatin.

In earlier publications Scott (1930) has supplied us with a detailed picture of the microincineration pattern of mineral salts during the different phases of mitosis. An outline of his observations is presented here:

Resting Cell:

    Nucleus—Clump Ash
    Nucleolus—Clump Ash
    Nuclear Wall Outlined

Prophase

    Elongated Nuclear Ash
    The ash retains a form similar to the chromatin in prophase.

Metaphase:

    The ash deposits seem to follow the form of the chromosomes in metaphase, but there is a fine dispersion of some of the salts.

Anaphase:

    Dense ash in center of cell. There is no apparent organization. Sometimes fine threads can be observed at the periphery of the ash masses.

He comments that there was scarcely any ash not associated with chromatin. Our previous discussion of the location of morphogens (Chapter 3) mentions that they are most intensely concentrated in the chromatin where they are synthesized. This report is particularly significant because of the demonstration of fine threads at the periphery of clumps of ash. We have postulated that protomorphogens have a remarkable tendency to polymerize, and the appearance of fine threads in the ash of incinerated cells suggests that the minerals were associated with elongated polymerized molecules. Also significant is the observation of a fine dispersion of mineral elements during the metaphase. This leads us to speculate that this phenomenon may indicate a degree of dispersion of morphogens at the time of chromosome splitting.

In an excellent review of this subject Scott (1943) suggests that the minerals of a cell are linked in some manner with its structural integrity. He mentions the outstanding presentation of Bernal (1940) which offers a new conception of the importance of electrolytes in protein structure, particularly that of elongated molecules. Both of these investigators recognize the importance of minerals in the organization of living molecules, which is the fundamental thesis of the morphogen hypothesis. We consider Scott's observations of incinerated cells in particular a significant demonstration of the postulated determinant effect of the morphogens.

"Organizers" in Embryonic Development.—Paul Weiss' comprehensive review of amphibian organization (1935) furnishes an extensive correlation of experiments which are further suggestive that the morphogens are the primitive organizers of form. In order for us to clearly review the evidence, it will be necessary to digress for a time from a strict discussion of the primary organizer and its reaction within the cell during mitosis.

There is one corollary in genetics that must be constantly kept before us. *Upon the relative complexity of differentiation of the organism will depend the complexity of its biological determinants.* We have discussed the influence of protomorphogen over the biological specificity of the protein molecule. The mode of activity of a chromosome, gene or cytomorphogen over the morphology of an organism still remains relatively unclarified. It is one thing to organize the biological specificity of a single molecule and quite another to organize a complete cell or an organism consisting of thousands of cells each bearing a distinct pattern and position in relation to the whole.

We suggest that it is the function of the chromosome during a certain period of embryonic development to determine the region of the embryo in which the various genes and cytomorphogens shall exert their spheres of influence. The relative necessity for this function would of course depend upon the complexity of differentiation of the adult organism. If our suggestion is to receive serious consideration it is important to analyze embryonic development in an attempt to define the periods in which the various distinct forms of *determinant* differentiation occur.

"Fields" of Organization.—Of great significance in embryological development is the occurrence of various distinct "fields" of organization. Weiss (1935) offers the following definition "A 'field' is a system of patterned conditions, the pattern of which is not pieced together by individual contributions of independent constituents, but is the expression of the dynamic activity of the whole system, the whole pattern tending to retain its typical organization beyond changes involving its parts." Experimental evidence establishes the center of an embryonic field as existing in the centers of the areas from which the normal development of an organ will arise.[1] The intensity of the organizing power of the field decreases with the distance from its center.

The insignificant self-differentiating ability existing in a portion of the blastula is shown by segregating a bit of tissue from the influence of the embryo, allowing it to develop by itself. Weiss has reported such experiments with amphibian embryos in the blastula stage of development. By referring to "maps," the portions of the blastula from which various types of cells would normally arise is determined. When such a portion is segregated and cultured *in vitro* outside the influence of the blastula its differentiation is inconsequential. These experiments demonstrated that cells which would have normally developed into specific differentiated tissue underwent chaotic multiplication.[2] There is no trend towards differentiation during the blastula stage. We may conclude from the series of experiments of which the above are typical that the cells of the blastula are highly dependent upon the

1 Spemann, H.: *Zool. Jahrh.*, 32:1, 1912. Harrison, R. G.: *J. Exp. Zool.*, 25:.413, 1918; 41:349, 1925. Kaan, H. W.: *J. Exp. Zool.*, 46:13, 1926.

2 Bautzmann, H.: *Naturwiss.*, 17:818, 1929; Kusche, W.: *Roux' Arch. f. Entwmech.*, 120:192, 1929.

embryo as a whole for further differentiation.

The gastrula and neurula stages off er significant information leading to an hypothesis of morphogen function. Weiss mentions that the sector of the blastula which, during gastrulation, is invaginated around the lips of the blastophore into the interior thus forming the ento-mesoderm, differentiates when isolated and cultured independently of the embryo. He compares the immutable differentiation characteristic of this tissue with the flexibility of other parts of the embryo as transplants and concludes that the conditions directing development crystallize first in this location. Roux:1 calls this ability of tissue to persist in its predetermined development, even when separated from the embryo, "self-differentiation." Weiss states that during gastrulation the "fields" undergo further subdivision into areas of more specific organizing ability. A "head" field, for instance, may differentiate into eye, ear and gill fields, etc. The "field" concept should be clarified so that the term embryonic "field" shall only include those areas which can effect self-differentiation when separated from the rest of the embryo. With this definition in mind it can be stated that there are few embryonic "fields" in the blastula; they develop and differentiate during gastrulation and the neurula stage.

During the neurula stage those parts of the organism which would develop chaotically when transplanted *in vitro* away from the influence of the blastula now will self-differentiate strictly according to their normal bent when so separated.

As Weiss points out, something of extreme import has occurred between gastrulation and the neurula stage. We postulate that the differentiation of the biological determinants is fairly well completed at this stage. The various fields of organization then develop into the neural groove, notochord and primitive segments. Each part is now more or less independent of the embryo, since its future is well mapped by its own biological determinants.

Chromosome Differentiation.—The gradual acquisition of self-differentiating capacity by various segments of the embryo during development from the blastula to the neurula stage would seem to suggest that during these periods the chromosome is gradually

1 *Terminologie der Entwicklungsmechanik,* Roux, W. Berlin. 1912.

"unwinding" and exerting its influence successively in organizing the whole structure by locating groups of cytomorphogens in a patterned position, both as to space and time.

The beginning of the development of "fields" in the blastula very likely constitutes a limited initial "segregation" of various specific genes into that part of the embryo where they are to exert their influence. During gastrulation it is apparent that these fields differentiate and in turn locate specific morphogen groups in that part of the embryo from which a detailed tissue is to develop.

It is significant that when presumptive tissue is isolated from the blastoderm and cultured *in vitro,* it does not differentiate into patterned structures but develops chaotically. Self-differentiating ability at this stage of embryonic development is confined to definitely localized "organizer" tissue. As the developing embryo reaches the neurula stage, cultivation of sections will result in differentiation along the strict lines of normal development. Our contention is that by the time the embryo has reached the neurula stage, the local "organizer tissue" has transferred genetic groups of cytomorphogens to the neighboring presumptive cells of the field, thus imparting self-differentiating ability to these cells.

One would assume that the differentiating powers of a fragment of tissue would be extremely limited until the segregation of the gene, followed by its splitting into groups of cytomorphogens, is complete. Such seems to be the case.

Many experiments have shown the existence of these embryonic "centers of organization" which exert a morphogenic influence on neighboring tissue by means of the influence of a hypothetical "organizer." Bautzmann has located an organizing center in the "gray crescent" of the amphibian egg which seemingly does not exert influence until gastrulation.[1] Ruud and Spemann[2] have determined that the organizing center of newt's eggs is in the dorsal half. Weiss reviews experiments indicating that the medullary plate receives its differentiation and morphology from the organizers of underlying mesoderm. It has been experimentally established that the roof of the archenteron organizes adjacent ectodermal layers into neural tissue.[3]

1 Bautzmann, H.: *Roux' Arch. f. Entwmech.,* 108:283, 1926.

2 Ruud, G., and Spemann, H.: *Roux' Arch. f. Entwmech.,* 52:95, 1922.

3 Marx, A.: *Roux' Arch. f. Entwmech.,* 105:19, 1925.

Needham (1931) has reviewed self-differentiation and organizer phenomena. He mentions that the organizing influence that radiates from the dorsal lip has been called the "organizer of the first grade," and some organs differentiating under its influence also exhibit, in turn, organizing influence on adjacent structures. This effect is termed "organizer of the second grade." Needham comments that some organs retain the ability for self-differentiation and others depend upon adjacent organs, but that only a few of these relationships have been completely studied. He concludes that although most of the experimental work on organizer phenomena has been conducted with amphibian material, it is probable that the information will be found to apply to all varieties of embryo.

Needham also reviews experiments which show that the degree of structure into which an embryonic fragment will differentiate when transplanted depends upon the age of the embryo. For instance, Hoadley[1] demonstrated that a piece of embryo 4 hours old transplanted to the chorio-allantoic membrane will produce an eye with pigment cells only; a piece from a 6 hour old embryo will produce pigment and retinal cells; a piece from an 8 hour old embryo will produce an eye with pigment and stratified retina; while a piece from a 10 hour old embryo in which the primitive groove is formed will produce a complete self-differentiating eye. Experiments such as this strikingly demonstrate the "unwinding" of the chromosome in the "center of organization" and consequent allocation of genes and cytomorphogens with potential organizing powers.

Countless brilliant experiments concerned with the transfer of grafts supply us with additional information on the existence of genic groups of cytomorphogens (the hypothetical "organizer" referred to in literature) and their independent determinant influence over developing tissue. So far the conception of determinant mechanism in the embryo has been outlined herein as follows: (1) the chromosome begins to map "fields" of morphogenic influence in the blastoderm; (2) these "fields" are gradually endowed with genic groups of cytomorphogens from the chromosomes; (3) during gastrulation the "fields" complete their acquisition of organizing ability and in turn segregate more specific groups of cytomorphogens into locales of organizing activity; (4) in the neurula stage these localized genic groups continue their self-differentiation and exert organizing

---

1 Hoadley, L.: *J. Exp. Zool.*, 42:143, 1925.

effects over adjacent cells through the release of cytomorphogens.

Transfer of "Organizer" Material.—A study of the experiments concerned with grafting "organizer" material into hosts leads to the discovery of a puzzling paradox. Some investigators report that the transferred "organizer" material determines the differentiation pattern; others have observed that the pattern of the host is followed; and others have demonstrated that under transfer conditions both normal and abnormal differentiation may occur.

A few of these conflicting experiments are reviewed by Weiss (1935) in an excellent survey of the field. For detailed study we suggest that the reader consult this review and later publications by the same author (Weiss, 1939). Spemann[1] has shown that ectodermal cells which would normally develop into skin differentiate into the neural system when transplanted into that "field" during the gastrula stage. Mangold[2] demonstrated in a series of experiments that material from the same part of the embryo, when grafted into other locations, took on the form of the location into which it was grafted and underwent differentiation into the organ normally arising from that region.

Experiments of this class lend themselves to the explanation that they represent the same phenomena as the normal transfer of genic groups from the center of organization to an area in the embryo which, until this transfer, is unable to organize itself. These experiments, however, illustrate the complement of this reaction since the undifferentiated cells are transferred to an organizer region. It is evident that when tissue, not yet having received its organizer group, is transferred to another region of the embryo, it receives genic groups from that region and differentiates accordingly.

Occasionally grafts will not develop in this manner under the influence of the organizers in the new location. Thus Spemann and Schotte[3] demonstrated that if the presumptive epidermis of the frog Rana esculenta, from a region in the gastrula which would

---

1 Spemann, H.: *Roux' Arch. f. Entwmech.*, 43:448, 1918; 48:533, 1921.

2 Mangold, O.: *Roux' Arch. f. Entwmech.*, 100:198, 1923.

3 Spemann, H., and Schotte, O.: *Naturwiss.*, 20:463, 1932.

ordinarily develop into suckers, was transplanted into the region of T. taeniatus which would ordinarily develop into balancers, no balancers were formed, but functional suckers were differentiated. Spemann[1] has published other experiments which show that the morphology of tissue arising from grafts may take on the characteristics of the graft rather than those of the host.

Other experiments show that the presumptive organization of both host and transfer may be developed. Mangold,[2] for instance, has demonstrated that a graft from one species may take on the appearance of its new locality but also produce form characteristic to the donor species. A graft from the belly of urodeles possessing a pair of head balancers was transferred to the head region of a urodeles that does not possess head balancers. The material took on the form of its new location but in addition produced the typical balancers of the species from which it was derived.

A careful study of the investigation of the results of the transfer of organizer regions indicates that the experiments may be placed in three different classes: (1) those in which the differentiation follows the presumptive pattern of the explant; (2) those in which the differentiation follows the presumptive pattern of the host, and (3) those in which the differentiation exhibits a combination of the explant and the host.

Needham (1942) has supplied a comprehensive review of the literature on morphogenesis which no students of this problem can afford to overlook. He analyzes several factors which supply the basis for an explanation of the differences in transplant experiments which we have listed.

Needham stresses the "competence" of embryonic tissue as being equally important with organizing potentiality. The induction of patterned differentiation by an organizer can only occur in embryo tissue which is reactive to this induction; this state is known as competence. Embryo tissue passes through a competent state during a period of embryonic growth and during this period specific organizers may stimulate it into patterned differentiation. *After it has become thus "determined," no other inducing stimulus or organizer can alter or reverse this determination.* Organizing inductors have no effect on tissue either *before it becomes* competent or *after it*

1 Spemann, H.: *Roux' Arch. f. Entwmech.*, 123:389, 1931.
2 Mangold, O.: *Naturwiss.*, 19:905, 1931.

has passed this period, whether it has received a determinant during this period or not.

Needham mentions that Hoadley[1] has advanced the hypothesis that the early embryonic cell is capable of self-differentiation only into a generalized cell type of low organization. Braus[2] pioneered in the demonstration that the specific protein characteristics of embryonic tissue are considerably different from the adult of the same species. This work was followed up by other investigators reviewed by Needham (1931) who have shown that the immunological character of cell protein varies during the development of the organism.

Organizer Morphogens as a Virus System.—We postulate that the organizer system, starting with the original germ chromosome complex and breaking down in an orderly manner into genes and finally cytomorphogens and protomorphogens, is a patterned living system of self-duplicating viruses, which differentiate in the medium of the generalized embryonic cell and exert their morphological influence on these cells. Wright (1941) has reviewed the self-duplicating aspects of gene physiology and remarked about the analogous properties of genes and the crystalline viruses. This concept stresses the separate but dependent relationship between the genic organizers and the basic low-organization embryonic cell. The reader will recall discussions in the first chapter which compares the elementary morphogens to a specific virus strain. The influence of new genes in altering the nature of similar low-organization cells has been reviewed by Manwaring (1934) in his discussion of the virulent gene that may be separated from virulent typhoid bacilli and added *in vitro* to a non-aggressive strain of typhoid, causing them to assume the same virulence as the original organism. These experiments strikingly demonstrate the influence of separate genic material over the morphology of low-organization cells. The whole concept of bacteriophage aptly set forth by d'Herelle (1924) (1926) (1930) is based upon the phenomena of virus transfer to, and influence over, receptive cells. This influence is demonstrated by considerable experimental evidence.

1 Hoadley, L.: *Biol. Bull.,* 46:281, 1924; *J. Exp. Zool.,* 43:151, 1926; *Arch. de Biol.,* 36:225, 1926; *J. Exp. Zool.,* 48:459, 1927.
2 Braus, H.: *Roux' Arch. f. Emwmech.,* 22:564, 1900.

The period of competence of embryonic cells is in some manner determined by their environment. Much investigation is needed to determine the exact nature of the factors influencing competence. The further progress of morphogenetic knowledge must await the results of such study.

Investigation of the factors regulating competence should bear in mind several possibilities suggested by anomalous physiological phenomena. First, there is a possibility that competence is directed by a hormonal "activator" which would probably arise from the chromosome or its genic groups. Second, there is the possibility that the period of competence is an inherent characteristic of cells which occurs only during a period of their cycle as a developing unit. Third, it is wise to keep in mind d'Herelle's comment (1924) that viruses are more likely to enter or leave the cell during the period of mitosis.

Further Comments on Organizer Phenomena.—Some understanding of the paradoxical nature of the results of graft experiments is now possible. If organizer is transferred to competent cells in the host, then the transplant pattern of differentiation may prevail. If, however, competent tissue is transplanted to an organizer region, the differentiation pattern of the host may prevail.

Weiss (1935) has commented that the intensity of the organizing influence of a graft seems to depend upon two factors. First, the size of the graft is in direct ratio to its organizing ability, and second, the closer the source of the graft to the center of the field, the greater its organizing influence. The hypothesis that there is an orderly cleavage of the chromosome, segregating virus-like genic groups into the center of the fields of organization where the cytomorphogens exert their influence over the morphology of neighboring cells, is supported by these comments on the conditions affecting the organizing potency of grafts.

The closer the source of the graft to the center of the field the greater amount of organizer or cytomorphogen it would be likely to contain. This is based upon the assumption that the center of the field of organization is the position of the functional genic group from which cleave smaller groups and eventually cytomorphogens to effect the neighboring low-organization embryonic cells into a patterned morphology. It is also obvious that the larger the graft the more

131

organizer it would transfer to its host and consequently the more powerful morphogenic effect it would exhibit.

Weiss considers that a graft exerts two quite separate influences on the host. First, if it contains organizers, it influences the morphology of the host to the limit of the strength of these organizers. Second, it may initiate differentiating activities in the host whether it directs them or not.

We have discussed this first influence and attempted to point out the various conditions which effect the final differentiation. We have mentioned that the host's cells must be competent to react with the graft organizers if the latter are to exert any effect. If the host's tissue has not reached the period of competence or has passed through it and is already "determined" by its own organizers, it will not be affected by the graft organizers. There seems to be some experimental evidence that in some cases there is a competition between the host organizers and the graft organizers, the more intense of the two exerting the more pronounced morphogenic action. An example of a mixing of organizer characteristics is seen when the graft of an ectodermal flap of one species is transferred to the gill region of another. The graft participates in gill formation, but at a rate characteristic of the donor species rather than of the host. Rormann[1] has reported occurrences of this phenomenon. We suggest that the graft cells had received some cytomorphogens from their own embryo which had succeeded in establishing at least the metabolic characteristics of the graft but more complex genic groups had not yet been influential in stabilizing the graft morphology, preventing it from participating in the differentiation of the gill region in the host.

A survey of the field of morphogenetic studies leads us to speculate that low-organization embryo cells are first differentiated into histological groups, such as epithelial cells, and later organized into more detailed localized structures such as the duodenal lining, etc. It is possible that the graft cells described in the above experiment had been separated into such a generalized group with a "determined" metabolism and, as such, were acceptable material for the construction of more detailed structures, such as gills, by the host.

It is not within the scope of this presentation to exhaustively discuss all the experiments dealing with grafts and link them to our morphogen

[1] Rotmann, F.: *Roux' Arch.. f. Entwmecb.*, 124:747, 1931.

hypothesis. The interested reader should consult such experts as Needham (1942) and Weiss (1939) for adequate reviews of such work.

Weiss' comment that the graft, in addition to the direct influence of its own organizers, also initiates differentiation in the host, brings our discussion to a more detailed study of the nature of the organizer itself.

Induction of Differentiation by Other than Organizer Material.
Geinitz[1] discovered that the organizer is not species specific. **He** was able to demonstrate that grafts from very different species, genera or families, could be transferred, exerting a remarkable effect on their new environment. Following up Geinitz' work many investigators[2] were able to demonstrate that organizer grafts from non-related species would induce normal differentiation and sometimes the formation of a secondary embryo (twinning) in the host. It is interesting to note that the differentiation pattern stimulated by this process is usually that of the host and not of the species from which the explant was obtained. Genetic terminology calls the explant in this case an "inductor" because it induces differentiation in the host rather than effects it through donor organizers *per se.*

It remained for Holtfreter[3] to discover that this inductor effect is not exclusively embryonic but a property of all adult tissue of all phyla. He was able to demonstrate the effect of various non-related grafts in inducing twinning or the production of secondary organs in a developing embryo.

Holtfreter[4] and others have also demonstrated that dead tissue can exert this inductive effect, although the inductions resulting from dead tissue are more likely to be a histogenetic process rather than an organ building one. Inductions from living tissue, on the other hand, frequently give rise to the differentiation of patterned organs in addition to the histogenesis of cell types.

1 Geinitz, B.: *Roux' Arch. f. Entwmech.*, 106:357, 1925.

2 See: *Biochemistry and Morphogenesis*, Needham, J. Cambridge University Press, Cambridge. 1942.

3 Holtfreter, J.: *Naturwiss.*, 21:766, 1933; *Roux' Arch. f. Entwmech.*, 132:307, 1934.

4 ———— *Ibid*, 128:584, 1933.

Holtfreter[1] also was able to demonstrate that portions of an egg not possessing inductor ability acquire it after boiling. Some adult tissues, however, were shown by this author[2] not to require boiling to manifest inductive ability.

However, both embryonic and adult inductors were demonstrated to survive prolonged boiling. The effects were greatly diminished by heating at 135° C. and destroyed at 150° C. or by ashing. Chuang[3] on the other hand demonstrated that the inductor effect of mouse kidney and of newt liver was practically destroyed at 100° C.

Treatment of non-inductor tissue with various organic solvents such as alcohol, ether, etc., caused the appearance of inductor ability. Needham comments that possibly any treatment that denatures proteins will cause the power of inducing differentiation to appear.

Simultaneously several investigators prepared cell-free extracts with inductor activity. Spemann, Fischer and Wehmeier[4] found the inductor to be soluble in acetone. Needham, Waddington and Needham[5] and other investigators made active extracts using ether as a solvent.

Recently, Levander (1945) has reponed success in producing new cartilage in muscle injected with alcoholic extract of bone. This has been confirmed by other workers (Annersten, 1940, and Bertelsen, 1945) .

It would seem that the inductor substance has the following properties: (1) it is non-specific; (2) when prepared from dead tissue, it induces histogenetic differentiation, while from living tissue it can induce organ morphogenesis; (3) it is relatively heat stable; this property varying with the source and type of inductor;
(4) it is released from a "masked" position in tissue with non-inductor properties by denaturation, and (5) it is soluble in or activated by, ether and acetone.

All of these distinguishing characteristics are identical with those we

1 Holtfreter, J.: *Roux' Arch. f. Entwmech.,* 132:225, 1934.

2 *Ibid,* 132:307, 1934.

3 Chuang, H.: *Roux' Arch. f. Emwmech.,* 140:25, 1940; *Biol. Centralb.,* 58:472, 1938.

4 Spemann, H., Fischer, F. G., and Wehmeier, E.: *Naturwiss.,* 21:505, 1933.

5 Needham, J., Waddington, C. H., and Needham, D. M.: *Proc. Roy. Soc. Ser. B* (London), 114:393, 1934.

have reported previously as characteristic of the morphogens except for the fact that the inductors seem to be non-specific. This will be discussed further on in our review of the organizer problem. The evidence thus far strongly suggests that the organizers and inductors are identical and that they consist of cytomorphogen groups.

However, conflicting evidence on this contention has been presented in that agents that are not the product of life processes have been demonstrated to be capable of inducing organization. Needham (1942) reviews the various experiments in which various synthetic polycyclic hydrocarbons have induced differentiation and organization in embryos. Weiss (1935) has reviewed investigations in which implantation of such irritants as celloidin resulted in inductor phenomenon.

Needham dwells long on the possibility that some of the polycyclic hydrocarbons are the primary inductor substance. He reports the other view, however, advanced in particular by Woerdeman[1] who concludes that the host plays an essential part in the inductor phenomenon and the inductor stimulus is merely an activation of host organizers. Weiss (1935) states that this is the only safe stand one can take in view of the present state of experimental knowledge. Because no suggestions have been advanced as to the manner in which a single synthetic chemical substance can organize even the various histological aspects of differentiation (to say nothing of the multitude of morphological organ patterns), we prefer the conception advanced by Woerdeman that these inductor substances exert their effect through the release or activation of morphogens (organizers) in the host.

This conception is further substantiated by the fact that many of the synthetic inductor chemicals are carcinogens (Shen, 1939). Attempts were made to prove that these factors did not produce their activity through release of host organizers by demonstrating that the inductor effect was not in linear proportion to the amount of the inductor.[2] We do not consider this adequate evidence since the metabolic effect of the morphogens is not in direct proportion to their concentration, but varies with it, and it is not unlikely that this property may complicate the organizing effects of morphogens also.

1 Woerdeman, M. W., *Proc. Kon. Akad. Wentenskap. Amsterdam,* 39:306, 1936.

2 See: *Biochemistry and Morphogenesis,* Needham, J. Cambridge University Press, Cambridge. 1942.

The catalytic nature of these synthetic inductors is amply shown by a comparison of the amount of synthetic inductor necessary for a neural induction as compared to the amount of nucleoprotein. (In our previous chapter we emphasize that nucleoprotein contains an intense concentration of morphogens.) This ratio is 1:250,000. The comparatively minute amount of synthetic. inductor leads us to suggest that it acts as a catalyst to release morphogens in the host, while the effects of the nucleoprotein depend upon the activity of its content of morphogens *per se*. In the absence of an inductor catalyst it is likely that comparatively little organizer is made available from the nucleoprotein, accounting for this tremendous difference.

We suspect that the effects of organizers themselves, as inductors, can be due to their action in releasing or activating the host's organizers (morphogens). Needham is aware of this possibility for he recognizes that the liberation of inductors is linked to the process of determination and histological differentiation. We have shown in previous chapters that intense concentrations of morphogens cause lysis and general disintegration of cells. This fact may be linked with the fact that non-active organizers or inductors may be released by boiling or by denaturation of their protein complexes.

There is one more problem in this study of grafts that Needham is careful to emphasize. That is the difference between regional organ differentiation and cell histological organization. The synthetic evocators and "dead" organizers usually have been shown to evoke only histogenesis and to be incapable of elucidation of the complex pattern of an organ. "Live" organizers, on the other hand, are capable of evoking differentiated organs as well, although the patterns may be in a different species and of a different organ than those which would have been determined by the organizer in the embryo from which it is taken. In this case the organizer from live tissue is acting as an inductor, not as an organizer *per se*.

Organizers or inductors from living tissue which has not been boiled are, in some manner, able to activate and stimulate the orderly mechanism by which the chromosome "unwinds" and exhibits a patterned cleavage into regional differentiation. The net result is the determination of the structure of a complete organ. On the other hand, organizer substance which has been boiled, or synthetic organizer, merely releases cytomorphogens from the host in a disorderly manner which promotes histogenesis only. Needham recognizes the link between chromosome cleavage and the characteristics of "live" organizers when he discusses the fact that the production of "twins" in embryos by such inductors is a similar process to the normal production of twins in the human, which is known to be controlled by inherited chromosome characteristics.

To sum up the review of graft experiments we might list the various possible paths by which differentiation may be effected as a result of graft transfer: (1) an organizer center of the donor embryo may be included in the graft and act upon competent tissue in the host to effect organ differentiation of a nature peculiar to the donor species; (2) the organizer center of the host may be "activated" by the addition of morphogens in the graft to effect organ differentiation peculiar to the host species, but this can only happen if the graft morphogens are of a more complex nature and not broken down by heat or denaturation; (3) the graft or cell-free graft extracts may induce donor type histogenetic organization in the competent cells of the host because of the donor cytomorphogens present in the transfer, or (4) the cell-free graft extracts, "dead" graft inductor, or synthetic inductors may "release" and "activate" the host cytomorphogens resulting in histogenesis of a nature peculiar to the host species.

It is obvious that we are dealing with a very complex picture in which various experiments will stimulate paradoxical conclusions unless all the possibilities are kept in mind. We have found that all the graft and inductor experiments we have been fortunate enough to study can be placed in one of the above four categories. The two problems in this study which require further elucidation are: (1) the nature of the influences that determined competence, and (2) the reason behind the evocation of organ determining effects by "live" inductors (the reason that more complex "higher" morphogen groups are able to stimulate the organized cleavage of the host chromosome while the less complex "lower" cytomorphogens do not have this ability). There is much fertile ground for further investigation in these fields in particular.

Needham concludes his review of morphogenesis with a discussion of the fibers appearing between the cells of the developing embryo. These fibers arise in the cell-free "caryolymph" which fills the cavities of the embryo organism during its early stages. This homogenous substance appears to be secreted by the cells and provides them with a natural culture medium in which development and migration may take place. This material is present in relatively large amounts during the early stages of embryonic development. It is the precursor of adult connective tissue.

Wavy fibrils appear in this cell-free substance and cells migrate into it later.[1] The experimental work of Weiss[2] strongly suggests that the matrix formed *by* these fibrils orients the developing embryonic cells into an organized pattern. He demonstrated that fibroblast cells cultured *in vitro* will migrate along the axes of tension created in a membrane of blood plasma placed in frames of various geometrical forms. Needham emphasizes the importance of these fibers in directing the developing embryonic cells into an organized pattern.

Baitsell has shown this fibril formation to be analogous to the formation of clot material in wound healing. We refer our reader to the following chapter where we attempt to identify the thromboplastin released from platelets with protomorphogens. Particularly significant is the observation that a disintegrating platelet secretes macromolecules which leave a trail of a fibrin thread behind them as they travel through the fibrinogen-containing serum. We have discussed the thread-forming properties of protomorphogens previously.

We postulate that the chromosome releases protomorphogen molecules into the intercellular "caryolymph" where they form fibrils along which developing cells migrate, thus orienting the whole embryo into its characteristic predetermined pattern. The work of Hardy and Nottage[3] has led Hardy[4] to propose a theory of trophic action of nervous phenomena

---

1 Baitsell, G. A.: *Proc. Soc. Exp. Biol. & Med.,* 17:207, 1920; *Proc. Natl. Acad. Sci.* (Washington), 6:77, 1920; *Am. . Anat.* 28:447, 1921.

2 Weiss, P.: *Am. Naturalist,* 67:322, 1933.

3 Hardy, W. B., and Nottage, *1* I. E.: *Proc. Roy. Soc. A.* (London), 118: 209, 1928.

4 Hardy, W. B.: *J. Gen. Physiol.,* 8:641, 1927; *Colloid Symposium Monogr.,* 6:7, 1928.

based upon the possibility that molecules are oriented by a nerve fiber. This is a brilliant attempt to extend a slight clue concerning the mechanism by which the "caryolymph" fibrils are patterned into a matrix, thus effecting the overall embryo morphogenesis. We suggest that these fibrils are patterned into a matrix under the influence, either of some physical orientation, or of the polarized field surrounding the cells whose chromosomes secrete the protomorphogen giving rise to the fibril. Burr[1] has performed a series of illuminating experiments on the electrical field surrounding various organisms. Of special interest is his comment that the field surrounding an egg may determine the pattern of organization of the developing organism. It is not unlikely that future experiments may demonstrate the influence of this electrical field in guiding the patterned formation of the "caryolymph" fibrils. Further experimental investigations of the factors which influence the organization of this fibril matrix will enable students of morphogenesis to understand more clearly this phenomenon which is the essence of embryonic organization.

In discussing the influence of the field surrounding the ovum, it may be of interest to study the conclusions of Rothen (1945). This investigator has demonstrated that the sphere of influence of antibodies is so vast (possibly hundreds of Angstrom units) that they may even react across thin biological membranes.

It would be inappropriate to conclude our brief discussion of the relationship of morphogens to differentiation without recognizing the outstanding work of both Paul Weiss and Joseph Needham. Practically all the bibliography in this chapter has been supplied by these two investigators in the various reviews that they have published. The reader will realize that this subject of morphogenesis must be expanded into volumes in order to be adequately treated. Any more than a brief mention of the links between the morphogen hypothesis and the specialized study of morphogenesis is beyond the scope of this presentation. We are indeed indebted to these two men in particular for their candid reviews which made this task possible.

1 Burr, H. S.: "The Meaning of Bio-Electric Potentials." *Yale J. Biol. & Med.*, 16:353-360, 1944.

Summary of Morphogens in Embryonic Differentiation

Summing up the hypothesis outlined in this chapter, we postulate as follows: (1) the chromosome is cleaved into genic groups which are segregated during gastrulation into the areas of the embryo from which the major adult structures will arise; (2) these genic groups are the functional centers of the various "fields" of organization; (3) these genic determinants of major adult structures are oriented in respect to each other by means of fibrils in the intercellular "caryolymph"; (4) these fibrils are formed by protomorphogens released by the chromosome which produces fibers in the "caryolymph" in the same manner as thromboplastin (protomorphogen) forms fibers from fibrinogen; (5) these fibrils are organized into a patterned matrix through the influence of either a physical strain or electrical polarity determined by the chromosomes; (6) during the neurula stage the genic groups cleave further into organizers which have the power of organizing more specialized components of the major adult structures; (7) during this stage, and also in some particular cells during gastrulation, the genic groups release cytomorphogens to neighboring cells which thereby are organized into specialized differentiated tissue; (8) at this point the general embryo cell is a low-organization individual incapable of differentiation until it receives the cytomorphogen from the genic center; (9) there is only a limited period in the life of these low-organization cells during which they are capable of receiving the organizing cytomorphogen from the gene; this period is known as "competence," and (10) the remarkable pattern of mineral distribution during embryonic development and during mitosis strongly suggests that the cytomorphogens of the gene and the genes of the chromosomes contain organized assemblages of mineral-linked protomorphogens.

The student of morphogenesis will recognize that this hypothesis seems to contain some of the contentions originally advanced by Weismann. Weismann held that there occurs an orderly segregation of the chromosome by which the portions which determine various organs eventually find their way to the point in which those organs will arise. Since Weismann, however, experimental evidence has shown that at each cell division the whole chromosome is duplicated and transmitted to the daughter cells. The modern viewpoint maintains that every cell of the body contains the sum total of the genic material carried by the original chromosomes[1].

It is difficult to reconcile this viewpoint with the demonstrated specific effects of the morphogens present in embryo transplants. The general contention is that most of the genic material is "masked" in some manner and only the portions selected for a certain organ or cell are "activated." Early experimenters with regeneration evolved their hyotheses around the contention that the genic material of the chromosome has a primitive specificity for the cell it is to determine. It has been shown, however, that genic material, when transferred to a new location, may enter into determinant activities of a different nature than it would have in its previous location. While the cytomorphogen determinants retain a semblance of tissue specificity (epithelial, cartilaginous, etc.), overall organ differentiation is controlled by the intact chromosome through its patterning of cytomorphogen histogenetic influences. It is therefore conceivable that genic material or cytomorphogens with more or less universal histogenic potentialities could participate in widely different differentiations determined by different chromosome matrix patterns. By crude analogy, the cytomorphogens might be compared to bricks which may be organized into a number of different buildings, depending upon the architect's plans.

From our standpoint it makes little difference whether the total genic chromatin is present in every cell, with most of it "masked" or whether, as Weismann speculated, only the significant portions are present. Our hypothesis is primarily concerned with establishing the assumption that the cytomorphogens are the organizer material from which the genes are composed and that the genes effect cell differentiation by releasing cytomorphogens, which are picked up by neighboring competent cells, whose morphology is subsequently "determined" by the absorbed cytomorphogen. This hypothesis simply links morphogen phenomena with chromosome differentiation, and the details must be worked out by further experimental investigations.

1 Morgan, T. H.: "Mendelian Heredity in Relation to Cytology." Section XI, *General Cytology,* Edited by E. V. Cowdry. University of Chicago Press, Chicago. 1924.

Metabolic and Determinant Morphogen Cycles.—In Chapter 3 we presented our hypothesis covering the metabolic reactions of the morphogens from the time of their synthesis in the nucleus to the point where they are secreted from the cytoplasm as waste products.

The morphogens are synthesized into chromatin material at the nuclear wall. The protomorphogens released from cytoplasmic proteins consequent to their dynamic state are used as substrate in this synthesis. As a result of nuclear chromatin metabolism, protomorphogens are discharged from the nucleus under the protection of a fatty envelope and re-utilized as part of the substrate for the synthesis of cytoplasmic proteins at the cell surface boundary. When the cytoplasmic protein molecules release the protomorphogen components near the nuclear wall as substrate for chromatin synthesis, the cycle is complete.

We mentioned another morphogen cycle that is concerned with organizing the structure of the cell. The schematic diagram (Fig. 5) indicated this simply as macromolecular particles secreted by the nucleus into the cytoplasm for the purpose of governing the histogenic organization of the cell. These particles become a part of the cytoplasmic proteins, and protomorphogen components are released by them, consequent to their dynamic state. Thus released, they are utilized again at the nuclear wall for chromatin synthesis.

The determinant cycle occurs once with every mitosis. It is nor necessary except to organize the development of the daughter cells at division; herein lies one basic difference between the determinant and metabolic morphogen cycles. The metabolic morphogen cycle occurs constantly as a result of nuclear metabolism and is not dependent upon division. This is indicated by the observation that protomorphogen secretion into the media continues after mitosis ceases, indicating the persistence of the metabolic cycle.

We shall refer to these two cycles as the metabolic cycle and the determinant cycle of the morphogens. The discharge of protomorphogens from the nucleus in the metabolic cycle is a consequence of a constant dynamic activity of chromatin metabolism. At each cell division nuclear reorganization occurs and the chromosomes are "cleared" of "waste" protomorphogens which are discharged into the cytoplasm, probably under the protection of a fatty envelope. These protomorphogens have no effect in the cytoplasm other than a toxic one; they are further discharged into the medium. Cowdry (1939) mentions the reorganization of the macronucleus at each division and the elimination of portions of the macronucleus into the cytoplasm as "waste" material.

Jennings (1940) has reviewed the work of Richards (1917) and others describing a different method in which the nuclear morphogens come in contact with the cytoplasm during the determinant cycle. It should be emphasized that the protomorphogens in the metabolic cycle are toxic and should not come in contact with the cytoplasm; therefore, they are protected by a fatty layer. In the determinant cycle, however, it is necessary that the morphogens come into an intimate controlled contact with cytoplasmic material in order to exert their organizing influences. Richards (1917) has observed that in the egg of Fundulus the chromosome vesicle enlarges and engorges with cytoplasmic material. After absorbing much cytoplasmic material, the vesicles discharge it back into the cytoplasm thoroughly mixed with great numbers of chromatin particles; this action occurs before each cell generation. It is most conspicuous at the very start of egg development, just before mitosis begins. It is of interest to note that as early as 1902, Conklin observed similar interaction in Crepidula upon which he comments as follows "One might speak of these changes in the nucleus as systole and diastole, by means of which an exchange of nuclear and cytoplasmic materials is brought about."

Jennings (1940) has observed similar phenomena in paramecia and infusoria. Referring to the interaction of the macronucleus and cytoplasm in infusoria he comments as follows: "The transfer of so great a quantity of nuclear material into the cytoplasm must greatly affect the nature and physiological activity of the cytoplasm; some of its presumable results we shall see later in the genetics of these organisms." This interaction is observed at definite cycles, the length of which depends upon the characteristics of the individual. Calkins (1930) has also described the discarding of portions of macronuclear material (into the cytoplasm) during division. (This process may be a part of the metabolic rather than the determinant cycle. In fact, it is quite difficult to definitely assign results of this type to one cycle or another.)

Guilliermond (1921) has asserted that the rod-shaped structures of the cytoplasm known as mitochondria have an important function in the elaboration of products of cell activity. Cowdry (1924) has reviewed the speculations of the early workers on the functions of mitochondria. Of special interest is the early idea that mitochondria are concerned with histogenesis; this suggestion is discussed by Meves (1918). Mitochondria have recently[1] been reported to be high in fatty substances, particularly in nucleoprotein and phospholipids. We have mentioned previously that the morphogens are a part of the nucleoprotein molecule and likely to be associated with fatty or lecithin material in the cytoplasm. In this case we postulate that the morphogens in the mitochondria are concerned with histogenesis and are not waste products protected by a fatty envelope. The association of these factors with phospholipids in the mitochondria may be necessary to prevent the determinant morphogens from exerting their influence suddenly and to allow their organizing activities to proceed in a slow and orderly manner. Drew (1922) has suggested that the fatty globules appearing in the cytoplasm arise from the degeneration of mitochondria. We have already postulated (see discussion of the metabolic cycle, Chapter 3) that these fatty globules consist of "spent" morphogens extruded from the chromatin and wrapped in a fatty envelope to prevent lethal effects in the cytoplasm. It is quite possible that some mitochondria from which the morphogens are not completely utilized may degenerate into these fatty globules, or the fatty materials in the mitochondria may attach themselves to the waste morphogens from the chromatin as protector coverings after the mitochondria morphogens have separated themselves to act as cytoplasmic organizers. It would be of interest to investigate the connection, if any, between the mitochondria and the morphogens of the cytoplasm with a view to identify each as particles of histogenetic influence.

Mitochondria have been subject to extensive investigation. We present here a few pertinent observations on their biochemistry and physiology.

1 Bensley, R. R.: "Chemical Structure of Cytoplasm." *Biol. Symp.*, 10:323-334, 1943.

Horning (1927) comments that the aggregations of mitochondria at the outer surface of the cell and nucleus is due perhaps to their phospholipid nature. More recently Davidson and Waymouth (1943-44) report that the cytoplasm of the liver cell contains ribonucleic acid which they assume to be present in the mitochondria, microsomes and secretory granules. One of the most consistent comments on mitochondria is that they contain nucleoproteins associated with phospholipids and fatty materials. This is one of the experimental facts upon which we base our postulation that mitochondria are connected with morphogen phenomena. If mitochondria contain morphogens they would have to have an ash residue. The reports of Chargaff (1942) include the presence of 1.6 per cent ash in a typical preparation of mitochondria lipids from rabbit liver.

Some investigators have assigned enzyme activity to the mitochondria since they are found associated with food vacuoles. Horning (1928) has found them to migrate slowly in the vicinity of food vacuoles, but has not observed them puncturing the vacuole membrane, although they are apparently seen within the vacuole from which arises the postulation concerning their enzymatic nature and function in assimilation. His previous studies (1926) indicate that the mitochondria are extruded from the protoplasm into the food vacuole. He (1926) observes that engulfed food circulates in the protoplasm, comes in contact with mitochondria which adhere to it, finally a food vacuole is formed; this slowly dissolves, along with the mitochondria, indicating an absorption and digestion of food.

Horning (1926) also comments that no evidence exists to indicate that mitochondria are formed *de novo* from the cytoplasm, but that they always increase on binary fission. Nevertheless, he states (1927) that they are not traceable through the nuclear membrane, although they are found aggregated outside, next to it in many cells. The mitochondria, themselves, have been observed by this investigator (1926) to undergo binary fission in an heterotrichan infusorian and he has not observed a constant number of mitochondria to occur in various other organisms. Later (1929-30), he concludes that the disappearance and reformation of mitochondria is associated with their synthetic ability.

As a basic principle, we suggest that, whenever morphogens of any nature containing active linkages appear in the protoplasm or tissue fluids they are normally immediately enveloped in a wrapper of a fatty phospholipid complex and this wrapper prevents the active morphogen linkages from engaging in any uncontrolled biochemical activities.

The oft-reported aggregation of mitochondria at the nuclear wall leads us to suggest that as morphogens are detached from the chromatin and extruded into the cytoplasm (during mitosis or conjugation phenomena) they are enveloped immediately outside the nuclear membrane in such a lipoid wrapper. We feel that all such mitochondria are of nuclear source and this phenomenon indicates why the lipoid-like body itself has not been seen to pass through the nuclear membrane (Horning, 1927).

Once in the cytoplasm, the morphogen groups in the lipoid wrappers known as mitochrondria are free to exert their morphogenic influences over histogenesis in an orderly and controlled manner. The fission, disappearance and reformation, of such mitochondria indicates that they contain nucleoprotein protomorphogens of a complex nature which are slowly "unwinding" as they exert their morphogenic influences.

The apparent influence over the digestion and absorption of engulfed particles indicates another manner in which the morphogenic influences of the mitochrondria may be evidenced. Before such engulfed food particles can be used to synthesize new cytoplasm they must be enzymatically reduced to be sure, but free protomorphogens must be brought in contact with them to insure the synthesis of proteins homologous to the cell. We believe it is possible that, in addition to whatever enzymatic reduction takes place, the mitochondria also supply the substrate materials within the food vacuole with these necessary protomorphogens, fresh from the chromatin. It thus becomes apparent that our previous conception (Chapter 3) that synthesis of new protein only takes place at the cell surface boundary must be extended *in these special cases* to include the synthesis of new protein in the locale of the food vacuole associated with mitochondria.

The reader may wonder why we have postulated two morphogen cycles, the metabolic and the determinant cycle. An alternate hypothesis would envision the determinant effects of the morphogens as a consequence of nuclear metabolism. This hypothesis would suggest that the nuclear chromatin becomes a part of cytoplasmic structure during the nuclear-cytoplasmic interchange or re-organization period of mitosis. Later as a consequence of the dynamic state of cytoplasmic proteins, protomorphogens would be released into the media and also at the nuclear wall for new chromatin synthesis. There is indeed not enough experimental evidence available to establish either the separate cycle or single cycle hypothesis to the exclusion of the other.

Reasons for Postulating Two Morphogen Cycles.—We are inclined to favor the idea of two separate morphogen cycles for the following reasons: (1) it is likely that all new cytoplasmic proteins during growth are synthesized at the cell wall, for at this point only, all substrate material is available, therefore protomorphogens could only become a part of the cytoplasmic proteins at the surface boundary of the cell; (2) this is indicated by the observation that cell division cannot occur until a "threshold" amount of protomorphogen is present in the media; (3) the histogenetic organizing influence of the morphogens is only needed at cell division, not for repair, (4) after cell division ceases the accumulation of protomorphogens in the media still continues with eventual lysis (See Burrows, Chapter 2); it is difficult to conceive of the histogenetic influences of chromatin occurring except at cell division.

It appears that the fundamental difference between the metabolic and determinant cycles may be outlined as follows: *the metabolic cycle* results in the discarding of protomorphogens from the chromatin, through the cytoplasm into the media as a result of
metabolic activities. These protomorphogens are determinants for cytoplasmic *proteins* synthesized at the cell wall; *the determinant cycle* results in the extrusion of chromatin into the cytoplasm containing cytomorphogen fragments which act as determinants for *histogenesis.*

Regardless of the lack of reliable evidence in this conflict we are presenting our hypothesis around the conception of two morphogen cycles, the metabolic and determinant. This is partially for the sake of continuity, and thus we have also mentioned an alternate hypothesis. Only further experimental investigation can integrate this problem into a clear theory of biodynamics.

In our discussion (Chapter 3) of the metabolic cycle we mentioned the suggestion that the dynamic state of cytoplasmic proteins periodically supplies free protomorphogens for chromatin sythesis at the nuclear membrane. It is quite important from a standpoint of cell morphology that these protomorphogens be synthesized into chromatin and discharged again as waste material through the cytoplasm into the media.

In the first sections of this chapter we have discussed the morphogens as the "determinant" organizer substance of the genes which is distributed to various locales in the developing embryo, where it becomes a part of the cytomorphogen of the cells. We see that at each cell division there is a mixture of nuclear and cytoplasmic constituents which "implants" chromatin material in the cytoplasm. It is probable that the mitochondria consist of morphogens at a point in their determinant cycle. This, then, is the mode of histogenetic action of the cytomorphogen. We should now investigate the methods by which the morphology of the cell, once organized, is maintained against the influence of external agents.

Environmental Modifications of Structure.—We shall shortly discuss the influence of environment over the morphology of cytoplasmic structure. During the metabolic cycle, each protomorphogen released by the cytoplasmic proteins to the nucleus is "reworked" by the heterochromatin, and any environmental modifications are eliminated. We suggest that this is an important part of the mechanism tending to perpetuate the cell's morphological integrity.

The effects of environment over a cell may be considerable and the fact that these environmental modifications can be inherited indicates that they are associated with a change of morphogen structure.

This brings again to the biological forefront the question of inheritance of acquired characteristics. Probably no biological question has been the butt of so much controversy as this in years past. Jennings (1940) has presented an adequate review of this problem as it applies to infusoria. Environmental modification of protozoa has been reported by many investigators[1]. Jollos (1921) has pioneered in the investigation of

---

1 Neuschlosz, S. M.: "Uber die Gewohnung an Gifte. I. Chiminfestigkeit bei Protozoen." *Pfiueger's Arch. Ges. Physiol.,* 176:223, 1919. "Untersuchungen uber die Gewohnung an Gifte. II. Die Festigkeit der Protozoen gegen Farbstoffe. Ill. Das Wesen der Festigung von Protozoen gegen Arsen und Antimon." *Ibid,* 178:61, 1920.
Dallinger, W. H.: "The President's Address," *J. Roy Microsc. Soc.,* 1: 185, 1887.

this problem in infusoria. Morphological and functional changes can be induced in individuals by altering culture conditions. These modifications may be caused to persist for many generations after the induced stimuli has been removed, but eventually the cells return to normal. The modification may consist of acclimatization to various chemical and physical agents. The disappearance of the modification after elimination of the inducing stimuli led Jollos to present the theory that such modification is cytoplasmic in extent and not due to genic differences. Raffel (1932) has attempted to interpret these phenomena as gene mutations, but Jollos (1934) has objected to this explanation. Sonneborn (1941) has recently reviewed the literature of protozoan inheritance.

In a later chapter we present additional evidence which can explain the puzzling occurrences of apparent inheritance of acquired characteristics occasionally reported. The final chapter of this biological enigma is yet to be written. As in most fields, further experimental research is necessary before comprehensive conclusions may be arrived at.

Manwaring (1934) has supplied additional information concerning the influence of environment over the morphology of bacteria. He reviews experiments in which pedigreed bacterial strains were caused to "mutate" into wholly unrelated antigenic types or genera. The mutant may revert back to normal type under optimum culture conditions but can in some cases be cultured for many generations with no tendency to revert. The "mutations" may appear at any time but are favored by sub-optimum culture conditions. The mutants may be a new unconventional, stable, species. Of eleven species investigated, none were found which did not "mutate."

As fantastic as such a phenomenon appears, it nevertheless has been observed by many investigators and presents additional evidence that the morphological integrity of a cell is profoundly influenced by environmental stimuli.

Bastian[1] has reviewed many experimental demonstrations of bacterial mutations induced by various abnormal culture conditions; he calls this

1 *The Nature and Origin of Living Matter,* Bastian, H. C. J.B. Lippincott Co., Philadelphia.

phenomenon heterogenesis. He even lists experiments[1] in which the spontaneous generation of living forms were apparently derived from sterilized colloidal silica. His sealed tubes were exposed to temperatures of 130° C. for periods of 5 minutes and later examination indicated the presence of living forms, principally bacteria. In view of modern knowledge on the stability of certain spore forms and especially Turck's observation on the extreme thermostability of cytomorphogen, it is at once apparent that spontaneous generation of life did not occur in these cases, but the necessary genic groups were maintained intact or spore forms survived his sterilization process. Apparently these groups were present in the original colloidal silica he employed for it was important that it be obtained only from one particular source.

Avery, McLeod and McCarty (1944) have published studies on the nature of the substance responsible for inducing the mutation or transformation of certain pneumococcal types. This article reviews the comments on the nature of these transformations. They demonstrate that a highly purified product may be extracted from type III pneumococci which in small amounts under the proper conditions can induce the transfer of type II pneumococci into type III, the same as those from which the inducing product was extracted. This transformation was first demonstrated in animals by Griffith (1928) and *in vitro* by Dawson and Sia (1931).

Avery and his co-workers found that this active product contained no demonstrable protein, unbound lipid or serologically active polysaccharide. It consists principally, if not solely, of a highly polymerized viscous form of desoxyribonucleic acid which exhibits no conclusive serological reaction with type III anti-pneumococcus rabbit sera. They conclude that this type of nucleic acid gives rise to enzymatic reactions which result in the demonstrated type transformation.

We postulate that the active agent is a special type of nucleic acid and that its influence over the energy reactions in chromatin nucleoprotein (see Chapter 3) changes them so that the active morphogen groups constructed and released in the chromatin are specific organizers for the transformed, type III, pneumococci. This could conceivably be accomplished by a change in the period associated with morphogen

1 *The Origin of Life,* H. C. Bastian. G. P. Putnam's Sons, New York. 1911.

morphogen synthesis and its preparation for organizing activities.

Stanley, Doerr and Hallauer (1938) find an analogy between the agent responsible for the type transformation and a virus. The conclusions of Avery et al make this analogy difficult to accept by strict interpretation. We feel that here, as in the problem of nonspecific inductors of tissue organization (reviewed earlier in this chapter), the inducing agent is not directly responsible for the change; rather it makes possible certain activities in the morphogen group which *per se* are directly concerned with the organization or reorganization of structure. These morphogens are, as we have postulated throughout, closely associated with, or actually, virus nucleoprotein molecules. Murphy (1931) (1935) perhaps has supplied certain evidence supporting this contention since he has specifically termed the hypothetical transforming agents in cancer as mutagens, to distinguish them from virus'. Carcinogens are not virus' nor are they, as we have discussed earlier in this chapter, directly responsible for embryonic induction or mutation in cancer cells. Rather, they make possible the induction or mutation under the direct control of the affected morphogens.

Pontecorvo and White (1946) have recently reviewed this problem and in particular have mentioned the cytoplasmic "self-duplicating" particles, plasmagenes. They also mention the recent discovery that an extract of yeast cells adapted to ferment a particular sugar will increase the speed of adaptation of non-adapted cells. The work on plasmagenes has proceeded for a great part on the problem of the "killer" reaction in paramecium. Wright (1945) has reviewed this problem in some detail. It seems that in the presence of a particular gene there is an extranuclear substance (kappa) which multiplies in the presence of this gene, but only persists for a few generations in its absence. It is apparent that cytoplasmic heredity may be produced by a cross between the paramecium containing kappa and one which does not. The presence of the gene synergist will not give rise to kappa, in spite of the fact that kappa will not persist for many generations in its absence. We might suggest that kappa is the protomorphogen and the particular gene necessary for its persistence is the locale in the chromatin where kappa is "reworked" in the nucleus during the metabolic and determinant cycles. In the absence of this chromatin "reworking" (the particular gene) kappa gradually disappears due to the absence of a part of its morphogen cycle.

In this case it would seem as if the kappa substance is a morphogen molecule capable of multiplying in the cytoplasm and being passed on to progeny through the medium of "cytoplasmic inheritance." The problems of cytoplasmic inheritance are stimulating and it is encouraging that they are right now receiving considerable attention, particularly in view of their relationship to the inheritance of morphogenic transformations or "acquired characteristics." More experimental evidence is necessary for a competent evaluation of the problem; we simply wish to emphasize the basic principle that morphogenic changes, inherited or not, may be induced either by the addition of a particular morphogen itself or through the medium of an inducing stimulus which acts indirectly through its influence over the morphogens. This phenomena appears throughout biology; in the study of embryonic induction and transplants, the characteristics of cultures, and the study of cancer. It should be kept constantly in mind in order to clearly evaluate the experimental reports as they accumulate.

We should not neglect the reports of Demerec (1946) on radiation induced mutations in Escherichia coli affecting their susceptibility to Tl. bacteriophage. The mutation rate remains high for several generations after the irradiation. Also the important and provocative reports of Lysenko (1946) should be carefully evaluated. Hudson and Richen's (1946) have published a critique of his work which seems to be most candid. Lysenko claims to alter heredity in plants by manipulation of the environment and his "new genetics" promotes many Lamarckian concepts. Hudson and Richens show much of his data to be inconsistent and unsupported by acceptable experimental evidence; they admit, however, that other geneticists have not taken the trouble to repeat his experiments. Without presuming to pass judgment we feel that Lysenko's work deserves the most careful investigation, and certainly his experiments should be repeated before criticism of his conclusions is presented. The orthodox concepts in genetics are being altered by recent discoveries, some of which are reported on these pages, and, under the circumstances, a most candid and open attitude is vitally necessary for the progress of genetic knowledge.

We shall now turn to the methods of tissue culture for still additional evidence of the influence of environment on cell morphology.

Drew (1923) has supplied some interesting information on the influence of environmental factors over the morphology of cells cultured *in vitro*. He has noted that when tissue from an animal is cultured for several generations and then transferred back to the donor it does not grow, but is reabsorbed and disappears. It is apparent that the differences in environment during this period have changed the specific nature of the animals cells and the transfers so that they are no longer compatible. Drew has reported in the same publication that kidney tissue *in vitro* can be caused to differentiate into tubules upon the addition of connective tissue; without this addition the kidney tissue grows into undifferentiated sheets.

Ebeling and Fischer (1922) have investigated this phenomenon and find that epithelial cells and fibroblasts may be cultured side by side and retain their integrity. The addition of connective tissue had little or no effect. This work had also been done at a previous date by Champy.[1] Landsteiner and Parker (1940) have supplied further evidence upon the stability of the morphology of cells in tissue cultures. They demonstrated that connective tissue from chicken kidney cultivated in rabbit serum not only retained its species specificity but caused the formation in the rabbit serum of serum proteins antigenically associated with the chicken.

It is apparent that connective tissue exerts an important influence over cell dynamics. Drew's observation that it caused differentiation of kidney cells indicates that it can function as the "evocator" of organizer function. We shall discuss connective tissue in more detail in our next chapter. It will suffice here to mention that we believe that connective tissue carries a concentrated amount of protomorphogens in a "bound" form.

We have thus briefly sketched four manners in which environmental factors have been demonstrated to influence morphology: (1) direct observations of morphological changes in infusoria subjected to conditioned media; (2) bacterial mutations arising from sub-standard culture conditions; (3) special agents, some carried in the cytoplasm, may exert an influence on cell characteristics which may be inherited, and (4) the influence of connective tissue over cells cultured *in vitro,* probably not in the same class as the first two examples. Rather this latter influence can probably be classified as an "evocator" response. That it

1 Champy, C.: *Bibliog. Anat.,* 33:184, 1913; *Compt. rend. soc. biol.,* 76:31, 1914.

does not universally occur is indicated by the work of Ebeling and Fischer showing that cells may be cultivated side by side; also by the work of Landsteiner and Parker demonstrating the remarkable stability of cells cultured *in vitro* to influences of added connective tissue.

Maintenance of Morphological Integrity.—In the face of the above evidence, the maintenance of morphological integrity against environmental influences requires special protective mechanisms. Delbruck (1945) has presented a theoretical study of bacterial populations with this consideration in mind. He mentions that for a stable strain to exist, forward mutations must be balanced either by reverse mutations or a high selection pressure against the forward mutation.

In certain species the cells rely upon a special mechanism known as conjugation to maintain the integrity of the species organization. Cowdry (1939) has reviewed this phenomena to an extent. He mentions that the macronuclei are concerned with the metabolic and organization reactions of the cell. The micronucleus, on the other hand, has little to do but to maintain itself and supply new macronuclei. The process of conjugation, and its complement, endomixis, consists essentially of elimination of the macronucleus and formation of a new one under the influence of the micronucleus. Calkins (1934) has reviewed the factors concerned with rejuvenation in protozoa. He notes that through the rejuvenating process of conjugation or of endomixis, protozoa are enabled to live through thousands of generations. Normally this rejuvenating process takes place in definite cycles of so many generations,[1] and without it the race dies out.[2]

A study of conjugation and endomixis does not properly belong in this discussion. It is mentioned simply to illustrate a method by which infusoria periodically re-organize themselves, are rejuvenated, and by means of replacement of the macronucleus from the micronucleus insure the consistency of the species integrity. This is another

1 Woodruff, L. L.: "Rhythms and Endomixis in Various Races of Paramecia Aurelia." *Biol. Bull.,* 35:51 56, 1917.

2 Calkins, G. N.: "Uroleptus Mobilis Engelm. II. Renewal of Vitality Through Conjugation." *J. Exp. Zool.* 29:121-139, 1919.

illustration of the necessity for a "reworking" of the morphogens at periodical intervals in order to perpetuate a specific morphology.

At the time Weismann evolved the germ plasm theory, evidence was unavailable to indicate the methods of chromosome synthesis. As a result Weismann assumed that the determinant substance was attenuated at each cell division. We see now that it is being constantly produced by the dynamic state of nuclear protein.

Spiegelman and Kamen (1946) propose a most interesting and highly significant concept of gene action. Their proposal seems to fit closely into the morphogen hypothesis. They state "Genes *continually* produce at different rates partial replicas of themselves which enter the cytoplasm. These replicas are nucleoprotein in nature and possess to varying degrees the capacity for self-duplication. Their presence in the cytoplasm controls the type and amounts of proteins and enzymes synthesized." They continue, reviewing the probability that these gene fragments compete with each other with the outcome that the enzymatic character of the cytoplasm is thereby determined. They recognize that altered environment could change this competitive result.

It is now possible to sum up the suggestions contained in this chapter. We have reviewed on page 138 the hypothesis covering the mechanism by which the chromosome morphogens find their way to the nuclei of the various cells in the developing embryo. Up to that time, however, we had given no hint concerning the influence of the cytomorphogen in the determinant cycle.

Review of the Morphogen Determinant Cycle.—We now re- view the hypothesis covering this determinant morphogen cycle: (1) at each mitosis there is a re-organization of the macronucleus and a nuclear-cytoplasmic interchange; (2) this leaves chromatin material in the cytoplasm where it organizes cell morphology; (3) chromatin (morphogens) may also influence cell morphology through the medium of the mitochondria; (4) the cell is influenced in various manners by its environment which induces limited inheritable variations in character; (5) the fact that all protomorphogens find their way through the chromosome in chromatin synthesis before they appear again for protein synthesis at the cytoplasm boundary aids in the maintenance of cell integrity, and (6) in some cells the process of conjugation or endomixis re-organizes the nuclei and assists in maintaining organization integrity.

The whole problem of morphogenesis is as intricate and inviting as any field of biology. It is a field that is not by any means thoroughly explored. It offers the most interesting of uncharted problems and probably most in the satisfaction of solving the deepest mysteries that baffle the mind of man.

BIBLIOGRAPHY

ANNERSTEN, C.: Experimental Examination of the Biochemistry of Bone Formation." *Acta Chir. Scand.,* 84:269, 1940-41.

AVERY, O. T., McLEOD, C. M., and McCARTY, M.: "Studies on the Chemical Nature of the Substance Inducing Transformation of Pneumococcal Types." *J. Exp. Med.,* 79:137-158, 1944.

BERNAL, J. D.: "Structural Units in Cell Physiology." *The Cell and Protoplasm,* Pub!. Am. Assoc. Adv. Sci., No. 14, p. 199. 1940.

BERTELSEN, A.: *Acta Orthop. Scand.* 1945.

CALKINS, G. N.: "Uroleptus halseyi Calkins. II. The Origin and Fate of the Macronuclear Chromatin." *Arch. Protistenk.,* 69:151-174, 1930.

———— "Factors Controlling Longevity in Protozoan Protoplasm." *Biol. Bull,* 67:410-413, 1934.

CHARGAFF, E.: *J. Biol. Chem.,* 142:491, 1942.

CONKLIN: "Karyokinesis and Cytokinesis in the Maturation, Fertilization and Cleavage of Crepidula and other Gastropids." *J. Acad. Natl. Sci. Philadelphia,* 12:1, 1902.

COWORY, E. V.: *General Cytology.* University of Chicago Press, Chicago, Ill. 1924.

———— *Problems of Ageing.* Williams & Wilkins Co., Baltimore, Md. 1939.

DAVIDSON, J. N., and WAYMOUTH, C.: "Histochemical Demonstration of Ribonucleic Acid in Mammalian Liver." *Proc. Roy. Soc. Edinburgh* 62, Pt. I, 96-98, 1943-44.

DAWSON, M. H., and SIA, R. H. P.: "In Vitro Transformation of Pneumococcal Types; Technique for Inducing Transformation of Pneumococcal Types In Vitro." *J. Exp. Med.,* 54:681, 1931.

DELBRUCK, M.: "Spontaneous Mutations of Bacteria." *Ann. Missouri Bot. Gard.,* 32:223-233, 1945.

EMEREC, M.: *Proc. Natl. Acad. Sci.,* 32:36, 1946.

DREW, A. H.: "A Comparative Study of Normal and Malignant Tissues Grown in Artificial Culture." *Brit. J. Exp. Path.,* 3:20-27, 1922.

———— "Growth and Differentiation in Tissue Cultures." *Brit. J. Exp. Path.,* 4:46, 1923.

EBELING, A. H., and FISCHER, A.: "Mixed Cultures of Pure Strains of Fibroblasts and Epithelial Cells." *J. Exp. Med.,* 36:285-291, 1922.

GRIFFITH, F.: *J. Hyg.* Cambridge, England, 27: 113, 1928.

GUILLIERMOND, A.: "Elements of Microbial Cytology." *Microbiology,* Ed. by C. E. Marshall, P. Blakiston's Son & Co., Philadelphia. 1921.

D'HERELLE, F.: *Immunity in Natural Infectious Disease.* Williams & Wilkins Co., Baltimore, Md. 1924.

————— *The Bacteriophage and Its Behavior.* Williams & Wilkins Co., Baltimore, Md. 1926.

————— *The Bacteriophage and Its Clinical Applications.* Charles C. Thomas, Publisher, Springfield, Ill., Baltimore, Md. 1930.

HORNING, E. S.,"Studies on the Mitochondria of Paramoecium."*Austr. J. Exp. Biol. & Med. Sci.,* 3:89-95, 1926.

————— "Observations on Mitochondria." *Ibid,* 3:149-159, 1926.

————— "On the Relation of Mitochondria to the Nucleus." *Ibid,* 4:75-78, 1927.

————— "On the Orientation of Mitochondria in the Surface Cytoplasm of Infusorians." *Ibid,* 4:187-190, 1927.

————— "Studies on the Behaviour of Mitochondria Within the Living Cell." *Ibid,* 5:143-148, 1928.

————— "Mitochondrial Behaviour During the Life-Cycle of a Sporozoon (Monocystis)." *Quart. J. Microsc. Sci.,* 73:135-143, 1929-30.

HORNING, E. S., and SCORR, G. H.: "A Preliminary Study of the Distribution and Changes in the Inorganic Salts During Embryonic Development of the Chick." *Anat. Rec.,* 52:351-366, 1932.

HUOSON, P. S., and RICHENS, R. H.: *The New Genetics in the Soviet Union.* Imperial Bureau of Plant Breeding and Genetics, Cambridge. 1946.

JENNINGS, H. S.: "Chromosomes and Cytoplasm in Protozoa." *The Cell and Protoplasm,* Publ. Am. Assoc. Adv. Sci., No. 14, p. 44. 1940.

JOLLOS, V.: "Experimentelle Prostistenstudien. I. Untersuchungen uber Variabilitay und Vererbung bei Infusorien." *Arch. f. Protistenk.,* 43:1-222, 1921.

————— "Dauermodifikationen und Mutationen bei Protozoen." *Arch. f. Protistenk.,* 83:107-219, 1934.

LANDSTEINER, K. L., and PARKER, R. C.:"Serological Tests for Homologous Serum Proteins in Tissue Cultures Maintained on a Foreign Medium." *J. Exp. Med.,* 71:231-236, 1940.

LEVANDER, G.:"Tissue Induction."*Nature,* 155:148-149, 1945.

LYSENKO, T. D.: *Heredity and Its Variability.* King's Crown Press, Columbia University, New York. 1946.

MANWARING, W. H.: "Environmental Transformation of Bacteria." *Science,* 79:466, 1934.

MEVES, F.: "Eine neue Stutze fur die Plastosomentheorie der Vererbung." *Anat. Anz.,* 50:551-557, 1917-18.

————— "Die Plastosomentheorie der Vererbung. Eine Antwort auf verschiedene Einwande." *Arch. f. mikr. Anat.,* Bonn. 92:41-136, 1918.

MURPHY, J. B.: "Discussion of Some Properties or the Causative Agent of a Chicken Tumor." *Trans. Assoc. Am. Physic.,* 46:182, 1931.

————— "Experimental Approach to the Cancer Problem. I. Four Important Phases of Cancer Research. II. Avian Tumors in Relation to the General Problem of Malignancy." *Bull J. Hopkins Hosp.,* 56:1-31, 1935.

NEEDHAM, J.: *Chemical Embryology.* 3 Vols., Cambridge University Press, London. 1931.

————— *Biochemistry and Morphogenesis.* Cambridge University Press, London. 1942.

PONTECORVO, G., and WHITE, M. J. D.: "Heredity and Variation in Micro-Organisms." *Nature,* 158:363-364, 1946.

RAFFEL, D.: "The Occurrence of Gene Mutations in Paramecium Aurelia." *J. Exp. Zool.,* 62-63: 371-412, 1932.

RICHARDS, A.: "History of Chromosomal Vesicles in Fundulus and Theory of Genetic Continuity in Chromosomes." *Biol. Bull.,* 32:249, 1917.

ROTHEN, A.: "Forces Involved in the Reaction Between Antigen and Antibody Molecules." *Science,* 102:446-447, 1945.

SCORR, G. H.: "The Disposition of Fixed Mineral Salts During Mitosis." *Bull. d'Histol, App.,* 8:251, 1930.

————— "Mineral Distribution in Cytoplasm." *Biol. Symp.* 10: 277-289, 1943.

SHEN, S. C.: *J. Exp. Biol.,* 16:143, 1939.

SONNEBORN, T. B.: "Inheritance in Ciliated Protozoa." *Cytology, Genetics and Evolution,* University of Pennsylvania Bicentennial Conference. 1941.

SPIEGELMAN, S., and KAMEN, M. D.: "Genes and Nucleoproteins in the Synthesis of Enzymes," *Science,* 104:581-584, 1946.

STANLEY, W. M., IN DOERR, R., and HALLAUER, E.: "Handbuch der Virusforschung." Julius Springer, Vienna, 1:491, 1938.

WEISS, P.: "The So-Called Organizer and the Problem of Organization in Amphibian Development." *Physiol. Rev.,* 15:639-674, 1935.

————— *Principles of Development.* Henry Holt & Co., New York. 1939.

WRIGHT, S.: "The Physiology of the Gen ." *Physiol. Rev.,* 21:487-527, 1941.

————— "Physiological Aspects of Genetics."*Ann. Rev. Physiol.,* 7:75-106, 1945.

CHAPTER 5

MORPHOGENS IN THE HIGHER ORGANISM

Introduction

In the preceding chapters the morphogen hypothesis was developed with respect to the protein molecule, the vitality of the cell, and embryonic morphogenesis.

In these chapters we have attempted to outline an hypothesis with an exactitude consequent to a detailed study of the specialized problems involved. It was our original intention to survey the literature with the object of orienting an intelligently planned program of experiments on the physiology of growth factors.

As we linked the facts together, we became conscious that our working hypothesis was becoming more and more comprehensive until it appears that no field of biological study is immune from the ideas suggested by the morphogen concept. Study and speculation in related fields is tempting, but we must restrain our enthusiasm at this time and confine our efforts to those studies which are pertinent to the original project.

Simply because our ideas in diverse fields are speculative and not developed in detail is no reason to withhold their publication. The morphogen concept leads to some obvious conclusions in various fields of biology. It is to the advantage of biological study if these conclusions are herein presented. They may stimulate useful criticism and experimentation by those who can apply specialized attention to the various problems suggested.

This chapter, therefore, is to be devoted to speculative surveys of widely divergent fields of biology in which the authors obviously cannot have expert knowledge.

Universal Aspects of the Morphogen Concept

The morphogen hypothesis herein presented basically entertains the following fundamental ideas: (1) chromosome fragments termed morphogens are in a dynamic state of synthesis and destruction as a part of the vital energy reactions of the cell; (2) the morphogen fragments accumulate in the cell fluids as the basic cause of senescence and death; (3) the morphogens from the chromosome are the determinants for every cell and every "living" molecule in biological structure, and (4) morphogens can only be synthesized in the chromatin material but must be present in the pericellular fluids for the synthesis of every biological protein molecule. They are, therefore, necessary for morphogenesis, mitosis and growth.

Plants.—We have confined our discussion of the metabolism of the morphogens to the animal kingdom. There is evidence however, that plant physiology as well is linked to the morphogen concept.

Turck (1933) was led to the study of protomorphogens as a result of his observations at the Lincoln Park Greenhouse of Chicago about 1893. He found that it was necessary to mix virgin Wisconsin prairie soil with exhausted greenhouse soil in equal proportions in order to obtain growth of plants. This in itself was not surprising. But he found that the addition of 5 to 10 per cent of the exhausted soil to virgin soil resulted in better growth than on the virgin soil alone.

Here is a phenomenon that lends itself readily to explanation by the morphogen hypothesis. The exhausted greenhouse soil was not alone sterile because of an exhaustion of foodstuffs. It was "poisoned" by an excess concentration of plant protomorphogens which inhibited and prevented further growth. The addition of large quantities of virgin soil not only replaced depleted foodstuffs but also diluted the protomorphogen of the sterile soil from an inhibitive to a stimulative concentration.

We might interpose the comment that this condition would only occur in an artificial environment such as a greenhouse. In nature and natural farmlands, it is unlikely that the soils would become supersaturated with poisonous concentrations of protomorphogens, due to the natural or artificial rotation of crops. It will be recalled that only homologous protomorphogens are toxic; heterologous protomorphogens may be utilized by other species, in fact are valuable in nutrition. Probably the natural rotation of crops is due to this phenomenon, each crop saturating the earth with its own protomorphogens and eliminating itself; but these protomorphogens are fertilizer for the next species.

Turck's observation that the addition of 5 to 10 per cent of exhausted greenhouse soil to virgin soil resulted in optimum growth is interpreted as the result of the addition of small amounts of protomorphogen and the consequent beneficial effect. The value of the addition of heterologous as well as homologous to the soil is receiving much attention by agricultural science in the judicious use of compost. Sir Albert Howard is the first exponent of this idea in his book, "An Agricultural Testament." Many other excellent expositions have recently appeared[1].

Turck devotes a chapter of his work on "Cytost" to a discussion of plant physiology. He interprets much of the work on plant growth and injury hormones as a protomorphogen effect. This work, however, must be carefully reviewed as many, if not most, of the plant growth hormones responsible for various tropisms cannot be classified as protomorphogens.

However, there have been observations of plant growth hormones which can be so classified. Most auxins (plant growth hormones) exert a stimulating effect only. A protomorphogen, however, must exhibit the characteristic of stimulating growth in dilute amounts and inhibiting growth in concentrated amounts. Such an"auxin"seems to be described by Stewart, Bergren and Redemann (1939). This factor, which they term an inhibitor, is extracted from the cotyledons of radish plants and exerts an inhibiting property roughly proportional to concentration, dilute amounts having a slight stimulating effect. This property is illustrated in the accompanying figure 6. It is strikingly different from the standard curve obtained with the usual plant auxin which does not inhibit in any degree of concentration.

The various morphogen concepts which we have developed in this thesis referring to nuclear metabolism, changes in cell boundary potential, etc., are obviously not to be adapted as such to the plant kingdom. There are,

---

1 *An Agricultural Testament,* Sir Albert Howard. Oxford University Press. 1942.

The Living Soil, E. B. Balfour. Faber & Faber, Ltd., London. 1945.

Bio-Dynamic Farming and Gardening, E. Pfeiffer. Anthroposophic Press, New York. 1943.

The Wheel of Health, G. T. Wrench. Reprinted by Lee Foundation for Nutritional Research, Milwaukee, Wis. 1945.

however, indications such as we have just reviewed, that the fundamental biology of plants revolves around morphogens much the same as does that of animals. This, however, is an independent and separate field of research.

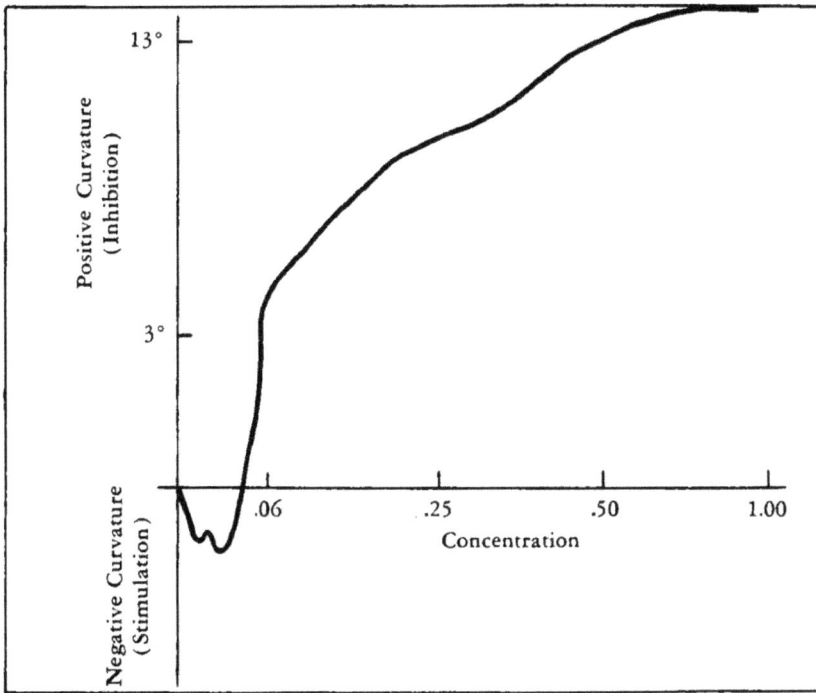

FIGURE 6

Effect on plant growth of applications of an inhibitor extracted from cotyledons of radishes. (Derived from Stewart, W. S., Bergren, W., and Redemann, C. E.: *Science*, 89:185-186, 1939.)

Cold and Warm Blooded Animals.—There are basic differences in the morphogen concept between cold blooded and warm blooded animals. Some of these differences may be listed as follows:

1. Senescence seems to be unknown among the cold blooded animals. No degenerative changes with old age have been reported with these groups.
2. There is no creatine-creatinine metabolic cycle in cold blooded animals.
3. Cold blooded animals do not have blood platelets; thromboplastic activity being regulated to various tissues, notably the spindle cells.

4. Cold blooded animals may be able to regenerate parts which are lost by injury.
5. Growth of cold blooded animals is not affected by the growth hormone of the anterior pituitary.[1]
6. Warm blooded animals have a constant rate of metabolism, whereas in cold blooded animals the metabolic rate is secondary to ambient temperature.

Of course, there are other basic differences between warm and cold blooded animals, but we have listed only those which may be linked with the differences in morphogen metabolism.

The reader will recall that our hypothesis envisions two morphogen cycles: (1) the metabolic cycle concerned with the dynamic reactions in the nuclear chromatin and associated with phosphagen and nucleoprotein energy reactions, giving rise to protomorphogens in the tissue fluids which in turn restrain and probably regulate the vital activities in the cell, and (2) the determinant cycle concerned with the organization of the cell and of the morphogenesis in general.

The enumerated differences between warm and cold blooded animals indicate a difference in both these cycles.

The determinant cycle may be considerably different in view of the ability of cold blooded animals to regenerate injured limbs. Either portions of the chromosomes are delivered *intact* to various locales by the cold blooded animal and *not intact* in the warm blooded animal, or there are constraining forces in the warm blooded animal preventing a duplication once morphogenesis is complete. It is also possible that the cells of cold blooded animals, being farther down the evolutionary scale, are able to revert to their primordial embryonic state and become "competent" to receive the inducing stimulus.

The metabolic cycle of morphogens is quite different in cold blooded animals. The absence of senile changes in this group indicates that protomorphogens released from cells do not accumulate in the cell fluids and inhibit the vital activity of the cells. The metabolic cycle of protomorphogens is linked with phosphagen (dipotassium creatine hexosephosphate) metabolism in warm blooded animals. The absence of creatine in cold blooded animals indicates a fundamental difference in this cycle. The lack of anterior pituitary growth hormone effect also suggests

1 Collip, J. B.: "Some Recent Advances in the Physiology of the Anterior Pituitary." *J. Mt. Sinai Hosp.*, 1:28-71, 1934-35.

a difference in the mechanism for the regulation of mitotic activity in cold blooded animals.

We shall suggest later that the blood platelets are an important link in the elimination of protomorphogens. Their absence in cold blooded animals indicates a considerable difference in the mode of protomorphogen elimination between cold and warm blooded animals.

We had thought that a careful survey of the biochemistry of cold blooded animals would suggest diverse approaches to the problem of senescence in mammals. However, unfortunately, the difference between the protomorphogen metabolism of the two groups seems at this time to be too far-reaching to bridge with analogy.

Morphogens and Connective Tissue

Single celled life and, in a sense, many of the lower organisms live in a surrounding medium which serves as "tissue fluid" for the living unit. There is little difficulty in preventing an excess accumulation of protomorphogens in the surrounding media. Most natural environments provide for a constantly changing media, and many organisms are able to move about into a fresh environment.

In the mammal an entirely different picture is presented. Each cell is surrounded by tissue fluids which require special mechanisms to supply food and eliminate wastes. In a sense, each mammalian cell may be said to exist in a tissue culture whose media consists of the surrounding tissue fluids. Were the fluids surrounding each cell not constantly supplied with foodstuffs and voided of protomorphogen and other wastes, the mammalian cell would soon die from auto-intoxication and starvation.

The bloodstream and numerous other physiological systems are a part of the mammalian solution for the problem of constantly supplying the cells with food and eliminating their wastes. In this chapter we shall attempt to outline various routes for the elimination of protomorphogens from pericellular fluids and other diverse aspects of morphogen metabolism in the mammal.

If the reader will refer to the schematic diagram of the morphogen cycles in Chapter 3, figure 4, he will note that protomorphogens discharged as waste from nuclear metabolism appear in the fluid surrounding the cell. We have outlined an hypothesis which suggests that an accumulation of these protomorphogens in the cell fluid lowers the vitality of the cell and eventually causes cytolysis.

Assuming that the individual cell morphogen cycle is represented thus, it now becomes our problem to pick up the thread of continuity at this point and suggest methods by which these extracellular protomorphogens are disposed of in the mammal.

Morphogens and the Precipitation of Fibrin.—Over fifty years ago in the laboratory of Prof. W. W. Podvysotski in Kieff, Dr. Galin (1889) made some experiments which lend a clue at this point. (It is fitting that the first clue to this hypothesis should originate in Russia, which in this modern day is reporting some of the most advanced work in biology.) Galin concluded that connective tissue can be saturated by poisonous substances floating in the blood and tissues, even when normal toxin eliminating functions (viz., kidney) are performing satisfactorily.

It was commented at that early date that experimental evidence suggests the elastic tissue as a depot for the storage of various poisons, more especially when the kidney function is disturbed.

Burrows (1926) has supplied us with a useful conception of the manner in which tissue "toxins" or protomorphogens are "bound" into the connective tissue fibers of the organism. He advanced the theory that all cells liberate a substance which he termed "archusia." In low concentrations in the media this substance is beneficial to the health and growth of cells. In high concentrations of archusia growth ceases and cytolysis occurs. We consider this archusia to be the same substance we have discussed in this report under the term "protomorphogen" and we shall refer to it as such in our discussion. We might mention that Burrows observed that archusia is soluble in saline solution, a property of nucleoproteins previously mentioned. It may be of interest to note Burrows' report that archusia is only produced in the presence of oxygen. Since archusia (protomorphogen) production is a function of the metabolic morphogen cycle, it may be of interest to speculate that there is an oxygen cycle concerned here which may be the most important one for tissue growth. Fisher (1942) has analyzed the oxygen consumption of yeast and concluded that of two or three cycles there is only one that is important to normal cell division. This one is probably concerned with morphogen metabolism.

Under certain concentrations of archusia in the media Burrows demonstrated that the cells secrete a fat-like substance which he terms "ergusia." He has shown that this ergusia lowers the surface tension of the cell boundary. This effect has been discussed in Chapter 3 in respect to the manner in which the vitality of cells is lowered and cytolysis occurs when the media is not replaced at periodic intervals. Burrows (1926) has presented a lucid explanation of cell movement based upon the effect of ergusia in lowering surface tension on a local area of the cell surface boundary.

Burrows and Jorstad (1926) have supplied evidence that the substance archusia is very similar to vitamin B, and ergusia is similar or identical to vitamin A. (We shall comment on these conceptions later in our discussion of the nature of ergusia as it appears to us.)

Burrows (1926) has discussed the coagulation that occurs on the addition of cells to plasma *in vitro*. This coagulation appears as two distinct types. One type, occurring immediately upon the addition of the cells, has the nature of a gelatinization without the appearance of fibers. The other type, called "secondary" coagulation by Burrows, occurs only after a latent period (the time of which depends upon the nature of the added tissue) and results in the production of true fibers of fibrin. Both types occur universally with all types of cells except lymphocytes and phagocytes.

It is with the so-called "secondary" coagulation that we are concerned. Burrows has demonstrated that this is a result of the ultimate effect of ergusia on fibrinogen. Ergusia is then an active blood coagulant resulting in the precipitation of fibrinogen into fibrin and is a universal thromboplastin present in all tissues. The latent period before "secondary" coagulation is the time necessary for the cells to release sufficient ergusia to initiate this coagulation phenomenon. Drinker (1942) mentions that scar tissue formation is a similar phenom non, being the result of excess fibroblastic growth due to the accumulation of substances ordinarily removed by the lymph (this explains callus formation from constant irritation).

In Chapter 3 we have suggested that Burrows' ergusia is protomorphogen "wrapped" with lipoid "envelopes," preventing the lethal effect of intense concentrations of "raw" protomorphogen. In a review of structural proteins Astbury (1945) comments that electron microscope investigation of fibrous proteins implies a fibrous structure built up from elementary patterned molecules, or "templates." We believe that the active protomorphogen linkages supply the basis for the "template" action of these primary molecules of fibrous tissue.

Protomorphogen is, therefore, the initiating stimulus to the formation of connective tissue in the organism. The white connective tissue is formed by the precipitation of fibrin from fibrinogen in the presence of protomorphogen. The synthesis of a protein requires a substrate and a determinant for the protein structure. The substrate for white connective tissue consists of fibrinogen, and it is likely that we can consider the thromboplastin or protomorphogen the determinant. The possibility that thromboplastin may be a determinant in some cases is strongly hinted by the observation that filaments of fibrin can be seen radiating from a platelet disintegrating in plasma.[1] This would explain the local specificity of connective tissue proteins since their determinant protomorphogens arise from the metabolism of neighboring cells.

Burrows (1927) has demonstrated that newly formed fibrin stains as white fibrous tissue when cell extracts containing protomorphogen are added to the fibrin. He concludes that white fibrous tissue is, therefore, fibrin with ergusia (protomorphogens) adsorbed or attached to the molecules.

In this respect the connective tissue as a depot for electrolytes must not be overlooked. Manery, Danielson and Hastings (1938) have observed that dense connective tissue has a high electrolyte content resembling serum more than any other tissue. The reader will recall our comments in Chapter 1 discussing the importance of various minerals in the protomorphogen molecule. Protomorphogen is probably always associated with mineral matter which links the specific determinant into a predetermined pattern.

Drew (1923) has supplied additional evidence concerning the presence of protomorphogen in connective tissue. In Chapter 4 we discussed the effect of protomorphogens as inductors of differentiation. Drew demonstrated that the addition of connective tissue from any organ induced

---

1 *Starling's Principles of Human Physiology,* E. H. Starling. Ed. by C. L. Evans, Lea & Febiger, Philadelphia. 1936.

the differentiation into typical tubules of undifferentiated kidney cells. Landsteiner and Parker (1940) have mentioned that connective tissue can organize proteins identical with or closely related to serum proteins, indicating the presence of protein determinants in connective tissue. The conversion of mononuclear leucocytes into fibroblasts *in vitro* has been demonstrated many times. Fischer (1925) noted that when placed in contact with homogenous muscle tissue the leucocytes are transformed into true fibroblasts after the fourth transplant. This may indicate the effect of the morphogen factors on the morphology of the adjacent leucocytes.

Wolfe and Wright (1942) have recently suggested that products of cell metabolism may be released into the cell fluids and penetrate into connective tissue fibers, playing an important role in their staining reaction.

It seems probable, therefore, that all white connective tissue is formed in a manner similar to a blood clot, namely, by the precipitation of fibrin from fibrinogen under the control of a thromboplastic agent. In addition, it appears that the newly formed fibrin adsorbs protomorphogen molecules and in time becomes sufficiently saturated with them to stain as connective tissue. It is likely that the specific adsorption of protomorphogens by connective tissue may be classed as chemisorption rather than a pure physical adsorption.

Thus, the idea presented in 1889 by Dr. Galin receives experimental and theoretical substantiation. And, for the purposes of the morphogen hypothesis we must keep in mind the thromboplastic activity of protomorphogen and the conception of connective tissue as a local storehouse for discharged protomorphogen. Connective tissue may be considered as having a powerful affinity for all protomorphogen molecules.

Thromboplastin.—At this point it is pertinent to review a few of the facts relating to the coagulation phenomenon and particularly to the thromboplastic activity of various tissues.

Quick (1942) has recently published a comprehensive review of the field of hemorrhagic diseases which supplies a modern version of the coagulation processes.

The fundamental reaction in the coagulation of blood is the conversion of fibrinogen to fibrin, which is a fibrous substance forming the basis of white connective tissue. This reaction is precipitated in the bloodstream of the mammal by the breakdown of the blood platelet.

When lysis of these platelets is prevented, no clot can occur except when tissue juice is present.

The substance in platelets responsible for this action is termed *thromboplastin.* Intact platelets will cause syneresis or clot contraction. Burrows (1926) has demonstrated that ergusia will also effect syneresis. The thromboplastic activity of tissue juice, platelets, spindle cells and ergusia leads us to suggest that protomorphogen is responsible for this activity and, indeed, that protomorphogen is the primary thromboplastin.

There has been some argument about the exact position of thromboplastin in the clotting cycle. The concensus of opinion seems to be that it is the factor which is essential for the conversion of prothrombin to thrombin. Thrombin is the active agent, probably of an enzymatic nature, which converts soluble fibrinogen into insoluble fibrin. It appears, however, that thromboplastin is the "trigger" which initiates these reactions by forming thrombin. In spite of the presence of fibrinogen and prothrombin, no clotting occurs until thromboplastin is released.

The identity of protomorphogen with thromboplastin is further indicated by the fact that frog plasma filtered through charcoal will not coagulate (thromboplastin in the spindle cells being adsorbed) except upon addition of tissue extract.[1] There is much further evidence on the identity of protomorphogen and thromboplastin. Quick (1936) has noted three properties of thromboplastin which are particularly significant: (1) thromboplastic activity is diminished by extraction with lipoid solvents; (2) thromboplastic activity is diminished but not destroyed by heat indicating a relative thermostability of the product, and (3) thromboplastin possesses a limited degree of species specificity. The reader will recall previous evidence that protomorphogen is: (1) relative thermostable; (2) extracted by lipoid solvents, and (3) relatively species specific. In a recent review, Chargaff (1945) also states that thromboplastic substances are relatively thermostable and species specific.[2] He suggests that thromboplastic substances from various

1 Tait, J., and Green, F.: "The Spindle-Cells in Relation to Coagulation of Frog's Blood." *Quart. J. Exp. Physiol.,* 16:141-148, 1927.

2 Chargaff, E.: "The Coagulation of Blood." *Advances in Enzymology* 5:31-65, 1945.

tissues constitute "an entire family of conjugated proteins whose similarities and dissimilarities will have to be determined by detailed chemical and immunological studies." We postulate that the family of conjugated proteins referred to here is in reality that class of protomorphogens from different tissues, all of which are thromboplastic, but differ in various respects, i.e., solubilities, molecular size, specificity, etc. Bunting (1932) has reviewed the evidence indicating that platelet material suggests chromosome origin, staining with Janus green indicating the presence of mitochondria. The Morphogen Hypothesis holds that protomorphogens from cells are products of the cell's chromatin, and that they may appear in the mitochondria of cells (mitochondria being high in fatty substances, nucleoprotein and phospholipides).[1]

The chemistry of platelets is very suggestive of chromosome (morphogen) origin. Hammersten (1914) reviews early evidence indicating that platelets consist of a combination of nucleins and proteins as a derivative of the cell nucleus. Gruner (1914) has also mentioned that platelets contain nucleoproteins, lecithin and cholesterin. Wright (1932) reviews that the active thromboplastin from tissue extract is soluble in ether and alcohol and is of lipid character, probably a cephalin. This evidence was supplied by Howell who isolated this substance which seems to be an ether soluble cephalin. The reader will recall that protomorphogens are almost universally accompanied by lecithin and seem to be extracted with lipoid solvents.

Further evidence of the chromosome origin of thromboplastin has been supplied by Marshak and Walker (1945). They demonstrated that a chromatin derivative of liver has a hemostatic action when locally applied to bleeding wounds. Ungar (1944) comments on reduced coagulation time as a consequence of injection of serum from traumatized animals, indicating that tissue products released by trauma exert thromboplastic activity. Chargaff, Moore and Bendich (1942) found thromboplastic protein from lung to exhibit marked phosphatase activity. (Phosphatase is an enzyme found in the cell nucleus particularly concentrated in chromatin.) These investigators separated this protein from lung by means of saline extraction and ultracentrifugation. (Saline extraction has recently been reported most

1 Bensley, R. R.: "Chemical Structure of Cytoplasm." *Biol. Symp.*, 10:323-334, 1943.

efficient in extracting nucleoprotein from the nucleus.) More recently Maltaner (1946) has discussed the thromboplastic properties of some antigens and the fact that thromboplastic cephalin fixes complement in direct proportion to its thromboplastic activity. Our morphogen hypothesis recognizes that protomorphogen is responsible for antigenic activity.

The studies of Cohen and Chargaff (1940) on the thromboplastic protein from lungs have demonstrated the antigenic activity of thromboplastic protein, but the antibody-antigen complex was more potent than the antigen alone. This indicates that the thromboplastic linkages are not "covered" by antibody reactions. They found that if the protein was freed of all phosphatides, it was devoid of thromboplastic activity. However, their treatment removed all the phosphorus containing components including nucleic acids. We suggest that either the protomorphogens were removed from the protein (the protein lost its antigenic properties after extraction with boiling alcohol-ether) or the protomorphogen linkages which were thromboplastically active were irreparably damaged by these manipulations.

We might emphasize at this point our contention that, although all physiological thromboplastin is either protomorphogen or protomorphogen degradation products, this does not necessarily mean that all protomorphogen is active thromboplastin. Thromboplastic activity probably depends upon special active linkages which may or may not be available in a protomorphogen molecule, depending upon its location, complexity and associated factors.

The hypothesis that the primary thromboplastin is protomorphogen suggests two fields of further speculation. One, the presence of protomorphogens in platelets indicates a means of blood stream transfer to a position for convenient excretion. And two, the identity of protomorphogen with thromboplastin provides a convenient landmark as an indication of locales of intense protomorphogen activity. We shall discuss the latter first.

Lung tissue is recognized as being a potent source of thromboplastin. Chargaff, Moore and Bendich (1942) have demonstrated that macromolecules from lung are carriers of strong thromboplastic activity. Howell has prepared an active phospholipide from pig lung exhibiting extraordinary thromboplastic potency (Copley, 1945). Shapiro and co-workers (1944) have also demonstrated the intense thromboplastic activity of lung extract. Mason and Lemon (1931) have mentioned that three drops of fresh lung tissue extracts injected into the circulation of the rabbit will produce coagulation of all the intravascular blood in 20 seconds. Dyckerhoff and Grunewald (1943) have used cooked lung extract as a source of thromboplastin for their experiments. Chargaff, Moore and Bendich (1942) isolated a high molecular lipoprotein from beef lung which exhibited a very high thromboplastic activity. Chargaff (1945) reviews the experimental evidence on lung thromboplastin and concludes that it may be classified as a lipoprotein of a very high molecular weight and probably of cytoplasmic origin. Protomorphogen is a high molecular weight nucleoprotein associated with lipoid, secreted from the cytoplasm.

Widenbauer and Reichel (1941) have concluded that there are two classes of substances with thromboplastic activity rather than a single lipid-protein complex. As evidence they submit the fact that some of the thromboplastic lipids must be treated with fat solvents in order to demonstrate thromboplastic activity, while some tissue extracts do not require this treatment. We submit that there is an excellent possibility that some of the thromboplastic protomorphogen is more efficiently "sheathed" with lipoids than other and therefore some extracts will require fat solvent treatment in order for thromboplastic activity to become manifest. Supporting this contention is the report by the above investigators that some cephalin preparations lose their thromboplastic activity by standing, heat or acetone treatment, and then become coagulation inhibitors. This can be interpreted as an expression of the protomorphogen binding powers of such lipoids, impairing thromboplastic activity.

The existence of a high degree of thromboplastic activity in lung tissue is highly significant. Turck (1933) reports that one of the most universal effects of administration of homologous protomorphogen (termed "cytost" by Turck) is capillary stasis and inflammation of lung tissue. He observed that minute amounts of protomorphogen sprayed into the cages in which cats were kept resulted in lethal inflammation of lung tissue. (Ir is important to note that in all cases of protomorphogen toxicity and shock, the blood is highly viscous and has a diminished clotting time.) The normal presence of a high content of protomorphogen in lung tissue indicates that its protomorphogen ratio is close to the danger point and small excess amounts may cause inflammation.

The existence of a high degree of sensitivity of lung tissue to insufflation of homologous protomorphogen leads to some interesting speculations. Turck speculates that those British scientists who pioneered the entrance into the tombs of Egyptian mummies were exposed to homologous protomorphogen-loaded dust which eventually resulted in the death of those present.

We may diverse for a moment to speculate on another interesting possibility, that of the importance of argon in the air. Dr. Hershey of the University of Kansas discovered some years ago that rats die from fibrosis of the lungs in 5 to 6 weeks if there is no argon in the air. More recently Behnke and Yarbrough (1939) have reported the narcotic effect of argon to be higher than nitrogen at pressures of 4 to 10 atmospheres, although equal in this respect at normal air pressure. Of interest is their report that the oil-water solubility of argon is higher than nitrogen and the other rare gases tested, and the solubility of argon in oil is twice that of nitrogen.

It is pertinent to speculate upon the possibility that argon may be concerned with the insulation of protomorphogen molecules in view of the solubility of this gas in lipoids. It appears that argon, by reason of its greater solubility may displace nitrogen from lipoid phases. Argon is a most inert element and would lend itself ideally as a protective or sheathing agent. The reader will recall that protomorphogen is normally associated with lipoids and phospholipides. Also indicating such a relationship between argon and protomorphogen is the fact that argon has been particularly identified in brain tissue,[1] another locale of intense protomorphogen activity. The fibrosis of lung tissue reported by Dr. Hershey to occur in the absence of argon is significant since it has been demonstrated that protomorphogen promotes the formation of fibrous and hyaline structure. In the same vein is the report of Kairiukschtis (1931) that argon is more rapidly resorbed from the pleural cavity after

1 *The Chemistry of the Brain*, I. H. Page. Charles C. Thomas Publishers, Springfield, Ill. 1937. Pictet, A., Scherrer, W. and Helfer L.: *Compt. rend. soc. biol.* 181:236, 1925; also *Helvetica Chim. Acta*, 8:547, 1925.

injection than any other gas tested, including air. It may be that its resorption is concerned with this suggested argon function of protecting protomorphogens.

Nearly sixty years ago several experiments were conducted in an attempt to identify the toxic substance present in respired air.[1] It was proven that the toxic substance present in exhaled air was not CO_2 as suspected but rather a substance which could be removed by washing the air in H_2SO_4. It is interesting to speculate on the possibility that waste protomorphogens may be broken down in lung tissue and toxic volatile components excreted. This possibility is suggested by the observation of Franke and Moxon (1937) that volatile selenium and tellurium compounds appeared on the exhaled breath of rats receiving excess amounts of these trace elements in their diets. Dolique and Giroux (1943) conclude that the principal channel of elimination of selenium from a poisoned organism is through the lungs.

Recently Copley (1945) using Howell's methods, has prepared a highly potent thromboplastic substance from human placenta. The thromboplastic activity of placental tissue was first demonstrated by Sakurai (1929). The presence of potent thromboplastin in the placenta may have special significance.

It is likely that the developing embryo is producing and excreting protomorphogens at a rapid rate. We have seen that protomorphogens are secreted most rapidly by cells during active mitosis. Although the mother is protected by various means from this additional protomorphogen, the shortened coagulation time near the end of gestation illustrates the additional burden on her protomorphogen disposal system. Winternitz, Mylon and Katzenstein (1941) have noticed that administration of tissue extracts with thromboplastic properties leads to thrombosis in pregnant dogs, although innocuous to non-pregnant dogs. Pickering, Mathur and Allahab (1932) have also demonstrated an intense thromboplastin metabolism in pregnant animals. They have shown that the blood from the uterine veins is extremely hypercoagulable in pregnant animals; this indicates an active protomorphogen content.

In an investigation of tissue extracts and thrombosis, Winternitz, Mylon and Katzenstein (1941) have demonstrated that the thromboplastic potency of various tissue extracts can be listed in the following order:

1 Hamilton, J. B.: "Hygiene (Of Air)." *Ann. Univ. Med. Sci.,* 5:10, 1890.

crude testicle, lung, kidney, spleen, liver and muscle. We have already discussed the possible link between lung and disposal of volatile protomorphogen components. The testicle would, of course, have an exceedingly high concentration of protomorphogens since it is the seat of the germinal activity and chromosome assembly in the organism. Winternitz also found that acetone-ether extraction of these various tissues altered their potency in varying degrees, testicle being the least affected. This is to be expected since by the morphogen hypothesis lipoid solvents remove the protecting wrapper and "activate" protomorphogen in this manner. The protomorphogen in the germ cells of the testicle is a part of the intact chromosome, probably in "bound" or combined forms, and thus is not removed as easily by solvents affecting the lipoid wrapping which may serve to protect unbound or transit forms from premature attachment.

We shall later indicate evidence that the kidney and liver are concerned with the excretion of protomorphogen toxins and the spleen is concerned with the protection against protomorphogen by reason of its place as the center of the reticulo-endothelial system.

Dyckerhoff and Deschler (1943) have also reported that ether extractions of thromboplastic substances reduce their clotting activity, also that benzene destroys the clotting activity of plasma.

Cohen and Chargaff (1940) found that the protein component of lung thromboplastin did not exert clot accelerating activity after the removal of the phosphatides. The thromboplastic protein complex was found to act as an antigen. Apparently fat solvents release the active components of the protomorphogen-protein complex from the protein carrier.

Let us return and re-analyze the evidence establishing the archusia-ergusia hypothesis. Burrows has demonstrated that archusia is a substance soluble in saline, constantly secreted by cells which accumulates in the media. In lower concentrations it causes the cells to liberate a lipoid substance (coagulant). In higher concentrations there is a digestion of proteins and fat, synthesis of protoplasm and growth of the cell. In still higher concentrations the cells cease growth, become enervated, and undergo lysis.

Burrows (1926) has illustrated that fat droplets from the protoplasm have a thromboplastic effect, but fat droplets from the blood stream actually inhibit coagulation. This is to be expected since thromboplastic activity depends on protomorphogen content. The blood lipoid, if protomorphogen free, would no doubt act to improve the insulating fat envelope of the transit form of protomorphogen.

The reactions of archusia as outlined are substantially identical with those established in Chapter 3 for protomorphogen, viz. stimulation of growth and mitosis in small amounts and inhibition of growth with lysis in concentrated amounts.

The basic problem under discussion is the manner in which the organism controls and disposes of the protomorphogen constantly secreted by its cells. We have in the preceding pages presented evidence that the protomorphogen thus secreted: (1) is automatically "masked" to an extent by reason of its protective association with a lipoid complex; (2) is an active thromboplastin and as such the precursor of all white connective tissue by initiating the precipitation of fibrin from fibrinogen, and (3) is attracted to and combined with or adsorbed upon the fibrin thus precipitated. The connective tissue is therefore a great storehouse, possibly more or less temporary, for the protomorphogens secreted by all the living cells in the organism.

Elimination of Protomorphogen Toxins

We have thus far promoted the hypothesis that the protomorphogen which is secreted from all cells as a result of their metabolism stimulates the formation of connective tissue and also is adsorbed upon this connective tissue. The connective tissue forms a storehouse for the protomorphogen.

Unfortunately we cannot assume that the adsorption upon connective tissue marks the end of protomorphogen metabolism in the mammal. The protomorphogen adsorbed on fibrous tissue still vitiates the neighboring cells and, if allowed to accumulate would ultimately cause the dissolution of these cells. This is indicated by the work of Mayer (1935) who demonstrated the presence of this toxic factor in the marginal clot of tissue cultures. This is discussed more thoroughly in Chapter 3.

We must look further, therefore, for factors, endocrine or otherwise: (1) which break the bond between fibrous tissue and protomorphogen; (2) factors which prepare the released protomorphogen for transfer in the blood stream; (3) the mechanism of transfer in the blood stream, and (4) the ultimate mode of elimination from the organism. As we unwind the pattern of the protomorphogen disposal mechanism it will become apparent that it is a very intricate physiological system. Nevertheless, it is a most important system of the organism since impairment of its function apparently leads to most serious disorders such as cancer, premature senescence, arthritis and general degenerative changes. The cessation of this function leads to immediate and pronounced symptoms of shock, toxemia and death.

The consequence of impaired protomorphogen elimination will be discussed later under pathology. For the present, however, we shall confine ourselves to presenting the outline of methods of its elimination. Much of the evidence establishing this outline, however, is obtained from pathological conditions and they will be touched upon briefly from time to time, a more comprehensive discussion being withheld till later.

Factors Removing Protomorphogen from Connective Tissue

Elutogenic Factors.—The first step in the excretion of protomorphogen is its dissociation from connective tissue. It will be remembered that although some of the protomorphogen molecules have served as a determinant for the formation of connective tissue fibers and thus are an integral part of the connective tissue molecule, there is apparently a variable quantity adsorbed or loosely attached to the fibers.

We have studied a series of substances which promote the release of protomorphogen from connective tissue. This action, of course, may be a denaturing process. We are loath to group these factors under the term "denaturants," however, since there are many factors which denature proteins which do not do so in a physiological manner and are not a part of the normal metabolism of the organism. In order to restrict our discussion to those denaturants which are a part of normal physiology and have a specific purpose of releasing protmorphogen from connective tissue for elimination purposes we shall refer to them as "elutogenic factors." This word is derived from the term "elution" which refers to the removal of adsorbed substances. We consider the major part of protomorphogens in connective tissue to be associated in an adsorbed sense.

Epithelial Fibrinolysin.—Burrows (1926) mentions that epithelial cells secrete a lysin which dissolves the protomorphogen-induced clot so that it is reabsorbed and used for growth. Epithelial tissue is the only class of cells which secrete this lysin. It is significant that epithelial tissue is the only tissue in the mammal in which mitosis takes place after parturition (with the exception of the osteoblasts). It is apparent that this factor from epithelial cells not only separates the protomorphogen from the fibrous tissue so that it may be eliminated, but also reduces it (depolymerizes it?) to a form useful to the cell in synthesizing new proteins at the cell boundary.

Cowdry (1924) has reviewed similar experiments. When hyaline cartilage of chick embryols cultured *in vitro* with embryo extract, the hyaline substance disappears, spindle cells grow and multiply by mitosis. It is evident that the embryo extract has supplied a substance which causes a lysis of the hyaline material making it available for growth and mitosis. This is probably the same factor present in epithelial tissue. It is of interest to note that both embryo extract and epithelial tissue contain substances which stimulate mitosis as well. This problem will be discussed later under the subtitle "growth substances." For the present we are only interested in the effect of releasing protomorphogen from connective tissue.

Sex Hormones.—There may be a biological reason for activity of the sex hormones (estrogen and testosterone) in releasing the protomorphogens from connective tissue. The morphogen hypothesis envisions the chromosome of the germ cell as consisting of a stable framework whose physical structure and organization is characteristic of the species and the individual; this is an orthodox conception.

However, the morphogen hypothesis goes further and postulates that individual tissue and cell determinants are attached to this framework.

Only when these cytomorphogens are attached to the chromosome network will there be a functional and potential germ cell. We postulate that while the immortal chromosome framework is reproduced by the sex cells, the individual tissue determinants (cytomorphogens and protomorphogens) are reproduced and supplied only by the specific tissues they determine. They therefore afford some degree of somatic influence over the active chromosome.

Darwin proposed a similar theoretical mechanism. He supposed the determinants to be formed in all tissues and swarm by way of the blood stream into the testes or ovaries. They were supposed to be redistributed to the respective locales in the embryo. He called these determinants "gemmules."

Jacques Loeb (1906) inadvertently hinted at this possibility when he reviewed the evidence that the sex cells of the male hydrolyze blood proteins, incorporating hydrolysate factors in the spermatozoa. In the ovaries this hydrolysis does not occur, blood protein being used for the construction of the egg. In this respect it is interesting to note the observations of Fabre and Kahane (1938) that when animals are exposed to carbon dust, carbon is found in the liver and kidney and occasionally in the spleen and *testicles*. We shall discuss later the hypothesis that colloidal and macromolecular wastes are discarded through much the same elimination system as protomorphogens. Inasmuch as the liver, kidney and spleen are all involved in this system, it is not surprising to find carbon in these locales. Its presence in the testicle, however, is highly significant and indicates that it happened to be transported to the site of chromosome construction along with protomorphogen normally utilized for this purpose.

The orthodox idea that the complete chromosome with determinants is reproduced *in toto* in the germinal apparatus does not admit of the possibility of any inheritance of acquired characteristics. Indeed one of the major battles in biology has been waged around this conception. Witness Weismann's experiments in which the tails were cut from successive generations of mice; never. once did the tail of any offspring show recessive characteristics.

Nevertheless authenticated reports of the inheritance of acquired tendencies to various conditions are before us and cannot be disposed of with a few choice aphorisms.[1] Children of diabetic parents have a weakness for this condition. Indeed, if both parents are suffering from diabetes mellitus the chance of their children suffering from the same is greatly enhanced.[2] Einhorn and Rowntree[3] have succeeded in producing inheritable characteristics by removal of the thymus in successive generations, and the same[4] has been observed from administration of thymus extract to successive generations.

The morphogen hypothesis admits the possibility of such environmental effects producing inheritable tendencies. Although the basic structure of the organism is determined by the chromosome framework, the addition of released protomorphogen molecules to this framework makes possible the transfer to the chromosome of weakness in various organs since the protomorphogens released from the parent organs may be morbid or even totally lacking. The major question seems to be not the existence of the phenomenon, but a delineation between the influences of the morphogens as compared to the chromosome framework. The morphogen influence might be called somatic inheritance and is undoubtedly limited to the tendencies exhibited by particular tissues and not by organized. structures or characteristics. It is more of a "tendency" than a characteristic, and probably disappears after a few generations in the same manner as cytoplasmic inheritance disappears in unicellular life (See Chapter 4).

Let us return to the experiment in which the tails were removed in successive generations; this had no effect on the offspring. The morphogen hypothesis would fit this demonstrated fact. We believe the chromosome framework, whose integrity is not influenced by normal environment, carries the outline of the tail structure. This outline cannot, however, produce a tail without association with necessary protomorphogens ordinarily supplied by tail cells. If the tails of parents are removed from successive generations there is no effect on the chromosome since the cells of the spinal column are so closely associa-

1 *The Inheritance of Acquired Characteristics,* P. Kammerer. Transl. by A. P. Maerker-Branden, Boni & Liveright, New York. 1924.

2 *Endocrinology,* A. A. Werner. Lea & Febiger, Philadelphia. 1942.

3 Einhorn, N. H. and L. G. Rowntree: "The Biological Effects of Thymectomy." *Endocrinol.,* 21:659-669, 1937.

4 Einhorn, N. H.: "The Biological Effects of Thymus Extract (Hanson) on Thymectomized Rats." *Endocrinol.,* 22:325-341, 1938.

ted with tail cells that their protomorphogens may be the same and may be considered as derived from one complete organ from a protomorphogen standpoint.

Similarly the morphogen hypothesis can explain the inherited weakness towards diabetes of children of diabetic parents.[1] In this case also the chromosome *framework* is undoubtedly intact. However, the cells of the Islands of Langerhans of the pancreas are of such specialized nature that other protomorphogens cannot be substituted in the chromosome as successfully as they may for less well differentiated tissue. A deficiency of islet protomorphogen could result in the development of an incomplete chromosome and consequently the Islands of Langerhans in the embryo might fail to normally develop. A weakness towards diabetes mellitus is thereby inheritable, although the first appearance of the disease in the parent or the immediate cause in the offspring may be due to dietary abnormalities in either parent which are known to be involved in pancreatic disturbances.[2] It is obvious that if there is no islet of Langerhans tissue in the parent no such protomorphogen will be available. One small bit of healthy tissue, however, could supply enough to complete the chromosome.

An interesting sidelight on the possibility that morphogens transported in the blood stream are destined for attachment to the chromosome and genes in the germinal tissue is the occasional reports of bizarre observations which up to this point have had no conceivable explanation in the light of modem genetics and therefore have been dismissed as the workings of chance. We refer to the influence of previous conceptions on subsequent generations called telegony. It seems not altogether impossible that morphogens from an embryo may find their way into the mother's blood stream and ultimately become attached to chromosomes in the ova, thereby influencing the character of subsequent offspring.

1 Warthin, A. S.: "The Pancreas as an Endocrine Gland." *Endocrinology and Metabolism,* Ed. by L. F. Barker. Vol. 2, D. Appleton & Co., New York, 1922. John, H. J.: "The Diabetic Child; Etiologic Factors." *Ann. Int. Med.* 8:198-213, 1934. *(John suggests that the inheritable tendency to diabetes is a recessive Mendelian character; we are more inclined to place it in the category of somatic inheritance under our above definition).*

2 *The Vitamins and Their Clinical Applications,* W. Stepp, Kuhnau, and H. Schroeder. Translated by the Vitamin Products Co., Milwaukee, Wis. 1938. *Chemistry and Physiology of the Vitamins,* H. R. Rosenberg. Interscience Publishers, Inc., New York. 1942.

A famous report of telegony is the influence of the successful mating of a mare with a zebra. Subsequent successful matings with normal stallions resulted in offspring with striped markings, persisting for three generations.[1] Carefully conducted control experiments, however, indicated that this was due to chance, since mares not previously covered by a zebra sometimes gave birth to offspring with the same markings, when sired by the same stallions.

Nevertheless, the opinion among animal breeders is widespread that pure blooded females are ruined by successful mating with mixed-breed males. Expensive animals have been often sacrificed as a result of such accidents. While no carefully conducted controlled researches have tested these ideas, it is well to remember that animal breeders are practical men, and not prone to sacrifice thousands of dollars on superstition alone.

Therefore the morphogen hypothesis revives the term somatic inheritance but restricts it to the hypothetical influence of individual morphogens rather than the more significant influence of the genes and chromosomes. The Weismann concept that the gene is reproduced in the germinal apparatus and passed on unaffected by the environment is established upon solid experimental ground. The morphogen hypothesis would only modify it to the extent that the *framework* is strictly germinal and follows this law. This framework of course determines the general characteristics of the organ. But the individual morphogens, the active individual protein and cell determinants, are attached to this gene framework and they are reproduced only in the soma. What specific degree of influence over the embryo these somatic morphogens exert is yet to be determined.

To return to our primary discussion of elutogenic factors, inasmuch as our hypothesis postulates the necessity of a constant source of protomorphogen for attachment to the chromosome matrix, it seems likely that there are specific sex hormones concerned with their release from connective tissue. There seems to be evidence indicating such a function for two of the sex hormones. Bierbaum and Moore (1945) have illustrated the effect of estrogen administration on the bone marrow of dogs. Estrogen injection resulted in hyperplasia of the bone marrow followed by hypoplasia. There was a marked

1 *Horse Owner's Cyclopedia,* J. H. Walsh. Porter & Coates, Philadelphia, 1871.

decrease in megakaryocytes. We shall present evidence later that the reticulo-endothelial system is intimately concerned with the disposal of protomorphogens removed from connective tissue or discharged by cells. In particular the megakaryocytes engage m their phagocytic removal. This effect of estrogen, therefore, may be interpreted as the result of a release of protomorphogens from connective tissue storehouses and the temporary overloading of the reticulo-endothelial disposal mechanism (especially the megakaryocytes).

Nasu (1940) demonstrated that the velocity of blood coagulation was accelerated by extracts of male and female urines, the latter being more potent (as would be expected the protomorphogen of the male is diverted in greater quantities to the testicle for the greater output of germ cells in the male gonad). We have suggested that the thromboplastin in platelets is derived from the protomorphogen of the tissues. This indicates the release of protomorphogen by sex hormones, increasing the platelet thromboplastin or its rate of release from the platelets.

Birch (1930-31) (1932) has reported encouraging results in the treatment of hemophilia with ovarian substance. This is significant in view of her reports and also those of Quick (1942) that hemophilia is associated with an impairment of the release of thromboplastin by the platelets. Quick mentions that a similar situation appears in menstruating women, indicating the link with some sex hormone. He lists various investigators who report success with various female sex hormones but concludes that the negative results rule out the possibility of these factors becoming useful in therapeutics. Nevertheless, the fact that certain sex hormone principles are successful in some cases is evidence of the link between these factors and the elution of protomorphogens from bound or adsorbed forms. Obviously there is more than this involved in hemophilia, but even a few encouraging reports are significant in their support of the elutogenic activity of sex hormones, an activity which becomes more apparent as various disease syndromes are interpreted in the terminology of the morphogen hypothesis.

An interesting collateral activity of sex hormones is reported by Taurog and co-workers (1944). They conclude from their observations that diethylstilbestrol promotes the rate of formation of phospholipides in the liver. This may be a compensatory synergism in that phospholipides are necessary to properly "insulate" protomorphogens released into the blood and fluids.

Ziskin (1941) has presented evidence of the elutogenic effect of testosterone. Treatment of monkeys with testosterone propionate resulted in an improvement of the keratin layer, hyperplasia of the epithelium and connective tissue, and improved density of the connective tissue in the gingival and oral mucous membranes. The reader will recall that free protomorphogen promotes increased density of connective tissue and keratinization of surrounding tissue. Our provisional hypothesis, therefore, is that certain of the sex hormones, probably testosterone and estrogen, effectively promote the elution of protomorphogens from tissues where they are adsorbed. This activity is probably linked with the necessity of supplying morphogen elements from the tissues to the germ cells where they become attached to the chromosome network forming a complete and functional chromosome.

Guanidine is one of the most powerful denaturants of proteins known.[1] Guanidine is one of the end products of the oxidation of guanine from nucleoproteins (Robertson, 1924). It is possible, therefore, that other elutogenic factors, i.e., the sex hormones, maintain a guanidine (or methyl-guanidine) threshold in the blood which serves its purpose as an important elutogenic factor. Guanidine is discussed further under the discussion of creatine and the detoxifying cycles in this chapter, and also in Chapter 6 under the notes on eclampsia of pregnancy.

Thyroid Hormone.—Thyroid hormone is probably one of the most effective physiological denaturants and consequently an elutogenic factor. That thyroid denatures body protein, especially connective tissue, has been demonstrated many times. We suggest that the pyrogenic activity of the thyroid is a result of the release of tissue metabolites, one of which (pyrexin) is a specific pyrogenic factor.[2] It is likely that pyrexin is released by the autolytic breakdown of the protomorphogen molecule. The relationship of pyrexin to protomorphogen, however, will be dealt with in more detail later in a discussion of tissue exudates and inflammation.

---

1 Neurath, H., and Greenstein, J. P.: "The Chemistry of the Proteins and Amino Acids." *Ann. Rev. Biochem.*, 13:117-154, 1944.
2 Menkin, V.: "The Significance of Biochemical Units in Inflammatory Exudates." *Science*, 101:422-425, 1945.

The elutogenic influence of the thyroid hormone is further indicated by its influence over the growth of fibroblasts *in vitro*. Von Haam and Cappel (1940) have remarked that it stimulates mitosis when added in small amounts and inhibits in greater concentrations. Its elutogenic effect is probably proportional to its concentration. The release of a small amount of adsorbed protomorphogens in the media would stimulate growth while the release of much, without adequate. disposal would inhibit mitosis. Estrogen and testosterone, also elutogenic factors, have been shown to exert similar effects, although there is conflicting evidence on this point.[1]

The elutogenic effect of thyroid lends a plausible explanation to the mutual antagonism of this hormone and vitamin A. The antithyroid influence of vitamin A is satisfactorily established. The simultaneous administration of vitamin A with thyroxin injections, for instance, results in a lower BMR than would be obtained with thyroxin alone.[2] Assuming the pyrogenic effect of thyroxin to be due to the liberation of "pyrexin" as an end product of protomorphogen dissolution in the blood stream, we might suspect the catalytic action of vitamin A in promoting the phosphatide "sealing" of the released protomorphogens to inhibit their breakdown and consequently impair the release of pyrexin.

This effect is further indicated by the increase in muscle phosphatides as a consequence of thyroxin administration. Pasternak and Page (1935) remark that the phosphatides are formed under thyroxin influence. There is the strong possibility that they are formed and utilized in association with the protomorphogen released by thyroxin in order to prevent the lethal influence of "raw" protomorphogens which might undergo autolysis.

There is a possibility that the elutogenic effect of estrogen depends upon the stimulation of thyroid secretion by this hormone. De Amilibia, Mendizabal and Botella-Llusia (1936) have demonstrated the influence of folliculin in stimulating thyroid activity.

The existence of an anti-fever hormone in the thyroid has been reported by Mansfeld (1940). Injection of this factor drops resting metabolism 30-

1 Von Haam, E., and Cappel, L.: "Effect .of Hormones Upon Cells Grown in Vitro. I. The Effect of Sex Hormones Upon Fibroblasts." *Am. J. Cancer,* 39: 350- 353, 1940.

2 Rappai, S., and Rosenfeld, P.: "Gaswechselversuche bei mit Thyroxin und Vitamin A behandelten Teiren." *Pfluger's Arch. Ges. Pbysiol.,* 236:464-470, 1935.

50 per cent. This factor may exert its influence by preventing either the elutogenic effect of thyroxin or, what is more likely, preventing the autolytic breakdown of the protomorphogens or proteins released by this function. This thyroid factor deserves more extensive investigation particularly if it prevents the autolytic breakdown of tissue products.

Trypsin.—We have seen that protomorphogens are removed from connective tissue by various substances which we have classified as "elutogenic factors." Elutogenic factors of probable importance in normal physiology include the male and female sex hormones, a fibrinolysin from epithelial (and probably embryonic) cells, thyroid and guanidine.

There seems to be evidence that trypsin may act both as an elutogenic factor and also to split the protomorphogens that are released from connective tissue by other elutogenic factors. In Chapter 3 reference was made to the investigations of Simms and Stillman (1937) in which it was demonstrated that moderate tryptic digestion makes a growth inhibitor (protomorphogen) more soluble (depolymerization, shortening the culture lag period by extracting the protomorphogens from the protoplasm) and may even remove it from the culture media. This suggests that trypsin may function as an elutogenic factor, splitting protomorphogen from tissue and it also may depolymerize large protomorphogen molecules released by other elutogenic factors or split them into smaller, less complex components.

Trypsin is a normal component of the blood. It is probably prevented from splitting normal tissue beyond that necessary for its elutogenic action by a specific anti-tryptic antibody present in the blood. Ten Broeck (1934) has demonstrated the antigenic activity of trypsin. The variation in toxicity of different specific trypsins (Ten Broeck, 1934) (Weinglass and Tagnon, 1945), when injected into the blood, is a strong indication of the existence of a specific anti-tryptic substance maintained by the immune mechanism.

Eagle and Harris (1937) have demonstrated that the direct action of administered trypsin is to accelerate the coagulation of blood. Trypsin, *per se,* will not coagulate purified fibrinogen but accelerates coagulation probably by activating or releasing thromboplastin (Quick, 1942). Injection of large amounts of trypsin produces extensive intravascular coagulation and focal necrosis (Weinglass and Tagnon, 1945). The association of intravascular coagulation and necrosis from injection of trypsin indicates that the coagulation accelerative affect of trypsin is a consequence of release of thromboplastin (protomorphogen) from tissue. It is likely that normally the trypsin present in the bloodstream splits the protomorphogen removed from the connective tissue by other elutogenic factors.

It is quite possible that tryptic splitting is the first step in protomorphogen elimination. This activity may split off diffusible fractions to be removed via the kidney, and activate the macromolecular portions which then react with the natural tissue antibody by agglutination into a particle susceptible to reticulo-endothelial phagocytosis.

Paradoxically, injection of trypsin in moderate amounts results in a varying degree of incoagulability of the blood but Rocha e Silva and Dragstedt (1941) have called attention to the fact that the inhibition of coagulation is due to heparin released by the injected trypsin. This released heparin can overbalance the thromboplastin (protomorphogen) also released until the trypsin dosage is increased to a critical point where the heparin supply is inadequate to cope with the thromboplastin and acceleration instead of inhibition occurs. This is further demonstrated by the observation that the administration of heparin protects experimental animals against the necrotic effects of injected trypsin (Wells, Dragstedt, Cooper and Morris, 1945). In this case, the injected heparin adds to the effect of the trypsin-released heparin, so that a higher dosage of trypsin is necessary to reach the critical point where released protomorphogen can exert its influence.

Trypsin-Heparin System.—The release of heparin by trypsin is probably a part of the normal physiology of the action of this enzyme in the blood stream. The presence of the enzyme is possibly necessary for the splitting of protomorphogens and the concomitant release of heparin prevents the lethal influence of these split protomorphogens from becoming manifest.

There is much evidence concerning the anticoagulant influence of heparin and this action is well established. In his review of anticoagulants, Douthwaite (1944) states that heparin inhibits the release of thromboplastin (protomorphogen). Dyckerhoff and Grunewald (1943) seem to consider that the mode of heparin action is an antagonism to thromboplastin, administration of the latter overcoming to an extent the coagulation inhibition of heparin. Mellanby (1934) has emphasized the anti-heparin activity of administered thromboplastin.

Solandt and Best (1940) feel that the primary action of heparin is to prevent the initial clotting of blood, possibly by preventing the release of thromboplastin. They point out the difference between blood-clotting and thrombosis, the latter being initiated by an agglutination of platelets, the former by the release of thromboplastin. In small doses heparin only inhibits the clotting resulting from thromboplastin release, thrombosis not being affected. In larger doses both clotting and thrombosis were inhibited. Dyckerhoff and Marx (1944) seem to confirm this conception with their observation that thromboplastin administrations inhibit the first phase of heparin action, but are less effective in inhibiting the second phase. If the first phase of blood coagulation is inhibited by the action of heparin in preventing thromboplastin release, the administration of thromboplastin would overcome this influence. But the second phase in which larger administrations of heparin inhibit the agglutination of platelets seemingly would not be as effectively nullified by the mere administration of thromboplastin.

The primary influence of heparin, therefore, seems to be that of maintaining or stabilizing the integrity of the blood platelets, thus preventing the normal release of their thromboplastin (protomorphogen) by which coagulation is initiated. This contention is further substantiated by the investigations of Copley and Robb (1942) which brought forth no conclusive evidence that the platelet count is abnormally influenced by subcutaneous injections of heparin. Further, the work of Dragstedt, Wells and Rocha e Silva (1942) has brought into focus the influence of heparin in preventing the increase in plasma histamine as a consequence of administration of trypsin, antigens or proteoses. Quick's (1942) hypothesis, that the platelets remove released tissue histamine is pertinent here. If the effect of heparin is that of maintaining the integrity of the platelets, their ability to pick up released histamine should be augmented by heparin administration. Such seems to be the case.

The reports of Chargaff, Ziff and Cohen (1940) lead to the suggestion that heparin combines with the protomorphogen in platelets forming a complex more stable than ordinary. They demonstrated that heparin displaced the phospholipide portion of lung thromboplastin (protomorphogen) forming a stable heparin-protein complex. This combination of heparin with protomorphogen is apparently a specific reaction since Rigdon (1944) has reported that heparin has no effect on other types of agglutination such as bacterial or immune reactions.

Summing up the data, it would seem that: (1) the presence of trypsin in the blood stream is necessary both for its elutogenic action and its activity in splitting protomorphogens released by tissues into the blood via the lymph; (2) the lethal necrotic influences of this trypsin are inhibited by a specific anti-tryptic antibody, and (3) the sudden release of protomorphogen fragments by this trypsin is compensated for by its concomitant release of heparin which combines with the protomorphogen insuring a stable platelet structure.

Prandoni and Wright (1942) review evidence indicating that heparin is produced by the mast cells of Ehrlich. Its anti-protomorphogen influence is apparently quite beneficial to the organism and is further emphasized by Magerl's (1942) report that heparin administrations increase phagocytosis and, in general, stimulate the immune-biological system.

Maignan and Thiery (1942) have investigated the possible relationships of trypsin to alexin or complement. Their thought is that alexin is a complex of pancreatic trypsin and globulin. They report conflicting results on the values of blood alexin as a result of trypsin administration and of feeding or starvation (the latter increasing or decreasing the available pancreatic trypsin). It is possible that the heparin released by trypsin may influence the blood alexin in view of Magerl's (1942) report that heparin administration stimulates the immune-biological system.

Elutogens and Depolymerizers.—We have defined elutogenic factors as: (1) those substances which cause the elution of protomorphogen adsorbed on connective tissue, and (2) physiological denaturants releasing protomorphogen from the tissues of the organism where it is stored. Thus far it seems that epithelial fibrinolysin (including the embryo fibrinolysin), certain of the sex hormones, the thyroid principle

189

and blood trypsin fit into this class of substances. They all have important physiological functions concerned with the removal of protomorphogen from storage depots into the blood stream and lymph.

We have seen however, that the embryo factor has the property of promoting healthy synthesis of protein as well as of elutogenic activity. This property will be discussed in the next chapter under growth substances. Trypsin, however, also can be distinguished from other elutogenic factors in that it enzymatically reduces and depolymerizes protomorphogens, released either by itself or other elutogens.

It is important that we distinguish between depolymerizers and elutogenic factors, for each plays a cardinal role in protomorphogen disposal and the two activities are not identical.

We will recall that experimental evidence obtained with cultures of protozoa and of tissue *in vitro* demonstrates that accumulating protomorphogen vitiates the cell because there is a polymerization within and without the cytoplasm causing an accumulating concentration within.

It has been seen that small amounts of trypsin reverse this reaction, causing a depolymerization and consequent release of toxic protomorphogens from within the cell. This allows the cell to commence again its physiological activities of mitosis and repair. A local rejuvenation occurs, provided the protomorphogens in the media are summarily removed. In addition, it is likely that the protomorphogens from concentrated accumulations must be depolymerized before they can be utilized for the synthesis of new proteins and promote growth.

Briefly, therefore, depolymerization is the process by which the molecular aggregation of protomorphogen is reduced to the point where the individual units can be again utilized for growth of new tissue.

Elution, on the other hand, merely signifies the breakage of the bond, either chemical or physical, which binds protomorphogen intimately with connective tissue.

Elution may or may not be accompanied by depolymerization, depending upon the active agent. Normally it should be unless other means of insuring the disposal of the eluted protomorphogen are at hand.

If the eluted protomorphogen is not broken up at the physiological linkages by physiological depolymerizing substances, it may disintegrate into lethal moieties such as guanidine and necrosin, causing inflammation or more serious disorders.

The various physiological elutogens are variously matched with depolymerizing factors depending upon their role in the organism's economy. The sex hormones, for instance, are mildly elutogenic, but it is unlikely that their activity in this respect is accompanied by depolymerization of the eluted protomorphogen. The elutogenic activity of the sex hormones releases morphogens for the express purpose of supplying substrate material to the germinal apparatus for chromosome assembly. We believe it is transported to that locale under the protective influence of phospholipide sheaths and the specific activity of an internal secretion of the prostate. These will be discussed in detail further in this chapter. Uncontrolled depolymerization might reduce the morphogens to the point where they would no longer be useful to the germinal apparatus.

The thyroid elutogenic influence is of a different nature than that of sex hormones and probably it also requires phospholipide material to prevent the eluted protomorphogens from causing serious biochemical disruptions. In addition depolymerizing and constructive factors are probably cooperant with the thyroid in the normal physiological cycle accounting for the youthful influence of the thyroid when the other endocrines are functioning properly.

From the evidence reviewed the influence of trypsin would seem to be both elutogenic and depolymerizing. Its action in the physiological economy, however, is carefully controlled by various factors, for unhampered trypsin would certainly be fatal to any organism.

We may tentatively classify the elutogenic factors and depolymerizers which we suggest are a part of normal physiological activity as follows:

Elutogenic Factors	*Depolymerizers*
Sex Hormones	Allantoin
Thyroid	Urea
Epithelial and embryonic substance	Trypsin
Trypsin	Ultra-violet rays

Naturally, this whole hypothesis is presented as speculative material. Only the slimmest links of experimental evidence hold it together. As the reader will note, some of these factors appear under both classifications. This is to be expected since much experimental work must be performed before this problem can be properly evaluated. Furthermore, some of these factors, embryo substance in particular, consist of many different components which may be separated into those responsible for one function or the other.

We shall now briefly review some of the evidence which leads us to establish the depolymerizing activity of the factors included in that classification.

Allantoin and Urea.—Allantoin, found in the allantoic sac of certain mammals, has long enjoyed a reputation as a therapeutic adjunct in the promotion of healing. MacAlister (1936) has supplied an excellent review of this interesting principle. The relationship of allantoin to morphogen metabolism is indicated by the experiments on plants in which the presence of allantoin in the soil surrounding bulbs inhibited growth while the injection of allantoin into the bulb or stalk of the plant stimulated growth. Allantoin may appear in the soil as a consequence of plant metabolism.

This experiment suggests that allantoin has the function of depolymerizing protomorphogen molecules. The depolymerization of those surrounding the plant might, if not resulting in their disposal, have the irritating influence of concentrated protomorphogens in the media. However, depolymerization of the protomorphogens within the bulb would enable them to be excreted from the cells and the plant, lowering the internal concentration and stimulating growth.

Local administration of allantoin to wounds has proven it to be a healing agent of first importance. The depolymerization of protomorphogen molecules enables them to be disposed of through the proper channels and thus prevent inhibitory concentrations from stopping normal regenerative processes. Further, the depolymerized protomorphogens are again available as substrate for new protein synthesis.

Allantoin has been shown to be a growth promoter, possibly by catalyzing the synthesis of nucleic acids. This may be the result of depolymerization alone or the existence of another distinctly positive growth stimulating effect. We shall show later that some substances, i.e. embryo hormones, are elutogenic factors and depolymerizers, but have an additional positive effect in stimulating tissue repair and mitosis.

That this latter is not the only effect of allantoin is indicated by the observation that it does not stimulate tumor growth. Were it a growth promoter alone, it would do so. In our discussion of cancer later in this chapter, we shall present evidence that, in order for the pathological proliferation to succeed, there must be an excess accumulation of protomorphogens in the area of the neoplasm. By facilitating their removal the presence of allantoin would possibly inhibit tumor growth.

The presence of allantoin in the allantois also suggests the depolymerizing effect of this substance. The protomorphogens produced by the developing embryo must not be permitted to accumulate in the embryonic fluids or they will choke off growth. By depolymerizing them and preventing their adsorption on connective tissue, allantoin facilitates their removal through the placental wall. Closely related to allantoin and also a depolymerizing factor is urea; this substance substitutes for allantoin in the human species. Erickson and Neurath (1943) have revealed that urea denatures serum proteins changing the specificity of the molecules. Stanley (1940) has also discussed the denaturing of viruses by urea. We feel that urea has an important physiological function in human metabolism revolving around its mild denaturing effect and consequent depolymerizing influence.

Grunke and Koletzko (1939) has demonstrated that this denaturing action of urea plays an important part in accelerating clotting and fibrinolysis. The increased activity of protomorphogens depolymerized by urea may stimulate clot formation because of thromboplastic influence.

Urea has a healing influence similar to allantoin, and its influence in this respect has been appreciated for many years.

Rachmilewitz (1941) has demonstrated that uremic serum with a high content of urea stimulated the growth of fibroblasts *in vitro*. He reviews comment that the same effect of allantoin may be due to its hydrolysis and release of urea, although there is conflicting evidence on this point.

We might be tempted to include many other factors as physiological depolymerizers of protomorphogens. The activities of embryo substance and of anterior pituitary growth hormone, for example, lead us to suspect that they exert this influence. For instance, in a deficiency of vitamin A, anterior pituitary growth substance augments arthritic pains (due, we suspect, to the irritating influence of depolymerized protomorphogen and its end products such as guanidine and necrosin). These substances are discussed in detail in the next chapter under appropriate headings.

Nevertheless, we are confining ourselves at this point to a classification based upon the experimentally demonstrated activity of reducing lag period in cultures. We believe that this can be due only to depolymerizing activity since the lag period can be shortened only by reducing the time necessary for the protoplasmic protomorphogen concentration within the cell to be lowered to the level where mitosis may commence. (See Chapter 3)

Trypsin, ultra-violet radiation and, to an extent, allantoin, are the only substances which can properly be listed under this classification, if we use this criterion as the primary basis of selection.

In the next chapter we review and list several products, drugs and otherwise, which can be tentatively classified as either elutogens, depolymerizers, or both. Such a classification is based upon the pharmacological responses reported from clinical and experimental investigations of these products. They are not included in this discussion at this point, however, since we are attempting to confine ourselves in this chapter to a review of those mechanisms concerned with protomorphogen metabolism in the normal physiological cycles.

To recapitulate, protomorphogen is delivered into the pericellular fluids by all cells in consequence of their nuclear metabolism. (This nuclear metabolism is summarized by the schematic diagram in Chapter 3.) These protomorphogens are thromboplastic. They are determinants for the precipitation of connective tissue from fibrinogen and they are adsorbed on the connective tissue thus formed. This tissue becomes a storehouse for them. While thus adsorbed, much of their activity is masked, but it is believed that they still vitiate neighboring cells by their presence. They are released by elutogenic factors, sex hormones, thyroid and trypsin.

After their release there are three avenues of disposal: (1) use by the geminal mechanism for chromosome assembly; (2) excretion as waste products, and (3) utilization as substrate determinants for regeneration and repair. The latter route is probable if the other factors, nutritional and catalytic, are also present. (This is discussed further in the next chapter.) For the first two routes their transfer in the tissue fluids and blood stream is indicated. At this point we should concern ourselves with the methods of transfer which will be innocuous to the organism if their ultimate destination is excretion, and methods which will not damage the protomorphogens themselves if they are to be used for chromosome assembly.

The first and possibly most important subject to discuss in this respect is the protective association between protomorphogens and lipoidal substances.

Lipoidal Sheathing of Morphogens

We have presented considerable evidence in previous chapters that protomorphogens are associated with lipoidal or phosphatide molecules both in the protoplasm and in the tissue fluids. Robertson (1924) found his "allelocatalyst" soluble in acetone. Baker and Carrel (1925) demonstrated that washing cell extracts with alcohol-ether reduced the growth inhibiting effects of the extracts. Crile (1936) found that the determinant factor in autosynthetic cells was present in the lipoid constituent. Needham (1942) has reviewed considerable evidence which indicates that the inductor substances responsible for morphogenesis in the developing embryo are associated with lecithins. Fitzgerald and Leathes (1912) has reviewed evidence that ether soluble heat stable extracts of red blood corpuscles have antigenic properties. These antigens are probably protomorphogen fragments of nucleoproteins associated with phospholipides. The association of phospholipides with chromatin is also indicated by the experiments of Kaucher and coworkers (1945). They demonstrated that the nuclei of liver cells are significantly higher in phosphatides, mainly lecithin, than whole liver tissue.

We have shown that all of these substances have common properties under the morphogenic classification. That the morphogens themselves are not lecithins is illustrated in a striking manner by the experiments of Turck (1933), who demonstrated the allelocatalytic growth effects of tissue ashed at temperatures about 300 degrees C.

Protective Association with Lipids.—We have, therefore, presented the hypothesis that the morphogen molecule (consisting of nucleoprotein with significant mineral links) itself when not a component of a biological protein, is associated with phospholipides which surround the morphogen molecule, preventing it from exerting lethal effects on neighboring cells or protoplasm. We consider that the morphogen molecule has a tremendous affinity for phospholipide substance.

We consider that the substance Burrows calls "ergusia" is in reality identical with "archusia," but surrounded or associated with phospholipide. ote should be made that ergusia is "secreted" from the cell only after a medium concentration of archusia has accumulated.

In an investigation of macromolecular tissue lipoproteins Chargaff and Bendich (1944) discuss the x-ray evidence that these molecules occur as thin protein layers inserted between bimolecular lipid sheaths. They suggest the application of their lipid extraction technique to animal viruses consisting of high molecular weight lipoproteins. Burnet (1946) also discusses lipoproteins in respect to virus antigens with a phospholipide component, the latter being concerned with the surface activity of the molecule. Schmitt and Palmer (1940) has discussed x-ray diffraction studies of lipid-protein systems. It appears that monomolecular protein layers are bound between cephalin layers producing a stable system immobilizing the participants, and expelling the water from the cephalin. In their natural tissue environment protomorphogens are most often found as macromolecular nucleoproteins associated with lipids, probably in the above structural relationship. These protomorphogens are, in effect, specialized viruses.

Chargaff (1944) has reported the thromboplastic activity of phosphatide preparations from various tissues. He notes that purified lecithin preparations are inactive and that thromboplastic proteins are a thousand times more active than the most potent lipid preparation. Thus lipid or lecithin freed from protomorphogen does not exert thromboplastic activity by itself since this activity depends upon the protomorphogen.

Protomorphogen, therefore, may be considered to be both archusia or ergusia, the latter designation applying only to protomorphogen which is more concentrated and attached to a phospholipide sheath or envelope. The lipoid alone will not shorten coagulation time, in fact it has been demonstrated that the lipoids of the blood (not associated with protomorphogen) actually inhibit coagulation. This is by reason of the influence of lipoids in "sheathing" protomorphogens and impairing their thromboplastic potency.

The question immediately arises as to the extent to which the phospholipide prevents the protomorphogen from exerting other normal physiological effects. It is likely that the alteration of the protomorphogen potency is quantitative rather than qualitative. It appears that phospholipide-bound protomorphogen (as in the ergusia of Burrows) can act as a thromboplastin, and influence the growth or dissolution of the cell depending upon its concentration. It is likely that these effects are far less pronounced when it is associated with phospholipides, however, and in nature it is almost universally so combined. The evidence demonstrates that a smaller amount of lecithin-free protomorphogen is a more potent thromboplastin than the same concentration associated with phospholipide. It is also probable that large amounts of phospholipide can effectively "mask" small concentrations of protomorphogen, preventing any thromboplastic or other activity.

The question of the chemical nature of the lipid sheath of protomorphogens presents itself for our consideration. Before more detailed consideration we would like to suggest that the chemical nature of this material is probably slightly different in different tissues; brain cephalin, lipoid groups in association with lung macromolecules, and phosphatides from liver cytoplasm, for instance, all exhibit a variation in chemical properties. This probably accounts for the variation in solubilities of protomorphogen carriers reported by various investigators. (See Chapter 3.)

The complexity of the problem makes it impossible for us to accurately outline the chemistry of these sheathing substances. We can simply present several links with known products, leaving exact biochemical questions to further experimental investigation.

The weight of evidence lends support to the hypothesis that they are basically of a phosphatide nature. The association of protomorphogen with

197

these substances is apparent in solubilities and in the reports of morphogen activity in phosphatides, both in experiments on embryonic differentiation (Chapter 4) and in the investigations of thromboplastic substances.

Robertson (1923) (1924) has reported several experiments, covering feeding, cultures and injections, all of which indicate that lecithin inhibits growth in the early stages of mitotic activity and stimulates later. He recognizes that this is an expression of protomorphogen impairment since protomorphogens exert exactly the reverse influence. It is obvious that phospholipides are an important constituent of sheathing material.

We are not sure of the exact relationship of cholesterol to the sheathing of the morphogen molecule. Robertson (1924) comments that cholesterol: (1) increases the rate of neoplastic mitosis; (2) accelerates the rate of infusorian reproduction, and (3) inhibits the initial growth of white mice, with little effect on subsequent growth. He suggests that cholesterol may regulate protomorphogen by impairing the influence of lecithin. There are many antagonisms demonstrated between lecithins and cholesterol which lend credence to Robertson's suggestion (Foldes and Murphy, 1946).

Schulman (1945) reports that cholesterol adsorbs proteins strongly and non-specifically. They form a mixed monolayer when the ratio of protein to cholesterol is 4:1. Other workers have suggested that cholesterol is attached to a ring structure inside the protein molecule (Tayeau, 1941). Tayeau (1943) also suggests that serum cholesterol is linked with proteins in a complex not separable with ether, unless treated with bile salts. Macheboeuf and co-workers (1943) report the presence in the blood of cholesterol-phosphatide-protein complexes. They differ in the nature of the phosphatide, the fatty acid in the cholesterol ester and the protein. The relative proportions of phosphatides and sterides remain the same, however.

We are inclined to suspect that the sheathing material is a complex of cholesterol-esters with phosphatides peculiar to the specific tissue with whose protomorphogen it is associated. It is very possible that the phosphatide-steride-protein complexes being studied by Macheboeuf and his associates are protomorphogen molecules in protective association with lipids under transport in the blood stream. By its nature of adsorbing proteins (protomorphogen) forming a monolayer, cholesterol may activate the latter. This may explain Robertson's observations that cholesterol regulates protomorphogen in a manner opposing the influence of lecithin; for it is very likely that a lecithin sheath envelopes the mixed monolayer of cholesterol and adsorbed protomorphogen forming a stable lipoprotein layer such as those reported by Schmitt and Palmer (1940).

The recent studies of Tompkins (1946) augur the impression that the assemblage of this molecule occurs in the reticulo-endothelial cells. She suggests that these cells associate with cholesterol, forming the ester and effect the combination with proteins after which the whole molecule is excreted as such. The conception of protomorphogen activation by cholesterol due to its adsorption forming a mixed monolayer very likely may form a basis for an explanation of many of the high cholesterol affects. Excessive cholesterol has been reported to occur in many pathologies, and two of these in particular (arteriosclerosis, cancer) can be considered on the basis of untoward protomorphogen influence. Clinically phospholipides oppose these cholesterol influences. According to the morphogen hypothesis this is because the phospholipides afford "sheathing" layers between which the cholesterol and protomorphogen layer is held and thus rendered innocuous.

Vitamin A and Protomorphogen Protection.—Burrows and Jorstad (1926) have suggested that the substance they term ergusia is identical with vitamin A. They demonstrated that older cultures, in which the protomorphogen was associated with a rich amount of phospholipides, assayed high for vitamin A. They also demonstrated that the effect of coal tar applications on cells deficient in vitamin A was a hyaline degeneration similar to that observed when coal tar was applied to cells low in ergusia. When more vitamin A was present the degeneration resulting from coal tar applications was not as pronounced; this was also noticed with

cells high in ergusia.

It should be mentioned that coal tar and lipoid solvents have the effect of disturbing the association between protomorphogens and phospholipides, either dissolving out both or separating them so that the protomorphogen exerts a more potent or toxic effect on the neighboring cells. The degeneration in the above experiments was likely due to the increased effective potency of the surrounding protomorphogen which was separated from its lecithin associate. If the amount of phospholipide was low in proportion to the associated protomorphogen, degenerative changes could occur. On the other hand, if the ratio of phospholipide to protomorphogen was high, the toxic potency of the protomorphogen would not as easily be reached.

Because of the phospholipide nature of the lipoid associated with protomorphogen in "ergusia" and the large quantity involved, it is unlikely that it, in itself, is vitamin A. The above evidence, however, strongly suggests that vitamin A is concerned with catalyzing the protective association between protomorphogen and phospholipides and is an essential part of the lipoid complex.

There is much evidence that vitamin A is an important catalyst that must be present for normal association between protomorphogen and phospholipides. Burrows (1927) has shown that the presence of excess protomorphogen results in a coagulation of the protein base as a granular or hyaline mass. Atrophy precedes this hyalinization. X-ray has been shown to protect against vitamin A deficiency, at the same time causing hyalinization. Burrows interprets this x-ray effect as the result of removal of the ergusia from the local area, dispersing it into the organism. He also mentions the hyalinization of cells, which occurs in precancerous conditions, as being due to removal of ergusia.

We would interpret this removal of ergusia as a splitting of the association between the local protomorphogen and the protective lipoids, releasing more potent protomorphogen as a thromboplastic agent precipitating granular and hyaline tissue. It is interesting to note that atrophy and hyalinization of epithelial tissues are two of the most pronounced manifestations of vitamin A deficiency. An interference with normal protomorphogen metabolism and transport in avitaminosis A is indicated by the fact that the platelet count is diminished in vitamin A deficiency.[1] We shall show later that protomorphogen secreted by tissue cells may be carried in the platelets as a stage in the process of its elimination from the organism.

Rosenberg (1942) mentions that there is an increase in purines in A-depleted tissue upon the administration of vitamin A. He further states ". . . all primary and secondary symptoms of vitamin A deficiency can be explained on this basis." Purines are essential constituents of nucleo-

1 *Applied Physiology,* S. Wright. Humphrey Milford, Oxford University Press, New York. 1932.

protein and as such are components of the physiological protomorphogen molecule. This activity of vitamin A seems to be an expression of its fundamental catalysis of the sheathing phenomenon which prevents the toxic influence of "raw" protomorphogens and their end products, purines being a toxic end product of nucleoprotein degradation.

In their study of tissue changes in vitamin deficiencies, Wolbach and Bessey (1942) conclude that the epithelial changes in vitamin A deficiency are not a consequence of deranged endocrine function, but a direct influence upon the tissues themselves.

In discussing the influence of vitamin factors with respect to the protection of protomorphogen it is interesting to mention the importance of vitamin E as an indispensable factor in nuclear activities involving chromatin material. Mattill's review[1] of vitamin E discusses experimental evidence that there is a liquefaction of chromatin material in the germinal cells in a deficiency of this dietary factor. The influence of vitamin E is distinctly different from that of vitamin A, although a deficiency of each results in sterility, the former through impairment of nuclear and chromatin metabolism and the latter through the progressive degeneration and hyalinization of epithelial cells.

Burrows and Jorstad (1926) have suggested that archusia is a form of vitamin B because of its growth stimulating effects and similar activity to the vitamin B preparations of that day. We consider it unlikely the protomorphogen can be identified with vitamin B, although one of the B complex fractions (Bios) has been split into factors, one of which exerts an allelocatalytic effect on yeast growth. These investigators extracted protomorphogens (archusia) from a number of heterologous tissues including bacteria and noted that many different types of cells were stimulated by it. An axiom of the morphogen hypothesis states that protomorphogen will stimulate growth of heterologous cells but can only inhibit the growth of its own species. This is explained by reason of the nature of the reaction in each case. As a growth stimulant protomorphogen fragments may be nutritionally useful in the synthesis of new protein molecules. As an inhibitor, however, they can only react (polymerize)

1 *The Vitamins*. H. H. Mattill, *"Vitamin E."* Chapter 30, Symposium published by the American Medical Association, Chicago. 1939.

with more of the same kind within the cell protoplasm.

Even the fact that heterologous protomorphogens (archusia) were shown to prevent the lack of growth observed in cultures on a vitamin B deficient media is not sufficient to identify protomorphogen with vitamin B. Protomorphogen in small amounts is a powerful growth stimulator and might easily promote growth of cells in spite of the presence of sub-optimum amounts of vitamin B in the medium.

Biochemistry of the Sheathing Material

The biochemistry of the physiological systems concerned with the processing of the lipid sheathing material and the catalysis of its protective association with protomorphogen is a complicated and involved picture which cannot be discussed in detail at this time. It may be of interest to briefly survey some of the factors involved in these reactions and suggest some links with the thought of encouraging further experimental research in this field.

Thymus.—Crotti's review of the thymus problem (1938) reports the following conditions among those resulting from thymic extirpation: fever, asthenia, subcutaneous ulcers, fatty degeneration and autointoxication. The reponed results of thymus extirpation, however, have been most conflicting and inconsistent. All of the symptoms, nevertheless, are typical of those resulting from excessive protomorphogen accumulations in the absence of adequate protective association with lipids.

Some thymus principle may be closely concerned with the supply of lipoidal substrate for sheathing material. The so-called thymic complexion, smooth and juvenile, is witness to the effect of this organ in promoting a youthful epidermis due, possibly, to its activity in promoting the protective association of protomorphogens with lipoids. (Compare the aged and wrinkled complexion of persons in the desert areas of this country where excessive vitamin D is produced in the skin. Vitamin D is reponed to break down organic phosphorus compounds, phospholipins, and thus may readily impair their protective association with protomorphogens.) Thyroid, an elutogenic factor and one concerned with the sheathing cycle more in a destructive than constructive manner, is closely associated with thymus activity. A mutual inhibition has been demonstrated between these two factors (Crotti, 1938). Low (1938) has demonstrated that thyroid administration increases the cortical lymphocytes in the thymus, Bomskov and Brochat (1940) having suggested that the lymphocytes carry the thymus hormone. Low also claims that thyroid and estrogen administration cause thymic involution (estrogen is also an elutogenic factor).

The thymic involution associated with the "alarm reaction" to shock is evidence in point. The intense demands upon sheathing material consequent to the release of protomorphogens may result in this sudden involution due to excess release of thymus lipoid or demands upon this organ. It has recently been reported that choline deficiency results in prompt thymic involution.[1] This is of interest inasmuch as choline is a necessary substrate material for phosphatide synthesis, and apparently in its absence the thymus is overworked to the extent of involution. The regression of thymus tissue at puberty is an indication of the reduced necessity for lipoidal sheathing material consequent to the reduction in protomorphogen metabolism when growth is attained. The major degree of the regression occurs in the lymphoid tissue, the ratio of the secreting cells being increased as its result (Wolf, 1939).

Hanson (1930) has reported that the administration of a specific thymus extract caused the regression of carcinoma. Harrower (1933) reports Babes conclusions that there is an atrophy of the thymus in animals afflicted with tar cancer. In view of the suggestions (see Chapter 6 this volume) that cancer develops as a consequence of a decrease in available sheathing material, these reports are extremely significant and lend further credence to the possibility that the thymus is concerned with the metabolism of the sheathing substances. The action of both choline and thymus extract as anticarcinogens becomes more understandable.

The thymus is an organ which has enjoyed only moderate interest among investigators. There seems to be excellent indications that it is important in the economy throughout life, not only during the growth period as is generally considered. Further experimental studies of this tissue will no doubt be rewarded with far-reaching progress in our knowledge of physiology.

1 Christensen, K., and Griffith, W. H.: "Involution and Regeneration of Thymus in Rats Fed Choline-Deficient Diets."*Endocrinol.*, 30:574-580, 1942.

Methyl Donors.—The sheathing material is of lipoidal nature and any discussion of lipid biochemistry would be woefully incomplete without a mention of the vast importance of methyl donors, especially choline. However, many excellent reviews of the outstanding work on these lipotropic factors have recently appeared and any attempt at a detailed discussion in these pages would be out of the question.

It is extremely significant from a standpoint of sheathing lipids, however, to note that over 95 per cent of the plasma phospholipides in man contain choline (Taurog, Entenman and Chaikoff, 1944). Boxer and Stetten (1944) have demonstrated by means of choline containing heavy nitrogen that choline deficiency impairs the *rate* of choline introduction into phosphatides without altering the percentage composition. Perlman and Chalkoff (1939) have also supplied data which tend to support the contention that the effect of choline on fatty livers is due to its influence upon the *rate* of turnover of choline phosphatides.

Fishman and Artom (1944) have demonstrated that the decrease in liver phosphatides produced by a protein deficient diet is corrected by choline, but not by methionine, another important methyl donor. On the other hand, Vigneaud and co-workers (1940) have stated that choline cannot yield a methyl group directly to guanidoacetic acid to form creatine. Methionine can directly cause this methlylation, however.

It is evident that methylation includes several trans-methyl reactions, depending upon the end substrate acted upon. Important methylations are concerned with the formation of liver phosphatides with choline and the methylation of guanido-acetic acid to creatine under the influence of methionine. Apparently homocystine is an intermediary agent between these two important reactions.[1]

Recently[2] it has been suggested that choline is also intimately concerned with the important phosphatide turnover in the kidney. In fact, both the kidney and liver lesions of choline deficiency can be ascribed to a failure of phospholipide synthesis in those areas.

1 Vigneaud, V. du, Chandler, J. P., Moyer, A. W., and Keppel, D. M.: "The Effect of Choline on the Ability of Homocystine to Replace Methionine in the Diet." *J. Biol. Chem.*, 131:57-76, 1939.

2 "Choline and Phospholipid Synthesis." *Nature,* 158:630, 1946.

The formation of phosphatides and their rate of turnover is, of course, a vital part of the biochemical cycle of the sheathing material and therefore essential for normal protomorphogen metabolism. We shall see later that the methylation of creatine is an important link in another detoxifying cycle, that of disposing of the guanidine resulting from protomorphogen degradation.

Methyl donors and the methylation processes, therefore, are vitally important for proper protomorphogen disposal in two distinct fashions, choline being essential to one, and methionine to the other.

Unsaturated Fatty Acids (Vitamin F).—In chapter 6 we shall review the evidence that the toxic irritating factors produced in inflammation, burns in particular, are end products of protomorphogen degradation. At this point it is interesting to note that the local administration of cod liver oil salve is astonishingly effective in the treatment of burns of all kinds. Stepp and his co-workers[1] state that pure vitamin A preparations are not effective and mention that other cod liver oil constituents such as the unsaturated fatty acids may be the active ingredients. We have received clinical reports that the local administration of vitamin F and associated unsaturated fatty acids is singularly effective in reducing the pain associated with burns. Vitamin F is linked with the highly unsaturated fatty acids and, of course, with lecithin metabolism (Perlenfein, 1942).

This activity is apparently a consequence of promoting the sheathing of protomorphogen molecules with lipid substance, preventing their degradation and consequent release of toxic irritating factors. In this respect it is interesting to note the report (Rosenthal, 1943) that brain cephalin lowers the mortality rate from burn shock. This is probably due to the same mechanism.

The promotion of sheathing processes by the unsaturated fatty acids may be due to the relationship of these substances with vitamin A metabolism. Stepp and his collaborators[1] strongly suggest that vitamin A is associated with lecithin metabolism and probably that of the unsaturates. In studies on the conversion of

1 *The Vitamins and Their Clinical Application,* Stepp, W., Kuhnau, J. and Schroeder, H. Translation published by The Vitamin Products Co., Milwaukee, Wis. 1938.

carotene into vitamin A, Hunter (1946) has commented that the presence of unsaturated fatty acids is intimately connected with this reaction. It appears, therefore, that there is a strong relationship between vitamin A and vitamin F, particularly in respect to the biochemistry involved in the sheathing of the protomorphogen molecule.

Vitamin F is either intimately associated with the highly unsaturated fatty acids or consists of some of the specific isomers of these unsaturates (Perlenfein, 1942). Another channel through which the unsaturates influence sheathing material is the participation in the formation of lecithin complexes with cholesterol esters. We have discussed some of the researches of Macheboeuf and coworkers (1943) from which we postulate that a cholesterol-ester and phosphatide complex is the lipoidal constituent of the lipoprotein protomorphogen molecule. The cholesterol ester portion apparently consists of cholesterol esterified with the most highly unsaturated fatty acids in the plasma.[1]

Artom (1933) reports that in the liver, the major proportions of exogenous fatty acids are in the cholesterol-ester or acetone-soluble portion while in the blood they occur in the phosphatides or aceton-precipitable fraction. He concludes that the phospho-aminolipids are concerned in the transport of fatty acids in the blood. Schmidt (1935) has supplied evidence that, under the influence of thyroxine, the liver phosphatide fatty acids decrease, those in the tissues increase. It is obvious that the phosphatides and cholesterol esters are concerned with the metabolic transport of the fatty acids. But, there is also the possibility that the highly unsaturated fatty acids are concerned with the processing and transport of the specialized phosphatide and cholesterol esters which partake in the establishment of the protective association with protomorphogens.

As a matter of fact, it is this postulation of a separate lipid metabolism concerned with sheathing materials, but not with the ordinary transport of fatty acids, which makes it difficult to properly evaluate the experimental evidence. More knowledge is necessary before it will be possible to separate the characteristics of the two cycles or, for that matter, be certain they are separate and distinct.

1 Kelsey, F. E., and H. E. Longenecker: "Distribution and Characterization of Beef Plasma Fatty Acids." *J. Biol. Chem.*, 139:727-739, 1941.

A recent review of this problem[1] entertains the proposition that, although the main function of plasma phospholipide is most likely associated with fat transport, there is some reliable evidence that is in contradiction to this popularly held viewpoint. We believe that a close analysis of the evidence will uncover and emphasize the metabolic activities concerned with processing of sheathing materials.

Bloor's comment (1939) that there are two general classifications of phospholipins in the liver may be pertinent. One contains more of the unsaturated fatty acids and is linked with metabolic processes of wear and tear (this may include the sheathing mechanism); the other contains food fatty acids and is linked with the transportation and combustion of food fats (this activity is outside the scope of this discussion).

Robertson (1924) furthers the contention that the unsaturated fatty acids are concerned with sheathing material as he reports that phospholipides devoid of thromboplastic activity actually contain some of the most highly unsaturated links. We consider that phospholipins devoid of thromboplastin consist of sheathing material before its association with protomorphogen.

Liver Metabolism.—Robertson (1924) also reports that liver lecithins contain the most highly unsaturated linkages in the organism. Cahn and Houget (1936) suggest that the sterol fatty acids are desaturated in the liver and transferred to phosphatides. These are in turn distributed to the tissues where they are broken down again during metabolism, the fatty acids returning to the liver to participate anew in the synthesis of phospholipins. This is of interest in view of Artom's (1941) comment that the phospholipide content of muscle is in proportion to its activity. The more active a tissue, the more protomorphogens are produced, raising the requirement for phospholipide sheathing material.

We conclude that the liver is the center for the processing of sheathing material. Later discussions in this chapter will indicate that the bile is an important avenue of protomorphogen elimination. It appears that the liver produces an "active" phosphatide which is transferred to the tissues where it combines with cholesterol esters and protomorphogen as an innocuous

1 "Role of the Liver in Plasma Phospholipide Metabolism." *Nutrition Reviews,* 5:135, 1947.

molecule suitable for transfer.. This molecule is very likely one of the phospho-protein-sterides under study by Macheboeuf and co-workers (1943).

This molecule is acted upon in the liver, with the following consequences: the cholesterol ester is converted into cholic acid and excreted in the bile, its fatty acid component being desaturated and attached to the phosphatide which is thus "activated" and returned to the blood, and the protomorphogen component is also excreted in the bile. Tayeau (1943) has demonstrated that bile salts split this protein cholesterol complex and combine with the cholesterol component. Possibly the liver enzyme which dehydrogenates lecithin in the presence of hypoxanthine or xanthine is concerned with the biochemical processing of the phosphatide molecule in this cycle.[1]

This pattern offers a degree of explanation for some puzzling observations on the cholesterol ester: free cholesterol ratios in liver damage. The cholesterol ester remains normal in ligature of the bile duct and even after removal of part of the liver it returns to normal after a few days. Free cholesterol is increased considerably, however. If the liver cells are damaged (yellow atrophy, etc.), the cholesterol ester is lowered.[2] As far as the biochemistry of the sheathing material is concerned, it is apparent that the cells of the reticulo-endothelial system esterify the cholesterol; but this activity may be dependent upon the supply of other sheathing substrates, desaturated phospholipides in particular, which are not available in sufficient quantities if the liver is damaged. Biliary obstruction alone, then, would simply prevent the elimination of cholesterol causing its rise in the blood as free cholesterol with a concomitant compensatory rise in phospholipides. The quantity of cholesterol esters would not change since the activity of the reticulo-endothelial cells in production of the protomorphogen-phosphatide-cholesterol-ester molecule would not be impaired and the liver destruction of this molecule would proceed as normal.

1 Annau, E., Epcrjessy, A., and Felszeghy, O.: "The Biological Dehydrogenation of lecithins and fat acids," Z. *Physiol. Chem.*, 277:58-65. 1942.

2 See discussions in: Sinclair, R. G.: "Fat Metabolism." *Ann. Rev. Biochem.*, 6:245-268, 1937; "Biliary Tract and Pancreatic System," *The 1946 Yearbook of General Medicine.* Yearbook Publishers, Inc., Chicago, Ill. 1946.

In the case of liver cell damage, however, the supply of desaturated phospholipide substrate may be impaired, causing interference with the production of sheathing material. This would mean a drop in cholesterol esters (components of the sheathing molecule). All these suggestions are contingent upon the assumption that the sheathing biochemistry is a very important part of the phospholipide-cholesterol cycles. There is no direct data to substantiate such an assumption, metabolic fat transport being very significant.

Adequate activity of the liver in this processing seems to be very necessary for protomorphogen disposal. The recent report of an anti-burn-shock principle from the liver, different from the antianemic factor, is interesting,[1] since rapid disposal and protection of protomorphogens released by burns is essential if shock is to be avoided. A possible endocrine linkage is the presence in the liver of a detoxifying hormone, Yakriton (Sato and associates, 1926) or Anabolin (Harrower, 1933), which prevents the lethal effects from chloroform administration and is useful in eclampsia. The reader may recall Turck's experiments with chloroform administrations, indicating that lethal consequences are due primarily to the excessive release of tissue protomorphogens. In the following chapter we shall investigate evidence indicating that eclampsia is linked to a considerable extent with the toxic effects of excessive protomorphogen destruction.

It is interesting to note that this liver detoxifying hormone, Yakriton, prevents the tadpole metamorphosing influence of thyroid (Horiuti and Ohsako, 1934) and in general opposes the action of the thyroid hormone. This tadpole metamorphosis acceleration by thyroid may be a consequence of an increased rate of protomorphogen activation or sheathing removal in the embryonic tissue, thus accelerating the rate of differentiation (See Chapter 4). In fact, the thyroid appears to be an important elutogenic factor, removing protomorphogens from connective tissue, and from tissue itself, by reason of its protein denaturation influence. Yakriton may oppose this thyroid influence either by increasing the protomorphogen disposal in the liver, or by catalyzing the rate at which sheathing material is processed in this latter organ.

1 Prinzmetal, M., Hechter, O., Margoles, C., and Feigen, G.: "A Principle from Liver Effective Against Shock Due to Burns." *J. Clin. Invest.*, 23:795-806, 1944.

Thyroid and Iodine.—Clinically, the administration of vitamin F has been reported to raise the blood iodine content 300 per cent (Hart and Cooper, 1941). These observations also report a concomitant amelioration of prostatic hypertrophy. We suggest (Chapter 6) that the prostate is also intimately concerned in the protection of the protomorphogen molecule in the tissue fluids and blood stream although it serves a different end-purpose than the phospholipin protection now under discussion.

Meyer and Gottlieb (1926) review evidence that iodine administration results in a lowered viscosity of the blood. (We consider that a clinical indication of an overloaded protomorphogen disposal mechanism is an increased blood viscosity perhaps linked with thromboplastic activity. Such an increase is seen in senescence and in Addison's disease.) Excessive administration of iodine, however, results in irritation and congestion of the mucosa and epithelial tissue everywhere in the organism. McCarrison and Madhava (1933) report that when pigeons were kept in filthy cages, hyperplasia of the thyroid, adrenals and spleen, and reduction in the size of the thymus and testicle resulted. Filthy cages increase the demand upon protomorphogen disposal systems in a manner analogous to Turck's experiments in which concentrated protomorphogen was sprayed into test animal cages, resulting in deleterious effects on the inhabitants.

From these comments it appears that iodine in small amounts is utilized in the biochemical reactions concerned with sheathing the protomorphogens in the blood and tissue fluids, but that excessive amounts are damaging. This is further indicated by clinical reports showing that ferrous iodide administrations relieve some of the symptoms of vitamin A deficiency. Possibly a link may be discovered in the irreversible inactivation of mucinase by iodine.[1] As a provisional hypothesis we suggest that iodine in some physiological form is utilized in the fatty acid transfer reactions of the lipoprotein protomorphogen molecule in the liver which release the protomorphogen and cholesterol for excretion in the bile, and return desaturated phospholipin to be utilized further in a protective association with protomorphogen molecules. (The presence of large amounts of vitamin F potent unsaturated fats in the kidney indicate that this process may occur there also. We shall discuss the biliary excretion of protomorphogen later in this chapter.)

1 *Enzymes,* Sumner, J. B. and Sommers, G. F. Academic Press, New York, 1943.

Chidester (1944) cleverly recognized the clinical implications of this provisional hypothesis. He was the first to emphasize the fat iodine balance and its importance in the health and vitality of practically every tissue in the organism. He early recommended the use of iodized cod liver oil in many conditions and steadfastly emphasized the great clinical value of ferrous iodide. If iodine is a key link in the biochemistry of the sheathing material, then it may well be universally concerned with disease since impairment of protomorphogen disposal or protection is harmful to any associated tissue.

Ferrous iodide administrations may relieve vitamin A deficiency symptoms to some degree either by: (1) promoting the biochemical system concerned with the sheathing processes and protomorphogen disposal, or (2) releasing sheathing lipids allowing them to gravitate into areas where they are needed more critically (the vitamin A deficient areas). An analogous action is seen in Burrow's experiments in which moderate irradiation relieves vitamin A deficiency (by releasing "ergusia" for use elsewhere in the organism).

This postulated activity of iodine naturally presupposes an important function of the thyroid. It is interesting to note that the administration of thyroid hormone lowers liver phospholipins and increases their concentration in muscle (Schmidt, 1935). The thyroid apparently catalyzes the processing of sheathing material in the liver and its redisposition to the tissues as "activated" phosphatides for association with protomorphogens.

It is difficult to analyze the exact nature of the thyroid influence beyond our provisional hypothesis of the role of iodine in the transfer reactions of the protomorphogen lipoprotein molecule. The thyroid may be important in this respect in view of its activity in curing some forms of rickets (Stepp, Kuhnau and Schroeder, 1938). The beneficial influence on rickets could be a result of the splitting of the phosphatide-cholesterol-protein complex, making phosphorus available. Recently this has been suggested as the primary mode of vitamin D activity.[1]

Foldes and Murphy (1946) have studied the effects of thyroid disease upon blood cholesterol and phospholipide. They report evidence which definitely indicates the influence of the t hyroid in

1 Dam, H.: "Fat-Soluble Vitamins." *Ann. Rev. Biochem.*, 9:353-382, 1940.

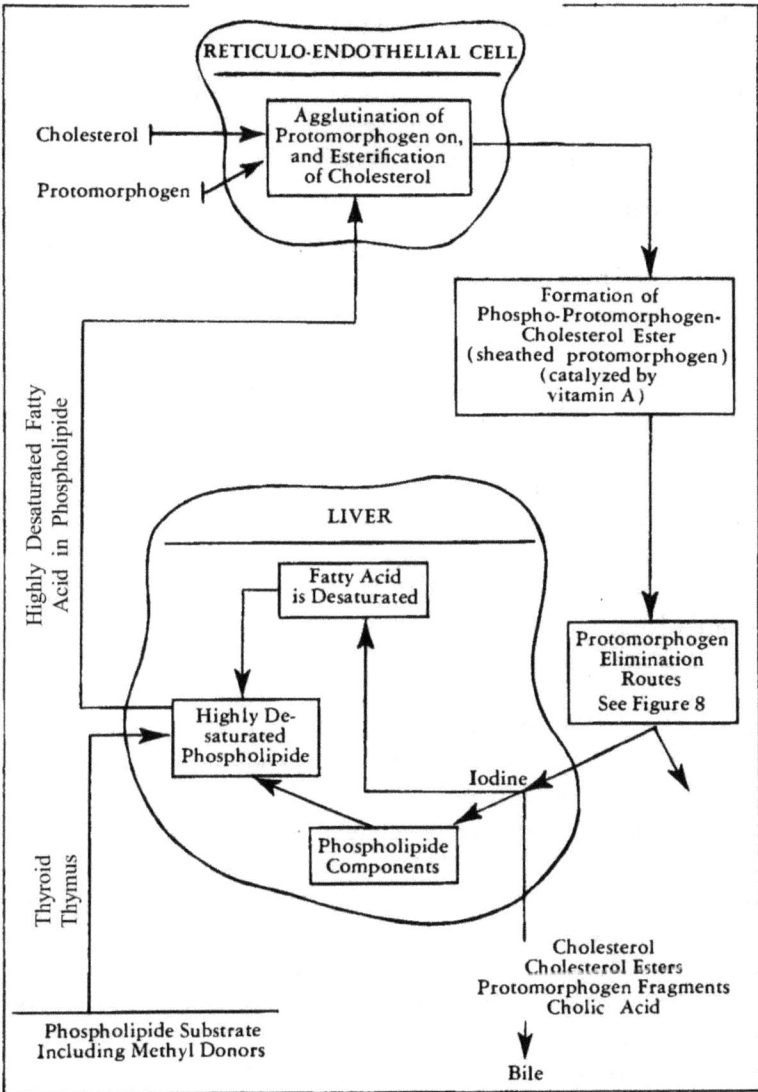

FIGURE 7

Biochemistry of the sheathing materials as suggested by available evidence. this refers to materials used in sheathing only and they cycle may or may not be different from the cycle involved in the metabolic transport of fatty acids.

promoting the deposition of phospholipides from the blood into the tissues. They review the new concepts advanced by Hoffmann and Hoffmann to explain the chemistry of hyperthyroidism. They suggest that the thyroid promotes the enzymatic breakdown of lecithin with the release of "lysolecithin," which is toxic in increased concentrations.

Creatine Formation.—More likely thyroid activity is more complicated, for we know that it is essential for the methylation of creatine (Stuber, Russmann and Proebsting, 1923). Creatine metabolism is of importance here since it seems to be a part of the physiological detoxifying system.

Guanidine, a potent toxin and protein denaturant, is one of the end products of nucleoprotein degradation. Consequently, it is released wherever protomorphogens are not properly disposed of or protected. Being a denaturant, it gives rise to the release of more protomorphogen from neighboring tissue, resulting in a potential vicious cycle of local irritation and inflammation and a general systemic toxicity.

Normally, it becomes guanido-acetic acid and is methylated into methyl-guanido-acetic acid. Methyl donors must be present for this conversion. Hence, methionine is normally necessary for this detoxifying effect. Glutathione may be a part of this picture since it spares cystine and methionine.[1] It is our thought that the creatine excretion of pregnant females and children may be related to these reactions since they are subjected to the intense protomorphogen metabolism of cell division not characteristic of the adult male.

Creatine is a precursor of phosphagen, discussed as a part of the cell energy system in Chapter 3 of this volume. Phosphagen is dipotassium hexose creatine phosphate. Potassium and phosphorus are substrate essentials for these reactions. The parathyroid is probably concerned with the phosphorus supply. Vitamin D has been reported to prevent tetany in parathyroidectomized test animals and in view of the recent report that this vitamin acts through its ability to make inorganic phosphorus available, it is possible that the parathyroid has a similar function. By rendering phosphorus available to fix methyl-guanidine as creatine, the

1 Stekol, J. A.: "Glutathione in Relation to Growth of Rats Maintained on Diets Containing Bromobenzene and Naphthalene." *J. Biol. Chem.*, 123:cxvi-cxvii, 1938.

well-known action of parathyroid as a guanidine eliminator is explained. It appears that the formation of creatine is an important avenue for the fixation of guanidine which may be released in the tissues as protomorphogen undergoes local splitting. What relationship, if any, these reactions may have with the sheathing processes we cannot suggest at this point. Rather, they are concerned with disposal of unsheathed protomorphogens.

To sum up the available evidence on the biochemistry of the sheathing lipids: (1) the thymus may be linked with the production of special lipoidal substances utilized in the sheathing processes; (2) methyl donors, of which choline is the most important, are necessary for phosphatide production and the reprocessing of the sheathing phosphatides in the liver; (3) the unsaturated fatty acids (vitamin F) are concerned with the processing and transport of the special phosphatides and cholesterol esters composing the sheathing molecule; (4) the sheathing (phospho-protein-steride) molecule is acted upon in the liver with the result that the cholesterol ester component is converted to cholic acid and excreted in the bile; its fatty acid moiety is desaturated and attached to the phosphatide component which is now available for re-formation of sheathing molecules, the protomorphogen component being also excreted in the bile; (5) the liver detoxifying hormone, Yakriton, may be concerned with these reactions, and also the liver anti-burn-shock factor; (6) iodine apparently is a carrier or "key" by means of which the sheathing molecule is dismantled in the above manner in the liver, the thyroid controlling this activity; (7) creatine formation from guanidine whose methylation is promoted by the thyroid, is also an avenue by means of which protomorphogen end products are made innocuous, and (8) the parathyroid is probably concerned with the fixation of creatine into phosphagen.

TRANSFER AND ELIMINATION OF PROTOMORPHOGENS

Relation of the Prostate Secretion.—We have put forward the hypothesis that, since the complete chromosome is a stable network with individual tissue determinants attached in significant locales, some morphogens released by tissues must be transported to the germinal centers where they become a part of the complete chromosome. We suggest that, although the chromosome network is reproduced only in the germ centers, the tissues are the only locale in which the individual morphogens can be reproduced. This presents the problem of their transfer from the tissues to the germinal centers. We predicted that the sex hormones are concerned with the elutogenic function of releasing the intact morphogens from connective tissue.

The prostate secretes a fluid which acts as a medium to carry the spermatozoa in semen. It might be advantageous to investigate the possibility that this secretion not only assists in maintaining the integrity of the determinants in the spermatozoan chromosomes in the seminal fluid, but also acts as an endocrine secretion in the blood stream to protect determinant fractions in transit before they are assembled as chromosomes in the gonad.

There is hint of some interesting ramifications, in the literature on prostate. Barker (1922) edited a review of prostate references up to that time. The acceleration of degenerative senile changes following prostate removal is significant. A too rapid hydrolysis of protomorphogens would unduly increase their concentration in the blood and tissue fluids, overloading the eliminative mechanism and exerting an adverse influence on the vitality of all cells. It is possible that the prostate secretes a hormone (both externally and internally) that hinders this hydrolysis.

Huggins and McDonald (1944) have reported an increase in a fibrinolysin in the prostatic fluid of chronic prostatitis. Note that this fibrinolysin is not a thromboplastin lysin (protomorphogen destroyer) but rather an enzyme which prevents the formation or hydrolyzation of the reaction products or end products of the coagulation system initiated by thromboplastin. Huggins and Vail (1943) report further information on prostatic fibrinolysin. They emphasize the stable nature of the coagulation-producing activity of prostatic fluid. This coagulation stimulation is slightly inhibited by heparin, a flocculated precipitate developing instead of a firm clot.

The clot producing influence of prostatic fluid may be due to the protomorphogen (thromboplastin) contained therein. The observation that this activity is exceedingly stable lends credence to our suggestion that prostatic fluid prevents the hydrolysis or breakdown of intact morphogens. The existence of a fibrinolysin in prostatic fluid further fits this system in that it prevents the formation of products which would react with the morphogens, precipitating them out by means of the coagulation mechanism. (Prostate fibrinolysin destroys plasma fibrinogen even in low dilution. (Huggins and Vail, 1943.))

Clearly the suggested influence of prostate in preventing the hydrolysis of blood protomorphogens is less important in females, as would be surmised. The demand for somatic protomorphogens for chromosome assembly is naturally infinitely less in the female. The female does contain in many reported cases para-urethral glands homologous to the male prostate. In some females, however, they seem to be absent. It is significant that this tissue in the female is depressed by castration (Huggins, 1945).

We feel that the whole prostate picture can be profitably re-investigated with these possibilities in mind. If the prostatic secretion is of importance in the metabolism of protomorphogen transfer, a new therapeutic weapon is available.

Relation of the Lymph.—Dustin (1933) emphasizes the intense outburst of cell division in the thymus and lymphoid tissue resulting from injection of proteins, arsenic and dyestuffs. This mitotic stimulation is alternated with periods of pyknosis probably due to the demand on the lymphoid tissue caused by the administration of the foreign substance. We wish to call attention to the observation that proteins, arsenic and dyestuffs all come under the general category of non-diffusible wastes, as do blood protomorphogens. These facts indicate that the lymphoid tissue is linked with their disposal.

Carlson and Johnson (1941) review evidence that dust and other exogenous particles collect in the lymph nodes and are engulfed by reticulo-endothelial phagocytes 1n this locale. Reinhardt, Fishier and Chaikoff (1944) have performed experiments which show the transfer of plasma phospholipides to the thoracic lymph even though they may pass across a capillary or sinusoid wall. The association of protomorphogen with phospholipides would lead us to the tentative conclusion that they also may be transferred into the lymph ducts from the extracellular fluids.

Drinker and Yoffey (1941) have published an excellent review of the lymphoid system. It is significant that the lymph flow is greatest in growing animals. Such an organism is producing protomorphogens at an accelerated rate due to active mitosis, and it would be expected that the excretory mechanisms would be augmented during this period. Such protomorphogen, as may be present in the lymph, is no doubt combined with a lipoid "wrapper" which would prevent the manifestation of thromboplastic or toxic activity. Such seems to be the case since the lymph has a varying thromboplastic activity, being weakest when it contains the most fat or phospholipides.

Concerning the disposal of lipoid "wrapped" protomorphogen deposited in the blood from the thoracic lymph duct, several avenues of investigation remain open. The "wrapped" protomorphogen may be transferred as. such in the blood to a point of elimination or it may be acted upon by some of several mechanisms (natural antibody, phagocytosis, kidney enzymes, etc.).

In conclusion it seems safe to assume that protomorphogens split from connective tissue by elutogenic factors may collect in the lymph where they associate with lipoids and are transported in this condition to the blood. They are handled in the blood by any number of mechanisms we shall review. It appears that exogenous particles also follow this course in the lymph although they may or may not have an affinity for phospholipides.

Relation of Natural Tissue Antibody.—The lymph nodes, due possibly to their content of reticulo-endothelial cells, are locales of antibody production. Dougherty, White and Chase (1944) conclude that the antibodies are concentrated in the lymphocytes, although McMaster and Hudack (1935) do not consider the lymph nodes as the sole site of antibody production. A recent editorial review of the problem in the Journal of the American Medical Association[1] regards it as definitely established that the lymphocytes either .store or produce antibody globulin, probably the latter. The locale of antibody production has not yet been established to the satisfaction of all investigators. Some emphasize the importance of the reticulo-endothelial cells (Cannon and co-workers, 1929; Sabin, 1939; and Jungblut, 1928, to mention a few) and some, interpreting the reticulo-endothelial origin theory differently, consider the lymphocyte and other cells to play an essential role (Ehrich and Harris, 1945). These same investigators later (1946) report their belief that the reticulo-endothelial cells may cause the liberation of smaller particles from the macromolecular antigen, each particle carrying the identical immunological characteristics as the original grouping. These

1 "The Cellular Source of Antibodies," *J. A. M. A.*, 128: 1232-1233, 1945.

are carried by the lymph to the nodes where antibody synthesis is stimulated.

Thus far we have not stressed the possible relationships of the immune-antibody system with the protomorphogen disposal problem. The existence of a blood antibody towards one's proteins is an idea which necessitates strong experimental support.

Such experimental evidence has been supplied by the investigations of Kidd and Friedewald (1942). They have detected an antibody in the serum of normal animals which reacts *in vitro* with a sedimentable constituent of normal tissue cells. The natural antibody has an affinity only for the extracts of normal and certain diseased tissue and will not react with various viruses. The natural antibody is heat labile, and the tissue antigen activity is injured at 56 degrees C. and progressively weakened and destroyed as the temperature is increased. The sedimentable constituent of normal tissue is, we believe, protomorphogen. This reacts with the immune system to produce a natural tissue antibody assisting in the elimination of, and protection against, discarded protomorphogen.

Burns, Scharles and Aitken (1930) have demonstrated that there is a serum principle which inhibits coagulation of blood initiated by thromboplastic tissue extracts. The inhibition was more pronounced with homologous tissue extracts. They also noted an acceleration of coagulation in some homologous tests which may have depended upon the amount of natural tissue antibody present. These experiments also indicate the existence of a natural tissue antibody which reacts with the thromboplastin (protomorphogen) from homologous tissue extracts.

Tocantins (1943) has also demonstrated a substance from normal human blood which reduces the thromboplastic effect of homologous brain tissue. This may likely be another indication of the ability of normal tissue antibody to react with thromboplastin (protomorphogen) from homologous tissue. Significantly, Tocantins found the amount of this substance to be higher in the blood of hemophiliacs.

v. Euler, Ahlstrom and Hesselquist (1945) have discovered that desoxyribonucleoproteins are split more readily by the serum of normal animals than by that of a sarcoma-bearing animal. This suggests the possibility of a specific hydrolytic enzyme (possibly associated with the natural tissue antibody) existing in the blood for the purpose of protecting against high blood protomorphogen concentrations. (Desoxyribo-nucleoprotein is a derivative of the nucleus and closely associated with, if not chemically a part of, the protomorphogen molecule.)

Kidd and Friedewald reported that the natural tissue antibody is not present in the newly born infant, developing only a few weeks after parturition. The fetus is protected from the accumulation of its protomorphogens by the placental system and has little need for a special antibody mechanism. This apparently develops after birth as the tissue protomorphogen reacts with the immune centers to produce the natural tissue antibody. The lower immune response of young newborn, however, is an oft-demonstrated phenomenon. Duran-Reynals (1945), for instance, reported a much longer lag before immunity to fibroma virus in the newborn over that of the adult.

To digress for a moment, the presence of the natural tissue antibody may suggest another reason for the importance of the adrenal cortical hormone to life and its influence in maintaining vitality and health. Dougherty, White and Chase (1944) have emphasized the relationship of the adrenotropic hormone and the adrenal cortical hormone to the activity of the immune centers. Injection of these extracts not only increased the activity of the immune centers but increased the antibody titre to injected antigen and continued administration of the hormones maintained elevated titres beyond the normal time. More recently (1946) these authors have confirmed their observations, reporting an antibody titre double that of controls in some experiments. The adrenal hormone, therefore, is undoubtedly closely linked with the elimination of protomorphogens. Inasmuch as the adrenal cortex is under the control of the anterior pituitary, this organ may be the key control of the natural tissue antibody. The primary importance of the immune-biological system in the protection against protein toxins may involve the anterior pituitary as a key organ in the disposal of protomorphogen.

The phagocytes apparently are concerned with the elimination of protomorphogen. Carrel (1924) found that macrophages rejuvenated stagnant fibroblasts *in vitro,* causing them to renew mitosis. The reader will recall that stagnation in a tissue culture is an inevitable consequence of gradually accumulating protomorphogen. The natural tissue antibody may be of importance in sensitizing phagocytes of the reticulo-endothelial system to protomorphogen. Bloom (1927) has found that certain histiocytic cells of the lungs are stimulated into phagocytic activity in the presence of a natural or acquired antibody. Burnet (1941) reviews that antibodies are produced in the reticulo-endothelial cells, and Gordon, Kleinberg and Charipper (1937) emphasize the importance of the reticulo-endothelial system in the production of antibodies to injected hormone products resulting in "anti-hormone" refractoriness. They demonstrate that this is not an exclusive function of the spleen, but of all reticulo-endothelial tissue, since as a consequence of splenectomy the remaining reticulo-endothelial tissue overcompensates with an "anti-hormone" refractoriness to injected hormones that is greater than that observed in the normal animal.

The evidence herein reviewed indicates that: (1) the protomorphogen released by elutogenic factors stimulates the reticulo-endothelial cells into the production of a natural tissue antibody which (2) quite possibly sensitizes the phagocytic elements of the blood to protomorphogen fragments, or (3) may either agglutinate or split them so that the phagocytic cells engulf the large particles and the kidneys excrete the diffusible residues, and (4) the adrenal cortex hormone, by reason of its stimulation of the antibody centers, is probably necessary for this action and consequently for the adequate disposal of protomorphogens.

Formation of Platelets.—The association of thromboplastin with protomorphogen leads us to a more detailed discussion of platelets and their formation.

The weight of evidence (Quick, 1942) suggests that the platelets are formed from pseudopodial processes of the megakaryocytes (mononuclear phagocytes found in the spleen and bone marrow). This indicates that the formation of platelets is linked with activity of the reticulo-endothelial system whence arise the megakaryocytes. The anti-toxic function of the reticulo-endothelial system is well established.

Many years ago it was reported[1] that the platelets exhibit different and specific staining characteristics in different infectious diseases. The specific dye affinities of platelet granules in different diseases is evidence that there may be bacterial end products in the platelets.

Evans (1932) has demonstrated that the platelet count may be doubled as a result of operations, fractures or parturition. This indicates that platelets may be concerned with the elimination of tissue toxins, including protomorphogens, which are released as a result of injury. The experiments of Holloway and Blackford (1924) are significant in this respect.-They report differential platelet counts of venous and arterial blood in the spleen, the carotid artery, jugular vein and femoral artery and vein. In all cases venous blood was found to contain more platelets than arterial, with an average differential of about 1.4:1. This also is in accordance with our suggestion that platelets carry away protomorphogen debris which accumulates in the tissue fluids as a consequence of morphogen metabolism. It has been reported that thromboplastin may carry antigenic specificity.[2] Such a property in platelets would be evidence of their protomorphogen content.

It is our provisional hypothesis that the platelets not only function as depots of thromboplastin to facilitate coagulation, but they are the vehicles by which the blood transfers non-diffusible wastes to the point of elimination. The major item among these wastes is the colloidal particles of homologous protomorphogen. In addition, the platelet may include bacterial fragments and endogenous wastes such as silica or carbon particles. In our study of the literature we have not seen the platelets referred to as vehicles for the disposal of endogenous non-diffusible wastes. This function will be discussed in greater detail later in this chapter.

Reticulo-Endothelial System.—The reticulo-endothelial system is probably the most important protective mechanism in the organism. This is emphasized by the work of Boone and Manwaring (1930) who found that an india ink blockade (overloading of reticulo-endothelial activity by

1 Eminet, P. P.: "Specifische Blutplattch und Die theorie der directen reactiven Aufeinanderwirkung." *Arch. f. Kinderheilh.*, 57:296-304, 1912.

2 Quick, A. J.: "On Various Properties of Thromboplastin (Aqueous Tissue Extracts)." *Am. J. Physiol.*, 114:282-296, 1936.

administration of india ink) reduced the rate of parenteral alien protein denaturation by 80 per cent. In a recent report Pokrovskaya and Makarov (1945) state that the reticulo-endothelial system is important for its defensive role in infection and in the elaboration of local tissue immunity. These activities play a decisive part in the processes of healing and tissue repair.

It would not be surprising to find the reticulo-endothelial system occupying a pre-eminent place in the list of physiological activities responsible for protomorphogen disposal. The whole problem of the defense against toxic products is in such a state of uncertainty that we shall not attempt to outline with any detail or sense of permanence the exact relationship of the reticulo-endothelial system to protomorphogen disposal. Rather we shall confine ourselves to generalizations introducing such evidence as may be pertinent without integrating the possibilities it may suggest.

We have already suggested that the reticulo-endothelial cells are associated with the formation of the phosphatide-protein-steride complex which, perhaps, represents protomorphogen in a protective association with lipins. Tompkins (1946) has demonstrated that the reticulo-endothelial cells absorb and esterify cholesterol, cause it to combine with protein, and excrete it in this combination. It has also been reported that cholesterol is esterified in the spleen.[1]

We have already discussed the formation of platelets from pseudopodial processes of the megakaryocytic cells of the reticulo-endothelial system. We have suggested that the platelets act as vehicles by which endogenous materials are carried in the blood stream to a point of elimination. The colloidal component of the nucleoprotein protomorphogen molecule split by blood trypsin may be among those eliminated in the platelets. The comments of Ehrich and Harris (1946) indicate that the reticulo-endothelial cells may split macromolecular molecules of protomorphogen, releasing smaller, antigenically active, component particles. These promote the formation of antibodies and may also be among those products carried in the platelets.

There are many protective and phagocytic cells in the reticulo-endothelial system which are not megakaryocytes and do not

1 Goreczkv. I.. and Kovats, J.: "The Distribution of Bound and Free Cholesterol in Spleen Reserve Serum." *Biochem. Ztscbr.*, 314:208-213, 1943.

therefore give rise to the formation of platelets. (The relationship of these cells to protomorphogen disposal will be discussed shortly.) There is some indication that the megakaryocytes are a specialized group of reticulo-endothelial cells which, by giving rise to platelets, allow the elimination of exogenous products which may not be susceptible to the lytic action of the phagocytic enzymes.

The intravascular administration of certain colloidal particles, such as carbon or various dyes, giving rise to what is called a "blocking" of the reticulo-endothelial cells has been mentioned. These colloidal particles are immediately engulfed by phagocytes and can be seen in the spleen, bone marrow and particularly lymph nodes where reticulo-endothelial cells are present. The particles temporarily overload the reticulo-endothelial system, whence the term "blocking."

Although the reticulo-endothelial system quickly compensates, they may remain in the node for some time, many years in fact, an observation that has led some to suggest the long life of reticulo-endothelial phagocytes. Apparently, however, this fact simply is a reflection of the intense phagocytic capacity of the reticulo-endothelial cells, the disintegration of a particle-filled phagocyte releasing particles which are immediately picked up by neighboring phagocytes, and so on ad infinitum.

It is an attractive idea to suppose that eventually in the process of phagocytic degeneration and re-engulfment of the particles the megakaryocytes chance to pick them up, in which case they are formed into platelets and released into the blood stream for elimination. All phagocytes could not form platelets whenever loaded with collodial particles; otherwise, in consequence of an overload of foreign macromolecules, the platelet count could rise to the dangerous extent that spontaneous coagulation and death might occur. We have suggested that the platelets serve two cooperating functions: (1) the blood stream transfer of exogenous wastes, and (2) thromboplastic agents. By holding the foreign particles in the reticulo-endothelial nodes, transferring them in the dynamic ephemerality of the reticulo-endothelial phagocytes, a dangerous rise in blood platelets can be avoided.

Recently Chargaff and West (1946) have demonstrated the existence of a thromboplastic protein in blood plasma similar to the thromboplastic protein from lung. This is apparently not associated with platelets, but is linked closely with the clotting time of the blood. These reports would indicate that while the platelets may be an important avenue of protomorphogen transfer in the blood stream, other avenues are also utilized. This is to be suspected in view of some of the problems met with in a complete analysis of these phenomena.

There is evidence from the study of pneumoconiosis which is linked with this hypothesis. Certain types of silica particles show white in dark field illumination and certain types do not. Cole and Cole (1940) have reviewed the evidence which shows that one type is eliminated from the nodes in the lungs; the other type remains in the area, eventually resulting in pneumoconiosis. Evidently there may be a selective affinity of megakaryocytes for one type of particle, these cells depositing the debris in platelets for transfer to the point of elimination. The elimination is likely through the bile, Boehm (1942) having noticed silica in the bile of glass workers.

Bile As An Elimination Route.—Other reports also indicate the importance of the bile as an elimination route of exogenous particles and colloidal macromolecules. Greenberg and Troescher (1942) have demonstrated that of an injected dose of labeled strontium or calcium, the bile accounts for about 25 per cent of the elimination. Greenberg, Copp and Cuthbertson (1943) have shown through experiments with labeled molecules that the bile plays an important part in the excretion of manganese, cobalt and iron in the order listed. Annegers and collaborators (1944) have demonstrated that the liver-bile route is of predominant importance in the excretion of arsenicals.

Our hypothesis of megakaryocytic engulfment with consequent platelet formation and elimination through the bile does not involve the biliary excretion of parenterally or intravascularly administered artificially radioactive isotopes such as investigated by Greenberg and co-workers. Only those isotopes which might find their way into protomorphogens or which might consist of non-diffusible particles are presumably engulfed by megakaryocytes, formed into platelets and excreted in the bile.

Sobotka (1937) has emphasized other biliary constituents which we believe suggest this avenue of protomorphogen elimination. Carnot and Gruzewsko (reviewed by Sobotka) have described a protein "cholenuclein" related to nucleoproteins, probably present in bile due to epithelial cell destruction. It has also been reported that proteins of exogenous origin may be excreted in the bile. Urea has been found to be a bile constituent in amounts always lower than serum but proportional to the bile solids. Its presence might be necessary because of its depolymerizing or denaturing influence, preventing the agglutination of polymerization of large molecules of protein, nucleoprotein or protomorphogen fragments.

Sobotka also mentions that hibernating animals accumulate a high ash bile, four times the normal content. This may be an expression of the accumulation of protomorphogen end products normally excreted through the biliary route. There seems to be considerable evidence that no bile circulates in the blood, but that which may be present is associated with the phagocytes. The blood bile salts have a tremendous affinity for serum proteins. Protomorphogens display these same characteristics.

Tashiro and co-workers (1931) suggest that some special cases of epithelial ulceration may be due to the removal of phosphatide and cholesterol which serves as a protection against bile salts circulating in the blood. Again the similarity to protomorphogens is manifest, these latter factors giving rise to ulceration or necrosis when separated from the cholesterol phosphatide protecting sheath. Bouchard (1894) has remarked that the greatest part of the toxic materials in bile resides in the pigments which may be adsorbed on charcoal. Bile decolorized by this method is much less toxic. He speculates that toxins removed by the liver are excreted in the bile. That these may be significant end products of metabolism is indicated by the fact that blood from an animal with a ligature of the portal vein is more toxic than normal blood. He has demonstrated that charred or boiled solid material from blood is exceedingly toxic when an aqueous solution is injected into an animal, causing convulsions and death. This links these poisons with the ashed cytost of Turck, described in Chapters 1 and 2, which we have identified as protomorphogen. Rehfuss and Williams (1941) have also investigated a toxic fraction and have observed that charcoal or kaolin removes most of the pigments and salts from bile. Bouchard identified the toxic fraction of bile with the pigments and mineral constituents, particularly potassium.

Rehfuss and Williams (1943) have identified a fraction in bile which is exceedingly toxic to various forms of animal and vegetable life. They purport to demonstrate that the improvement following removal of bile by duodenal drainage is a consequence of the degree of detoxification from this procedure.

There is therefore, a reasonable demonstration, admittedly by analogy only, which warrants the premise that these mysterious toxic factors in bile may be accounted for primarily by the presence of protomorphogen.

Bile Formation.—We have suggested that the reticulo-endothelial system engulfs non-diffusible and colloidal toxins and wastes; particularly, the megakaryocytes form platelets transferring the wastes to the point of elimination, probably the bile. The evidence we have reviewed intimates that protomorphogen wastes also are engulfed by the reticulo-endothelial cells and find their way into the platelets, appearing again in the bile. There is not a complete picture as yet of the path of elimination from the reticulo-endothelial cell to the platelet, thence to the bile. The evidence does establish the following:

1. protomorphogen and exogenous particles are engulfed by the reticulo-endothelial cells;

2. protomorphogen appears in the platelets as do exogenous particles, and;

3. protomorphogen toxins and exogenous particles appear to be eliminated in the bile.

The path, however, is not as clear cut as it might seem; not all reticulo-endothelial cells which engulf these particles can produce platelets, and the evidence concerning the destruction of platelets is ambiguous.

We shall not attempt to answer these questions, whose solution must be left to further experimental investigation. Some problems may be outlined as follows:

1. If all the reticulo-endothelial cells engulf protomorphogen and exogenous particles, and only the megakaryocytes are responsible for platelet production, then either: (a) there must be other means than the platelet for further elimination, or (b) during the life cycle of the phagocytes the material is re-engulfed by other reticulo-endothelial cells eventually being picked up and made into platelets by the megakaryocytes,

or (c) there is an endocrine or immune control by means of which the megakaryocytes are specifically "sensitized" to engulf protomorphogen in proportion to the demand for increase or decrease of the platelet balance in the blood.

2. We shall shortly present evidence concerning the destruction of the platelets by reticulo-endothelial cells. It is difficult to reconcile a hypothesis that platelets assist in protomorphogen disposal with the fact that they are in turn destroyed by the same group of cells which create them. The questions arise as to whether: (a) specific reticulo-endothelial cells create platelets (a demonstrated fact), and another specific group destroys them, carrying the protomorphogen into bile for elimination (the "destroyers" may be locally prevalent in the spleen and liver); (b) the platelets might exist only to maintain on tap a protomorphogen supply in the blood stream for thromboplastic purposes, although this point is debatable in view of the inclusion of exogenous toxins in platelets and their increase during periods of trauma, etc., or (c) the major destruction of platelets may occur in the liver or in a manner distinct from reticulo-endothelial action, their debris finding its way into the bile.

There is much evidence concerning the destruction of the platelets in the spleen and further evidence that bile is a product of spleen and reticulo-endothelial activity as well as of liver detoxification. Thrombocytopenic purpura, a hemorrhagic disease accompanied by a pronounced decrease in blood platelets, responds to splenectomy. For this reason the spleen has been regarded as the primary organ engaged in the removal of platelets from the blood. Quick (1942) mentions that Kaznelson is the chief proponent of the spleen-platelet destruction function, others ascribing this action to the whole reticulo-endothelial system. As mentioned above the chief support for these contentions is the marked rise in platelet (thrombocyte) count following the removal of the spleen in thrombocytopenic conditions.

Following up this line of thought, there is evidence that the reticulo-endothelial system (spleen in particular) is responsible for the formation of bile pigment (our suggestion being that certain of the bile constituents are macromolecular protomorphogen molecules and exogenous wastes collected by the platelets). Sacks (1926) reviews evidence that the reticulo-endothelial system is the primary site of bile pigment formation since among other things bilirubin is found in the blood of dehepatized dogs, indicating that the liver is not essential for its formation.

Goto (1917) has demonstrated that splenectomy results in a decreased formation of bile pigment. Cantarow and Wirts (1943) have demonstrated that liver injury does not influence either the volume of bile excretion, the amount of bile pigment or the biliary excretion of dye. However, they did diminish the biliary excretion of dye and bile pigments by blocking the reticulo-endothelial system with india ink. This is strong evidence that the reticulo-endothelial system excretes exogenous wastes and protomorphogen by means of the bile. The quantity of bile, however, was not altered by reticulo-endothelial blockade. The possibility that these wastes are transferred from the spleen to the liver in the portal circulation may be suggested by the evidence of a bilateral flow in the portal vein by means of which the splenic blood is mainly sent to the left side and mesenteric blood to the right side of the liver. Hahn, Donald and Grier (1945) have established this fact through the use of labeled phosphorus.

Many investigators have concluded that the spleen and the reticulo-endothelial system do not directly destroy blood platelets. The weight of evidence seems to support this contention. The most conclusive evidence in this respect has been supplied, we believe, by Holloway and Blackford (1924). Differential platelet counts on splenic arterial and venous blood failed to indicate a reduction in platelets in the splenic vein over those in the splenic artery. This experiment demonstrates that platelets are not destroyed or removed as a result of passage through the spleen.

The work of Hooper and Whipple (1917) strongly refutes any suggestion that the spleen destroys platelets and forms bile from the debris. These investigators determined that splenectomy in no way modifies the secretion of pigments in the bile. Their work is contradictory to the findings of Goto (1917) which we have reviewed. Burket (1917) determined that the diversion of splenic blood into the general circulation had no effect on the health, activity or bile production in experimental animals. This work also refutes any suggestion that the spleen mechanically removes platelets and deposits the protomorphogen and exogenous debris in the bile. This, however, does not eliminate the possibility that the liver removes protomorphogen and excretes it in the bile. The liver has been known as a detoxifying organ for many years, Bouchard (1894) reporting experiments that dogs die as a consequence of ligature of the portal vein.

The clue to platelet removal is probably supplied by Torrioli and Puddu (1938), who have concluded that the reticulo-endothelial system produces a substance (termed thrombocytopen) which stimulates megakaryocytes in low concentration and inhibits them in excess. Troland and Lee (1938) have confirmed this hypothesis by extracting a substance from diseased spleens which caused a marked reduction in the platelet count. This factor has since been reported by numerous investigators and denied by others.[1]

These reports indicate that the platelet production and balance may be regulated by a mechanism automatically controlling the level of specific megakaryocytic protomorphogen. So-called thrombocytopen meets some of the requirements for protomorphogen classification in that it specifically stimulates or inhibits megakaryocytic activity depending upon its concentration. It may, however, be a governor of the immune response centered in the spleen, described as the heart of the immune mechanism.[2] Antibody to protomorphogen can regulate mitosis depending upon its concentration; moderate amounts maintain optimum protomorphogen concentrations, while excessive amounts lower the protomorphogen concentration to a sub-optimum level and may even attack the protein of the tissue itself. (This conception of antibody action is derived from the experience with cytotoxins, reported in Chapter 6 of this volume.) Disorders in the immune system would, under this hypothesis, result in a disordered balance of megakaryocyte protomorphogen with consequent thrombocytopenia. Removal of the spleen, by interfering with the center of the immune response, could conceivably correct this condition. (Removal of the spleen has been reported to eliminate permanent acquired immunity to some diseases[2] indicating that the antibody producing center for some antigens may be localized in certain spleen cells.) The remaining reticulo-endothelial tissue may compensate for the removal of the spleen under these circumstances. Piney (1931) reports the experiments of Bedson (1926)

1 See Quick (1942) for a more complete discussion.
2 *The Spleen and Resistance,* Perla, D. and Marmorston, J. The Williams & Wilkins Co., Baltimore, Md. 1935.

which demonstrate that blockade of the reticulo-endothelial system with india ink causes an increase in blood platelets but not after splenectomy, indicating the spleen is the center of this control (in light of later information, the center of thrombocytopen production).

Further consideration of thrombocytopenic purpura may throw additional light on the problem of platelet removal and excretion. Quick (1942) has emphasized the observation that lack of platelets alone is not responsible for the hemorrhage or purpura in this condition. There is an excess of some principle which is responsible for vasodilation and a decrease of capillary resistance.

Quick advances the hypothesis that the platelets function as removers or carriers of histamine. Platelets are rendered more susceptible to agglutination and lysis when they contain histamine. He reasons, therefore, that an overproduction of histamine is the basic cause of hemorrhagic purpura and, incidentally, is responsible for the concomitant reduction in blood platelets. There is much evidence to support this viewpoint. Histamine is known as a vasodilator. Administration of histamine results in agglutination and reduction of platelets, as does anaphylaxis which liberates histamine. Rocha e Silva, Grana and Porto (1945) have also concluded from their experiments that histamine is bound to platelets.

To list the various factors entering into our hypothesis: (1) protomorphogen is secreted by the cells and immediately either catalyzes the formation of connective tissue or is adsorbed on neighboring connective tissue; it is stored in this manner; (2) elutogenic factors such as sex hormones, thyroid and trypsin release the protomorphogen from the connective tissue storehouse; (3) fragments of protomorphogen are seized by reticulo-endothelial cells and transferred into platelets by megakaryocytes, and (4) the platelets, being regulated by an autocatalytic mechanism, are destroyed; the protomorphogen debris finds its way into the bile where it is secreted.

This whole hypothesis is obviously too naive in its simplicity. It is presented simply as a start in the direction of unraveling an extremely complicated physiological mechanism. Inadequacies are obvious in that: (1) we have no evidence concerning the exact mechanism by which enough protomorphogen is formed into platelets to effect a noticeable excretion beyond the fact that protomorphogen is found in all platelets and the platelets are formed by megakaryocytes; (2) there is nothing to indicate whether the platelet route is an insignificant or the prime mode of protomorphogen excretion, and (3) although the platelet count seems to be regulated by an autocatalytic mechanism under reticulo-endothelial control, we can find no direct evidence concerning platelet destruction except that it apparently does not occur in the reticulo-endothelial system, and that platelet debris appears in the bile. Differential platelet counts of blood entering and leaving the liver would be significant. If indicative of destruction in the liver, this would offer a logical explanation of their final disposal. Some organ may exhibit a decreased platelet count in venous blood over arterial, and this information should supply the key to platelet disposal.

There are fragments of information that appear pertinent to this whole problem. We shall discuss these briefly under their own headings but make no attempt to fit them into the above hypothesis. It appears that there is simply not enough experimental evidence available at this time to establish a competent explanation.

Kidney Elimination of Protomorphogen.—Due to the large molecular structure of the protomorphogen molecule it is doubtful if it can be eliminated via diffusion through the kidney. There is the possibility that tryptic digestion, antibody lysis or enzyme destruction (these systems will be discussed shortly in relation to protomorphogen in the blood) may split protomorphogen into diffusible and non-diffusible components; the non-diffusible, being the more toxic (protein macromolecules), are excreted via the bile system just discussed, and the diffusible are excreted through the kidney.

The experiments of Pearce reviewed by Vaughan and co-workers (1913) indicate that exogenous proteins of a foreign nature accumulate in the kidney where the local enzymes split them into their components, which are either diffused out in the urine or transferred to other points of elimination, i.e., the bile. In these experiments it was found that rabbit kidney retained antigenic activity for four days after intravenous injection of foreign proteins. It is proven that this is due not to the absorption of the poisons in the kidney cells,

MORPHOGEN CYCLES IN THE ANIMAL ORGANISM

CELL SYSTEM

See Fig. 5—Chapter 3

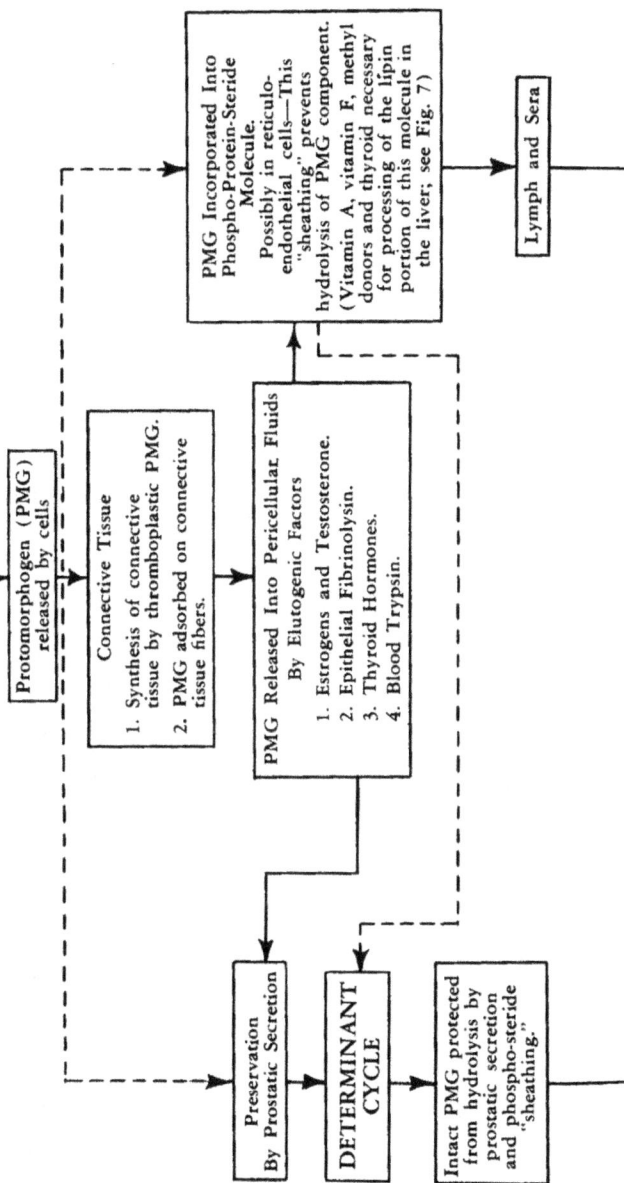

Protomorphogen (PMG) released by cells

Connective Tissue
1. Synthesis of connective tissue by thromboplastic PMG.
2. PMG adsorbed on connective tissue fibers.

PMG Released Into Pericellular Fluids By Elutogenic Factors
1. Estrogens and Testosterone.
2. Epithelial Fibrinolysin.
3. Thyroid Hormones.
4. Blood Trypsin.

PMG Incorporated Into Phospho-Protein-Steride Molecule. Possibly in reticulo-endothelial cells—This "sheathing" prevents hydrolysis of PMG component. (Vitamin A, vitamin F, methyl donors and thyroid necessary for processing of the lipin portion of this molecule in the liver; see Fig. 7)

Lymph and Sera

Preservation By Prostatic Secretion

DETERMINANT CYCLE

Intact PMG protected from hydrolysis by prostatic secretion and phospho-steride "sheathing."

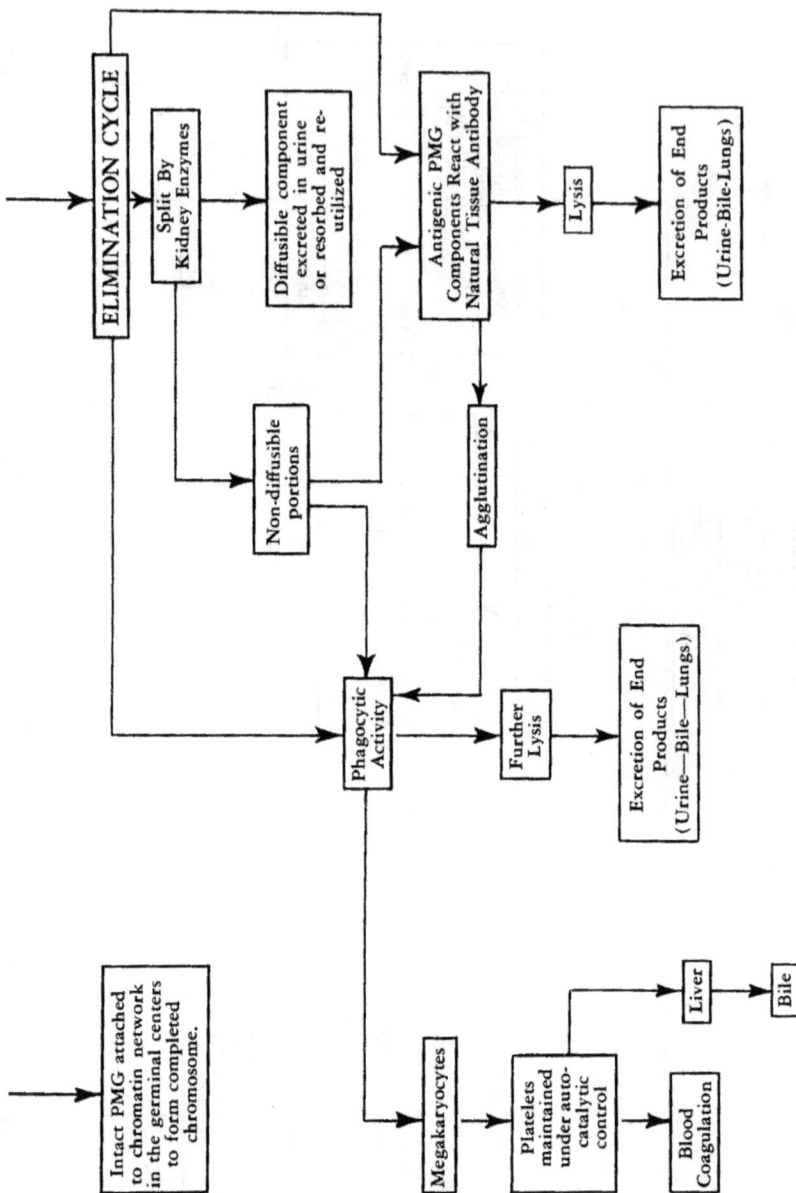

FIGURE 8

233

but rather to their collection in the vascular system of that organ, for washed kidney substance was not antigenic. The experiments of Rothen (1945) suggest a mechanism whereby enzymes locally present in the kidney cells may influence protomorphogen or foreign protein in the kidney capillaries without actual contact. He has demonstrated that an antigen may influence an antibody through impermeable layers a few Angstrom units thick and suggests that this may even occur through thin cell membranes. In view of Sevag's (1945) thesis that antigens are enzymatic catalysts, it is logical to suppose that enzyme activity may exert itself through a thin membrane. In this manner the enzymes of kidney cells may split proteins and protomorphogen in the kidney capillaries.

In experimental nephritis the elimination of foreign protein is delayed, indicating that the kidney enzymes play an important part in the denaturation and elimination of these toxic products. It is quite likely that protomorphogen debris is also thus affected by the kidney enzymes. Of special interest in these studies is the report of Greenstein and Chalkley (1945) that desaminases for nucleic acids are localized to a considerable extent in the kidney. We should not lose sight of the fact that protomorphogen is nucleoprotein debris. Platelet counts of blood entering and leaving the kidney would indicate whether this organ also destroys them and assists in the elimination of their debris.

Robertson (1924) doubts that his allelocatalyst (protomorphogen) is excreted by the kidney since it is non-diffusible. There is, however, a strong possibility that enzymatic destruction may occur, some or all of the by-products being excreted by the kidney. He states that the appearance of bile in the blood is accompanied by bile pigments in the urine. Bouchard (1894) lends credence to the possibility of protomorphogen excretion in the urine with his reports that the toxins in urine may be removed with charcoal, and that they consist primarily of mineral matter, especially potassium.

However, Robertson (1924) reported the presence of arginase in the kidney for which no function was then realized. The arginine of nucleoprotein (protomorphogen) may be split in the kidney system by this enzyme during the process of protomorphogen disposal. Further evidence of kidney protomorphogen excretion is given by the observation that the urine of typhoid patients can produce the local antigenic reactions described by

Shwartzman (1937). It is probable, therefore, that some antigenic molecules can be excreted in the kidney; the amount or character of these, however, would depend upon their relative molecular size.

Bouchard (1894) has discussed the resorption of bile as a consequence of intestinal putrefaction. This resorption results in the appearance of bile residues in the urine. He has also discussed the pathology of jaundice. In severe jaundice the bile pigments are adsorbed on connective tissue, giving rise to the alteration in skin coloration associated with this condition. The urine becomes intensely poisonous due to the large content of bile residues excreted by the kidney. The secretion of this poison eventually causes renal changes, decreasing the permeability of the kidney, resulting in renal inadequacy and uremia. Thus the kidney may be a second line of defense against protein toxins, the liver being the first.

Further evidence of the kidney participation in the excretion of protomorphogen is offered in Hammarsten's review (1914) of the non-dialyzable substances or adialyzable bodies appearing in the urine. These particles include nucleoprotein residues. They occur in normal urine but in increased amounts in the urine of pregnancy, febrile diseases, pneumonia, nephritis and eclampsia. All of these conditions will be reviewed in the following chapter in reference to the associated increased intensity of protomorphogen metabolism.

This evidence indicates that the kidney can excrete protomorphogen residues and thus assist the liver and bile in this function. The kidney elimination is quite different from the liver-bile method, however. Bouchard states that, although bile may be detoxified by adsorption on charcoal, urine from patients with severe jaundice retains its toxicity after charcoal decolorization. The difference of the kidney process is also indicated by the observation that administration of naphthalin to persons with injured livers results in the appearance of a different end product than the sodium naphthyl-sulphite normally observed. Bouchard has observed this phenomenon in several cases.

The evidence seems complete enough to suggest that the kidney takes part in protomorphogen elimination in conjunction with the liver and bile. Kidney enzymes split the protomorphogen molecules which collect in the renal capillaries; the diffusible residues thereby may appear in the urine, the non-diffusible being eliminated via the liver-bile route. In liver injury the kidney may be overtaxed into inadequacy by the demands of protomorphogen and other protein excretion normally taken care of by the liver-bile system.

Morphogens in the Biological System.—The universal existence of morphogen phenomena is apparent in: (1) plants, by the autocatalytic nature of fresh and used soil, and by certain plant growth factors which exhibit the morphogen phenomenon of stimulating growth in low concentrations and inhibiting in high, and in (2) cold blooded animals by reason of the information presented in Chapter 4 concerning the differentiation of the amphibian embryo.

Elimination of Protomorphogen.—This topic has been presented in some detail, although purposely with sketchy conclusions. It is best reviewed by means of a chart which we present in figure 8. The reader will note that we have carried through the morphogen "cycle" concept. The cell system of morphogen metabolism is shown in figure 5, Chapter 3. The morphogen released into the media or tissue fluids from this system is picked up and its disposal charted in figure 8.

The purpose of this chapter is to present reviews of known physiological activities in their respect to the morphogen hypothesis. No attempt has been made to establish a complete theory of morphogen elimination. The situation requires clarification by further investigation.

Quick (1942) has published a clear and complete discussion of hemorrhagic diseases which contain adequate reviews of the coagulation phenomenon and the related problems that we have touched upon in this publication.

In the following chapter we shall attempt a brief discussion of a few abstract physiological systems and their possible relationship to morphogens. We shall also re-analyze some pathological states in the light of the morphogen hypothesis in an attempt to suggest new therapeutic methods.

BIBLIOGRAPHY

ANNEGERS, J. H., SNAPP, F. E., IVY, A. C., and ATKINSON, A. J.: "The Effect of Bile Acids on the Biliary. Excretion on Neoarsphenamine and Mapharsen." *J. Lab. & Clin. Med.,* 29:853-862, 1944.

ARTOM, C.: "Sur le role des Phosphoaminolipides dans le Metabolisme des Graisses. I. Experiences avec Introduction Parenterale do Graisses lodeel." *Arch. Internat. Physiol.,* 36:101-128, 1933.

————— "Phospholipid Metabolism in Denervated Muscles." *J. Biol. Chem.,* 139:953-961, 1941.

ARTOM, C., and PERETTI, G.: "Sur le role des Phosphoaminolipides dans le Metabolisme des Graisses. II. Experiences avec Introduction Orale do Graisses lodees." *Arch. Internat. Physiol.,* 36:351-370, 1933.

ASTBURY, W. T.: "The Structural Proteins of the Cell." *Biochem. J.,* 39:lvi-lvii, 1945.

BAKER, L. E., and CARREL, A.: "Lipoids as the Growth-Inhibiting Factor in Serum." *J. Exp. Med.,* 42:143-154, 1925.

BARKER, L. F.: *Endocrinology and Metabolism.* D. Appleton & Co., New York, 1922.

BEDSON, S. P.: "The Role of the R. E. System in the Regulation of the Number of Platelets in the Circulation." *Brit. J. Exp. Path.,* 7:317-324, 1926.

BEHNKE, A. R. and YARBROUGH, O. D.: "Respiratory Resistance, Oil-Water Solubility, and Mental Effects of Argon, Compared with Helium and Nitrogen." *Am. J. Physiol.,* 126:409-415, 1939.

BIERBAUM, O. S., and MOORE, C. V.: "Marrow Depressant Effects of Estrogens in Dogs." *Fed. Proc.,* 4:144-145, 1945.

BIRCH, C. L.: "Hemophilia." *Proc. Soc. Exp. Biol. & Med.,* 28:752-753, 1930/31.

————— "Hemophilia." J. A. M. A., 99:1566, 1932.

BLAKESLEE, A. F., and AVERY, A. G.: "Genes in Datura Which Induce Morphological Effects Resembling Those Due to Environment." *Science,* 93:436-437, 1941.

BLOOM, W.: "Immune Reactions in Tissue Cultures. I. Reaction of Lungs from Normal and Immunized Rabbits to Pigeon Erythrocytes." *Arch. Path. & Lab. Med.,* 3:608-628, 1927.

BLOOR, W. R.: "Fat Transport in the Animal Body." *Physiol. Rev.,* 19:557-577, 1939.

BOEHM, E. E.: "A Spectrographic Study of Bile." *Proc. West Virginia Acad. Sci.,* 15:59-62, 1942.

BOMSKOV, C., and BROCHAT, F.: "Formation and Transport of Thymus Hormone in the Organism. Lymphocytes as Transporters of the Thymus Hormone." *Endokrinologie,* 23:145-161, 1940.

BOONE, T. H., and MANWARING, W. H.: "Parenteral Denaturization of Foreign Proteins. IV. Effect of Endothelial Blockade," *J. Immunol.,* 18:431-432, 1930.

BOUCHARD, C.: *Lectures on Auto-Intoxication in Disease.* F. A. Davis Co., Philadelphia. 1894.

BOXER, G. E., and STETTEN, DEW., JR.: "Effect of Dietary Choline Upon Rate of Turnover of Phosphatide Choline." *J. Biol. Chem.,* 153:617-625, 1944.

BUNTING, C.H.: "The Granular Leucocytes." *Special Cytology,* Ed. by E. V. Cowdry, Vol. II. Paul B. Hoeber, Inc., New York. 1932.

BURKET, W. C.: "Changes in the Peripheral Blood Consequent Upon the Diversion of the Splenic Blood Into the General Circulation." *J. Exp. Med.,* 26:849-863, 1917.

BURNET, F. M.: *The Production of Antibodies.* Macmillan & Co., Ltd., Melbourne, Australia. 1941.

———— "Vaccinia Haemagglutin." *Nature,* 158:119-120, 1946.

BURNS, E. L., SCHARLES, F. H., and AITKEN, L. F.: "Interaction Between Substances in Tissue Extracts and Blood Sera. Effect of Mixtures of These Substances on Coagulation of Blood." *Proc. Soc. Exp. Biol. & Med.,* 27:492-495, 1930.

BURROWS, M. T.: "Energy Production and Transformation in Protoplasm as Seen Through a Study of the Mechanism of Migration and Growth of Body Cells" *Am. J. Anat.,* 37:289-349, 1926.

———— "The Nature of Atrophy and Hyalinization of Cells and Tissues." *J. Cancer Research,* 11:72-85, 1927.

BURROWS, M. T., and JORSTAD, L. H.: "On the Source of Vitamin A in Nature," *Am. J. Physiol.* 77: 38-50, 1926.

———— "On the Source of Vitamin B in Nature." *Ibid.,* 77:24-37, 1926.

CAHN, T., and HOUGET, J.: As reviewed in *Ann. Rev. Biochem.,* 5:236, 1936.

CANNON, P. R., BAER, R. B., SULUVAN, F. L., and WEBSTER, J. R.: "The Influence of Blockade of the Reticule-Endothelial System on the Formation of Antibodies." *J. Immunol.,* 17:441-463, 1929.

CANTAROW, A., and WIRTS, C. W., JR: "The Effect of Dog's Bile, Certain Bile Acids and India Ink in Bilirubinemia and the Excretion of Bromsulfalein." *Am. J. Dig. Dis.,* 10:261-266, 1943.

CARLSON, A. J., and JOHNSON, V.: *The Machinery of the Body.* 2nd Ed., University of Chicago Press, Chicago. 1941.

CARREL, A.: "Tissue Culture and Cell Physiology." *Physiol. Rev.,* 4:1-17, 1924.

CHARGAFF, E.: "The Thromboplastic Activity of Tissue Phosphatides." *J. Biol. Chem.,* 155: 387-399, 1944.

———— "The Coagulation of Blood." *Advances in Enzymology,* 5:31-65, 1945.

CHARGAFF, E., and BENDICH, A.: "The Disintegration of Macromolecular Tissue Lipoproteins." *Science,* 99:147-148, 1944.

CHARGAFF, E., BENDICH, A., and COHEN, S. S.: "The Thromboplastic Protein: Structure, Properties, Disintegration." *J. Biol. Chem.,* 156:161-178, 1944.

CHARGAFF, E., MOORE, D. H., and BENDICH, A.: "Ultracentrifugal Isolation from Lung Tissue of a Macromolecular Protein Component with Thromboplastic Properties." *J. Biol. Chem.,* 145:593-603, 1942.

CHARGAFF, E., and WEST, R.: "The Biological Significance of the Thromboplastic Protein of Blood." *J. Biol. Chem.,* 166.189-197, 1946.

CHARGAFF, E., ZIFF, M., and COHEN, S. S.: "Studies on the Chemistry of Blood Coagulation. X. The Reaction Between Heparin and the Thromboplastic Factor." *J. Biol. Chem.,* 136:257-264, 1940.

CHASE, J. H., WHITE, A., and DOUGHERTY, T. F.: "The Enhancement of Circulating Antibody Concentration by Adrenal Cotical Hormones." *J. Immunol. Virus-Res. & Exp. Chemother.,* 52:101-112, 1946.

CHIDESTER, F. E.: *Nutrition and Glands in Relation to Cancer.* Lee Foundation for Nutritional Research, Milwaukee, Wis. 1944.

COHEN, S. S., and CHARGAFF, E.: "Studies on the Chemistry of Blood Coagulation. The Thromboplastic Protein from the Lungs." *J. Biol. Chem.,* 136:243-256, 1940.

238

COLE, L. G., and COLE, W. G.: *Pneumoconiosis (Silicosis). The Story of Dusty Lungs.* John B. Pierce Foundation, New York. 1940.

COPLEY, A. L.: "Studies on Human Placental Thromboplastin In Vitro and In Vivo." *Science,* 101:436-437, 1945.

COPLEY, A. L., and ROBB, T. P.: "Studies on Platelets. III. The Effect of Heparin In Vivo on the Platelet Count in Mice and Dogs." *Am. J. Clin. Path.,* 12:563-570, 1942.

COWORY, E. V.: *General Cytology.* University of Chicago Press, Chicago. 1924.

CRILE, G.: *The Phenomena of Life.* W.W. Norton & Co., Inc., New York, 1936.

CRORN, A.: *Diseases of the Thyroid, Parathyroids, and Thymus.* 3rd Ed., Lea & Febiger, Philadelphia, Pa. 1938.

DE AMILIBIA, E., MENDIZABAL, M. M., and BOTELLA-LLUSIAR, J.: "Ovarian Hormones and Thyroid Function." *Klin. Wchnschr.,* 15:1001, 1936.

DELSAL, J. L., and MACHEBOEUF, M.: "Phosphoaminolipides of Blood Serum. Nature of the Phospholipides Bound to Proteins of Horse Serum in the Form of Cenapses Precipitable by Acid." *Bull. soc. chim. biol.,* 25:358-360, 1943.

DOLIQUE, R., and GIROUX, J.: "Excretion of Selenium After Poisoning with Sodium Selenite and Sodium Selenate." *Bull. soc. chim. France,* 10:60-64, 1943.

DOUGHERTY, T. F., WHITE, A., and CHASE, J. H.: "Relationship of the Effects of Adrenal Conical Secretion of Lymphoid Tissue and on Antibody Titer." *Proc. Soc. Exp. Biol. & Med.,* 56:28-29, 1944.

DOUTHWAITE, A. H.: "Spoiled Sweet Clover and Other Anti-Coagulants." *Guy's Hosp. Gaz.,* 58:91-95, 1944.

DRAGSTEDT, C. A., WELLS, J. A., and ROCHA E SILVA, M.: "Inhibitory Effect of Heparin Upon Histamine Release by Tryspin, Antigen, and Proteose." *Proc. Soc. Exp. Biol. & Med.,* 51:191-192, 1942.

DREW, A. H.: "Growth and Differentiation in Tissue Cultures." *Brit. J. Exp. Path.,*4:46, 1923.

DRINKER, E. K.: *The Lymphatic System.* Lane Medical Lectures, Stanford University, California. 1942.

DRINKER, C. K., and YOFFEY, J. M.: *Lymphatics, Lymph, and Lymphoid Tissue.* Harvard University Press, Cambridge, Mass. 1941.

DURAN-REYNALS, F.: "Immunological Factors That Influence the Neoplastic Effects of the Rabbit Fibroma Virus." *Cancer Research,* 5:25-39, 1945.

DUSTIN, A. P.: "Nos Connaissances Actuelles sur le Determinisme de les Division Cellulaire chez l'Adulte." *Ann. et Bull. Soc. Roy. Sci. Med. et Nat. Bruxelles,* 1933:217, 1933.

DYCKERHOFF, H., and DESCHLER, W.: "Reduction of Coagulating Activity of Plasma and Thrombokinase by Extraction with Ether and Benzene." *Biochem. Zeitschr.,* 314:258-264, 1943.

DYCKERHOFF, H., and GRUNEWALD, D.: "Uber den Reaktionsmechanismus der Hemmung der Blutgerinnung durch einige seltene Erden und durch Heparin." *Biochem. Zeitschr.,* 315:124-138, 1943.

DYCKERHOFF, H., and MARX, R.: "Uber Heparinkomplement und eine Methode seiner vergleichenden Bestimmung im Blute." *Biochem. Zeitschr.,* 316:255-263, 1944.

EAGLE, H., and HARRIS, T. N.: "Studies on Blood Coagulation. The Coagulation of Blood by Proteolytic Enzymes (Trypsin, Papain)." *J. Gen. Physiol.,* 20:543-560, 1937.

239

EHRICH, W. E., and HARRIS, T. N.: "The Site of Antibody Formation." *Science,* 101:28-31, 1945.

ERUCKSON, J. O., and NEURATH, H.: "Antigenic Properties of Native and Regenerated Horse Serum Albumin." *J. Exp. Med.,* 78:1-8, 1943.

v. EULER, H., AHLSTROM, L., and HESSELQUIST, H.: "Cleavage of Nucleoprotein and Nucleic Acids from Liver and Jensen Sarcoma in the Presence of Serum of Normal and Sarcomatous Rats." *Nutrition Abstr. & Rev.,* 14:532, 1945.

EVANS, H.: As reviewed in *Applied Physiology,* S. Wright. Oxford University Press, New York. 1932.

FABRE, R., and KAHANE, E.: "Introduction to the Biochemical Study of Pneumonoconiosis." *Arch. maladies professionelles,* 1:21-28, 1938.

FISCHER, A.: "The Transformation In Vitro of Large Mononuclear Leucocytes Into Fibroblasts." *Compt. rend. soc. de Biol.,* 92:109, 1925.

FISHER, K. C.: "Cellular Metabolism." *Fed. Proc.,* 1:4, 1942.

FISHMAN, W. H., and ARTOM, C.: "The Relation of Diet to Composition of Tissue Phospholipids; Action of Choline and Choline Precursors in Weanling Rats." *J. Biol. Chem.,* 154:109-115, 1944.

———— "Relation of Diet to Composition of Tissue Phospholipids; Action of Choline, Vitamins, Amino Acids, and Their Combinations in Two Month-Old Rats." *Ibid,* 154:117-127, 1944.

FITZGERALD, J. G., and LEATHES, J. B.: "The Non-Antigenic Properties of Lipoids Extracted from Human Livers." *U. Calif. Publ.,* 2, 4:39-46, 1912.

FOLDES, F. E., and MURPHY, A. J.: "Distribution of Cholesterol, Cholesterol Esters and Phospholipid Phosphorus in Normal Blood." *Proc. Soc. Biol. & Med.,* 62:215-218, 1946.

———— "Distribution of Cholesterol, Cholesterol Esters and Phospholipid Phosphorus in Blood in Thyroid Disease." *Ibid.,* 62:218-222, 1946.

FRANKE, K. W., and MOXON, A. L.: "The Toxicity of Orally Ingested Arsenic, Selenium, Tellurium, Vanadium and Molybdenum." *J. Pharm. & Exp. Therap.,* 61:89-102, 1937.

GALIN: *Lancet,* page 755, 1889.

GORDON, A. S., KLEINBERG, W., and CHARIPPER, H. A.: "The Reticulo-Endothelial System and the Concept of the Anti-Hormone." *Science,* 86:62-63, 1937.

GOTO, K.: "The Relation of the Spleen to Blood Destruction and Regeneration and to Hemolytic Jaundice. XVI. The Influence of Splenectomy and of Blood Disintegration Upon the Production of Bile Pigment." *J. Exp. Med.,* 26: 95-804, 1917.

GREENBERG, D. M., COPP, D. H., and CUTHBERTSON, E.: "Studies in Mineral Metabolism with the Aid of Artificial Radioactive Isotopes. The Distribution and Excretion, Particularly by Way of the Bile, of Fe, Co, and Mn." *J. Biol. Chem.,* 147:749-756, 1943.

GREENBERG, D. M., and TROESCHER, F. M.: "Study with Radioactive Isotopes of Excretion of Calcium and Strontium by Way of the Bile." *Proc. Soc. Exp. Biol. & Med.,* 49:488-491, 1942.

GREENSTEIN, J. P., and CHALKEY, H. W.: "Desaminases for Ribonucleic and Desoxyribosenucleic Acids." *J. Nat. Cancer Inst.,* 6:61-72, 1945.

GRUNER, O. C.: *Biology of the Blood Cells.* Wm. Wood & Co., New York, 1914.

GRUNKE, W., and KOLETZKO, J.: "Studien uber die Blutgerinnung mit besonderer Berucksichtigung der Hamophilie; Einfluss des Harnes auf die kunstlich verzogerte (hamophilieanige) Gerinnung des Plasmas." *Ztschr. f. d. ges. exper. Med.*, 105:46-52, 1939.

HAHN, P. F., DONALD, W. E., and GRIER, R. C.: "The Physiological Bilaterality of the Portal Circulation. Streamline Flow of Blood Into the Liver as Shown by Radioactive Phosphorus." *Am. J. Physiol.*, 143:105-107, 1945.

HAMMARSTEN, O.: *A Text-Book of Physiological Chemistry.* John Wiley & Sons, New York. 1914.

HANSON, A. M.: "A Report of Four Cases of Inoperable Carcinoma Treated with Intramuscular Injections of Karkinolysin." *Minnesota Medicine*, 13:65-73, 1930.

HARRIS, T. N., and EHRICH, W. E.: "The Fate of Injected Particulate Antigens in Relation to the Formation of Antibodies." *J. Exp. Med.*, 84:157-165, 1946.

HARROWER, H. R.: *Practical Endocrinology.* Pioneer Printing Co., Glendale, Calif. 1933.

HART, J. P., and COOPER, W. L.: "Vitamin F in the Treatment of Prostatic Hypertrophy." *Report No. 1*, Lee Foundation for Nutritional Research, Milwaukee, Wis. 1941.

HOLLOWAY, J. K., and BLACKFORD, L. M.: "Comparison of the Blood-Platelet Count in Splenic Arterial and Venous Blood." *Am. J. Med. Sci.*, 168:723-728, 1924.

HOOPER, C. W., and WHIPPLE, G. H.: "Bile Pigment Metabolism. VIII. Bile Pigment Output Influenced by Hemoglobin Injection; Splenectomy and Anemia." *Am. J. Physiol.*, 43:275-289, 1917.

HORIUTI, I., and OHSAKO, H.: "Studies on the Detoxicating Hormone of the Liver (Yakriton)." *Tohoku J. Exp. Med.*, 24:505-515, 1934.

HUGGINS, C.: "The Physiology of the Prostate Gland." *Physiol. Rev.*, 25:281-295, 1945.

HUGGINS, C., and MCDONALD, D. F.: "Proteolytic Enzymes and Acid Phosphatase in the Prostatic Fluid in Chronic Prostatitis." *J. Urol.*, 52:472-474, 1944.

HUGGINS, C., and VAIL, V. C.: "Plasma Coagulation and Fibrinogenolysis by Prostatic Fluid and Trypsin." *Am. J. Physiol.*, 139:129-134, 1943.

HUNTER, R. F.: "The Conversion of Carotene Into Vitamin A." *Nature*, 158:257- 260, 1946.

JUNGBLUT, C. W.: "The Role of the Reticulo-Endothelial System in Immunity. V. Phenomena of Passive Immunity in Blocked and Splenectomized Mice." *J. Exp. Med.*, 47:261, 1928.

KAIRIUKSCHTIS, V.: "Versuche mit Neon-, Argon- und Heliothorat bei Kranken an Lungentuberkulose." *Zeitschr. Tuberk.*, 59:339-346, 1931.

KAUCHER, M., MOYER, E. Z., RICHARDS, A. J., and WILLIAMS, H. H.: "The Lipid Partition of Isolated Cell Nuclei of Dog and Rat Livers." *Fed. Proc.*, 4:157, 1945.

KIDD, J. G., and FRIEDEWALD, W. F.: "A Natural Antibody That Reacts In Vitro with a Sedimentable Constituent of Normal Tissue Cells. I. Demonstration of the Phenomenon." *J. Exp. Med.*, 76:543-556, 1942.

———— "II. Specificity of the Phenomenon: General Discussion." *Ibid*, 76:557-576, 1942.

LANDSTEINER, K., and PARKER, R. C.: "Serological Tests for Homologous Serum Proteins in Tissue Cultures Maintained on a Foreign Medium." *J. Exp. Med.*, 71:231-236, 1940.

241

LOEB, J.: *Dynamics of Living Matter.* Columbia University Press, New York. 1906.

LOW, F. N.: "Negative Effects of Endocrine Extracts on the Thymus of the White Rat." *Endocrin.,* 22:443-446, 1938.

MACAL1STER, C. J.: *Investigation Concerning an Ancient Medicinal Remedy and Its Modern Utilities.* John Bale, Sons & Danielsson Ltd., London. 1936.

MCCARRISON, R., and MADHAVA, K. B.: "The Effect of Insanitary Conditions on the Thyroid Gland and Other Organs of the Body." *Indian J. Med. Research,* 20:697-722, 1933.

MACHEBOEUF, M. A., DELSAL, J. L.. LEPINE. P., and GIUNTINI, J.: "Research on the State of Cholesterol Esters and of Phosphatides in Blood Serum. Study on the Homogeneity of the Phosphatide Steride Proteins by Ultracentrifugation and by Electrophoresis." *Ann. inst. Pasteur,* 69:321-333, 1943.

MCMASTER, P. D., and HUDACK, S. S.: "The Formation of Agglutinins Within Lymph Nodes." *J. Exp. Med.,* 61:783-805, 1935.

MAGERL, J. F.: *Zeitschr. lmnrunitatsforsch.* 101:1-12, 122-132, 168-177, 225-232, 1942.

MA1GNON, F., and THIERY, J. P.: "Influence of the Ingestion of Active Pancreatic Trypsin on the Amount of Blood Alexin in the Guinea Pig." *Compt. rend. soc. Biol.,* 136:645, 1942.

MALTANER, F.: "Significance of Thromboplastic Activity of Antigens Used in Complement Fixation Tests." *Proc. Soc. Exp. Biol, & Med:,* 62:302-304, 1946.

MANERY, J. F., DANIELSON, I. S., and HASTINGS, A. B.: "Connective Tissue Electrolytes." *J. Biol. Chem.,* 124: 359-375, 1938.

MANSFELD, G.: "Das wirksame Prinzip der oxydationshemmenden Schilddrusentatigkeit." *Arch. Exp. Path. u. Pbarmakol.,* 196:598-608, 1940.

MARSHAK, A.; and WALKER, A. C.: "Effect of Chromatin Derivative on the Healing of Skin Wounds." *Proc. Soc. Exp. Biol. & Med.,* 58:62-63, 1945.

MASON, E. C., and LEMON, C. W.: "Auto-Intoxication and Shock." *Surg., Gynec., & Obst.,* 53:60-64, 1931.

MAYER, E: "Experiments on the Limit of Growth in Tissue Cultures." *Skand. Arch. Physiol.,* 72:249-258, 1935.

MELLANBY, J.: "Heparin and Blood Coagulation." *Proc. Roy. Soc.* (London), B116 (796):1-9, 1934.

MEYER, H. H., and GOTTLIER, R.: *Experimental Pharmacology as a Basis for Therapeutics.* 2nd Edition, J. B. Lippincott Co., Philadelphia, Pa. 1926.

MIRSKY, A. E.: "Chromosomes and Nucleoproteins." *Advances in Enzymology,* 3:1-34, 1943.

NASU, M.: "The Effect of Sex Hormones on Blood Coagulation." *Okayama lgakkai Zassi (Mitt. med. Ges. Okayama),* 52:2783-88; 2789—German, 1940.

NEEDHAM, J.: *Biochemistry and Morphogenesis.* Cambridge University Press, London. 1942.

NORTHROP, J. H.: *Science News Letter,* Oct. 5, 1946.

PASTERNAK, L., and PAGE., I. H.: "On the Question of the Phosphatid Metabolism and the Influence of Thyreoidio." *Biochem. Ztscbr.,* 282:282, 1935,

PERLINFEIN, H. H.: "A Survey of Vitamin F." *Report No. 3.* Lee Foundation for Nutritional Research, Milwaukee. Wis. 1942.

PERLMAN, I., and CHAIKOFF. I. L.: "Radioactive Phosphorus as Indicator of Phospholipid Metabolism; On the Mechanism of Action of Choline Upon Liver of Fat-Fed Rat," *J. Biol. Chem..* 127:211-220. 1939.

PICKERING, J. W., MATHUR, S. N, and ALLAHAB, M. B.: "The Role of Tissue Juices in Thrombosis." *Lancet,* 2:387-388, 1932.

PINEY, A.: *Recent Advances in Haematology.* 3rd Edition, P. Blakiston's Son & Co., Inc., Philadelphia, Pa. 1931.

POKROVSKAYA, M. P., and MAKAROV, M. S.: "The Relation of the Wound Exudate to Healing." *Am. Rev. Sou. Med.,* 2:513-518, 1945.

PRANDONI, A, and WRIGHT, I.: "The Anti-Coagulants. Heparin and the Dicoumarin-3, 3' Methylene-Bis- (4-Hydroxycoumarin)." *Bull. New York Acad. Med.,* 18:431-458, 1942.

QUICK, A. J.: "On Various Properties of Thromboplastin (Aqueous Tissue Extracts)." *Am. J. Physiol.,* 114: 282-296, 1936.

―――― *The Hemorrhagic Diseases and the Physiology of Hemostasis.* Charles C. Thomas, Springfield, Ill. 1942.

RACHMILEWITZ, M.: "Effect of Uremic Serum and Urine on Growth of Fibroblasts In Vitro." *Arch. Int. Med.,* 67:1132-1143, 1941.

REHFUSS, M. E., and WILLIAMS, T. L.: "Study of the Third Bile Fraction." *Am. J. Dig. Dis.,* 8:407-415, 1941.

―――― "The Effect of the Liver Fraction of Duodenal Drainage on Certain Forms of Animal and Vegetable Life." *Ibid,* 10:435-440, 1943.

REINHARDT, W. O., FISHLER, M. C., and CHAIKOFF, I. L.: "The Circulation of Plasma Phospholipids: Their Transport to Thoracic-Duct Lymph." *J. Biol. Chem.,* 152:79-82, 1944.

RIGDON, R. H.: "Effect of Heparin on Phagocytosis. Observations on P. lophurae in the Chick." *J. Lab. & Clin. Med.,* 29:840-849, 1944.

ROBERTSON, T. B.: Chemical Basis of Growth and Senescence. J. B. Lippincott Co., Philadelphia. 1923.

ROBERTSON, T. B.: *Principles of Biochemistry.* Lee & Febiger, Philadelphia, Pa. 1924.

ROCHA E SILVA, M., and DRAGSTEDT, C. A.: "Liberation of Heparin by Trypsin." *Proc. Soc. Exp. Biol. & Med.,* 48:152-155, 1941.

ROCHA E SILVA, M., GRANA, A., and PORTO, A.: "Inhibitory Effect of Glycogen on Anaphylactic Shock in the Rabbit." *Proc. Soc. Exp. Biol. & Med.,* 59:57-61, 1945.

ROSENBERG, H. R.: *Chemistry and Physiology of the Vitamins.* Interscience Publishers, Inc., New York. 1942.

ROSENTHAL, S. M.: "Experimental Chemotherapy of Burns and Shock. III. Effects of Systemic Therapy on Early Mortality." *U.S. Pub. Health Repts.,* 58:513-522, 1943.

―――― "IV. Production of Traumatic Shock in Mice. V. Therapy with Mouse Serum and Sodium Salts." *Ibid,* 58:1429-1436, 1943.

ROTHEN, A.: "Forces Involved in the Reaction Between Antigen and Antibody Molecules." *Science,* 102:446-447, 1945.

SABIN, F. R.: "Cellular Reactions to a Dye-Protein with a Concept of the Mechanism of Antibody Formation." *J. Exp. Med.,* 70:67-79, 1939.

SACKS, B.: "The Reticulo-Endothelial System." *Physiol. Rev.,* 6:504-533, 1926.

SAKURAI, K.: *Sei-i-Kwai M. J.* 48:52, 1929.

SATO, A.: "Studies on Detoxicating Hormone of Liver; Preliminary Report. Effect of Detoxicating Hormone." *Tohoku. J. Exp. Med.,* 8:232-233, 1926.

SCHMIDT, L. H.: "The Phospholipid Content of Liver, Skeletal Muscle and Whole Blood as Affected by Thyroxine Injections." *Am. J. Physiol.,* 111:138-144, 1935.

SCHMITT, F. O., and PALMER, K. J.: "X-ray Diffraction Studies of Lipide and Lipide-Protein Systems." *Cold Spring Harbor Symposia on Quantitative Biology,* 8:94-101, 1940.

SCHULMAN, J. H.: "The Physical Chemistry of Lipoid Protein Association." *Biochem. J.,* 39:liv-lvi, 1945.

SEVAG, M. G.: *Immuno-Catalysis.* Charles C. Thomas, Springfield, Ill. 1945.

SHAPIRO, A., and others: "Plasma Clot Tensile Strength. Effect of Same Physical Factor, Anticoagulant and Coagulants." *J. Lab. & Clin. Med.,* 29:282-295, 1944.

SHWARTZMAN, G.: Phenomenon of Local Tissue Reactivity. Paul B. Hoeber, New York. 1937.

SIMMS, H. S., and STILLMAN, N. P.: "Substances Affecting Adult Tissue In Vitro. The Stimulating Action of Trypsin on Fresh Adult Tissue." *J. Gen. Physiol.,* 20:603-619, 1937.

――― "II. A Growth Inhibitor in Adult Tissue." *Ibid,* 20:621-629, 1937.

――― "Production of Fat Granules and of Degeneration in Cultures of Adult Tissue by Agents from Blood Plasma." *Arch. Path.,* 23:316-331, 1937.

SINCLAIR, R. G.: "Fat Metabolism." *Ann. Rev. Biochem.,* 6:245-268, 1937.

SOBOTKA, H.: *Physiological Chemistry of the Bile.* Williams & Wilkins Co., Baltimore, Md. 1937.

SOLANDT, D. Y., and BEST, C. H.: "Time-Relations of Heparin Action on Blood-Clotting and Platelet Agglutination." *Lancet,* 238:1042-1044, 1940.

STANLEY, W. M.: "The Structure of Viruses." *The Cell and Protoplasm.* American Association for the Advancement of Science. 1940.

STEPP, W., KUHNAU, and SCHROEDER, H.: *The Vitamins and Their Clinical Application.* The Vitamin Products Co., Milwaukee, Wis. 1938.

STEWART, W. S., BERGREN, W., and REDEMANN, C. E.: "A Plant Growth Inhibitor." *Science,* 89:185-186, 1939.

STUBER, B., RUSSMANN, A., and PROEBSTING, E. A.: *Biochem. Z.,* 143:221-235, 1923.

TASHIRO, *et al.:* "Studies on Bile Salts" *Med. Bull. Univ. Cincinnati,* 6:40-156, 1931.

TAUROG, A., ENTENMAN, C., and CHAIKOFF, I. L.: "The Choline-Containing and Non-Choline-Containing Phospholipides of Plasma." *J. Biol. Chem.,* 156:385-391, 1944.

TAUROG, A., LORENZ, F. W., ENTENMAN, C., and CHAIKOFF, I. L.: "The Effect of Diethylstilbestrol on the In Vitro Formation of Phospholipides in the Liver as Measured with Radioactive Phosphorus." *Endocrin.,* 35:483-487, 1944.

TAYEAU, F.: "Destruction of the Cholesterol-Protein Complexes of Blood Serum by Bile Salts." *Compt. rend. soc. Biol.,* 137:239, 1943.

――― "The State of Cholesterol in Blood Serum." Compt. rend. 212:575-576, 1941

TEN BROECK, C.: "The Differentiation of Trypsins by Means of the Anaphylactic Test." *J. Biol. Chem.,* 106:729-733, 1934.

TOCANTINS, L. M.: "Demonstration of Anti-Thromboplastic Activity in Normal and Hemophilic Plasmas." *Am. J. Physiol.,* 139:265-279, 1943.

TOMPKTNS, E. H.: "Reaction of the Reticule-Endothelial Cells to Subcutaneous Injections of Cholesterol." *Arch. Path.,* 42:299-319, 1946.

TORRIOLI, M., and GALEAZZI, M.: "The Biology of Megakaryocytes Surviving In Vitro. III. Action of Muscle and Liver Extract." *Policlinico (Sez. Med.),* 41:647-651, 1934.

TORRIOLLI, M., and PUDDU, V.: "The Biology of Megakaryocytes Surviving In Vitro. I. Technic." *Policlinico (Sez. Med.),* 41:245-248, 1934.

———— "II. Action of Spleen Extract." *Ibid*, 41:248-254, 1934.

———— "Recent Studies on the Pathogenesis of Werlhof's Disease." *J. A. M. A.*, III:1455-1456, 1938.

TROLAND, C. E., and LEE, F. C.: "A Preliminary Report on a Platelet-Reducing Substance in the Spleen of Thrombocyropenic Purpura." *Bull. Johns Hopkins Hosp.*, 62:85-86, 1938.

———— "Thrombocytopen: A Substance in the Extract from the Spleen of Patients with Idiopathic Thrombocytopenic Purpura That Reduces the Number of Blood Platelets." *J. A. M. A.*, III:221-226, 1938.

TURCK, F. B.: The Action of the Living Cell. The Macmillan Co., New York. 1933.

UNGAR, G.: "Endocrine Reaction to Tissue Injury." *Nature*, 154:736, 1944.

VAUGHAN, V. C., VAUGHAN, V. C., JR., and VAUGHAN, J. W.: *Protein Split Products (In Relation to Immunity and Disease)*. Lea & Febiger, Philadelphia. 1913.

VIGNEAUD, V. DU, CHANDLER, J. P., COHN, M.., and BROWN, G. B.: "The Transfer of Methyl Group from Methionine to Choline and Creatine." *J. Biol. Chem.*, 134:787-788, 1940.

VON HAAM, E. and CAPPEL, L.: "The Effect of Hormones Upon Cells Grown In Vitro. II. The Effect of the Hormones from the Thyroid, Pancreas and Adrenal Gland." *Am. J. Cancer*, 39:354-359, 1940.

WEINGLASS, A. R., and TAGNON, H. J.: "Effects of Chymotrypsin on Insulin and Blood Glucose." *Am. J. Physiol.*, 143:277-281, 1945.

WELLS, J. A., DRAGSTEDT, C. A., COOPER, J. A., and MORRIS, H. C.: "Observations on the Antagonism Between Heparin and Trypsin." *Proc. Soc. Exp. Biol. & Med.*, 58:57-59, 1945.

WIDENBAUER, F., and REICHEL, CH.: "Coagulation-Active Cell Substances." *Biochem. Z.*, 309:100-107, 1941.

———— "The Different Effects of Coagulation-Promoting Substances." *Ibid*, 309:415-417, 1941.

WINTERNITZ, M. C., MYLON, E., and KATZENSTEIN, R.: "Studies on the Relation of the Kidney to Cardiovascular Disease. III. Tissue Extracts and Thrombosis." *Yale J. Biol. & Med.*, 13:595-622, 1941.

WOLBACH, 'S. B., and BESSEY, O. A.: "Tissue Changes in Vitamin Deficiencies." *Physiol. Rev.*, 22:233-289, 1942.

WOLF, W.: *Endocrinology in Modern Practice*. 2nd Edition, W. B. Saunders Co., Philadelphia, Pa. 1939.

WOLFE, J. M., and WRIGHT, A. W.: "The Fibrous Connective Tissue of the Artificially Induced Maternal Placenta in the Rat with Particular Reference to the Relationship Between Reticulum and Collagen." *Am. J. Path.*, 18:431-454, 1942.

WRIGHT, S.: *Applied Physiology*. Oxford University Press, New York. 1932.

ZISKIN, D. E.: "Effect of the Male Sex Hormone on the Gingival and Oral Mucous Membranes." *J. Dent. Research*, 20:419-423, 1941.

CHAPTER 6

MORPHOGENS AND PATHOLOGICAL PROCESSES

THUS FAR in our discussion of the morphogen hypothesis, we have covered: (Chapter 1) the evolutionary basis for protomorphogens and experimental evidence for their existence as elementary protein determinants with minerals as essential parts of their structure; (Chapters 2 and 3) the relationship of protomorphogens to growth and senescence in single celled structures; (Chapter 4) the relationship of protomorphogens to embryonic differentiation; and (Chapter 5) the physiology of protomorphogens in higher organisms. In this chapter we will attempt to discuss various pathological conditions in which we believe there is evidence of a link with protomorphogens or their controls. The morphogen hypothesis seems to afford a new approach to many diverse pathological conditions in the animal. Where the thread of such a link exists, we shall discuss it herein hoping thereby to stimulate a reappraisal of the problems involved.

TISSUE INJURY AND INFLAMMATION

We have reviewed experimental evidence that small amounts of protomorphogen in the media are necessary for growth in tissue cultures, while concentrated amounts inhibit growth and result in cytolysis. It is interesting to study the regeneration of tissue in the animal organism from the standpoint of optimum protomorphogen concentration in the surrounding tissue fluids.

Protomorphogen and Healing.—Reissner (1941, 1942) has reported some interesting observations relating to organo-therapy in dentistry. His experiments were conducted with intramuscular, subcutaneous or submucous injections of homologous extracts of jaw bones and dental tissues. Rapid healing and regeneration of gingival tissue in gingivitis and pyorrhea have been reported. Oral administrations of the extracts were also added to the treatment. It is possible that these extracts were the means

247

of administering homologous protomorphogens or protomorphogen fragments which were utilized by the tissue in its regeneration. Reissner concludes that each organ operates under a characteristic protective or defensive mechanism and that he was supplying the "hormone" to stimulate this function. It is also possible that the administration of these homologous protomorphogens antigenically stimulated the animal's immune-biological system into the production of the natural tissue antibody which aided in the elimination of an *excess* concentration of protomorphogens in the local tissue area. (Refer to a later discussion on "Cytotoxins" in this volume for a more complete discussion of this possibility.)

The oral administration of various endocrine organ residues has been variously reported to exert remedial influences. Unpublished reports from many physicians have substantiated this conclusion. It is possible that this is the means of supplying, not only the hormone elicited by the organ in question, but also specific protomorphogens to aid in the regeneration of the secretory cells. (The reader will recall that protomorphogens exhibit an organ specificity which is out of proportion to their species specificity.) In view of this possibility we prefer to predicate, until more evidence is available, that Reissner's homologous tissue extracts supply specific protomorphogens to assist in the regeneration of gingival and other dental tissue.

Marshak and Walker (1945) have conducted extensive experiments on the influence of chromatin derivatives on the regeneration of injured tissue. They report a stimulation of the regenerative functions, and in some experiments labeled chromatin was located in the regenerating nuclei. The chromatin derivative was also found to exert a hemostatic action on bleeding wounds. These reports indicate that the morphogen content of the chromatin material was used in the synthesis of new chromatin, enhancing cellular regeneration.

Embryonic Growth Promoting Factors.—The interpretation of such phenomena as due to the administration of protomorphogen must be made with care. There are many growth promoting substances which do not come under the protomorphogen classification and these, of course, exert their influence in a manner quite apart from supplying morphogens as substrate material for chromosome synthesis. Indeed, probably the great majority of substances reported in the literature as stimulating agents for healing and regeneration cannot be classified thus.

The embryo growth promoting factors are probably those most commonly reported in the literature as stimulating to mitosis. Fischer (1940) has reviewed material on the embryo growth factor, commenting that it has no thromboplastic effect; this eliminates it as a possible protomorphogen substance.

The embryo growth factor probably is active as a catalyst in chromatin synthesis, its thermolability and enzymatic nature being indicative of this function. Sperti, Loofbourow and Lane (1937) have conducted a lengthy study of the growth promoting "injury factor" released by tissues as a result of injury, either by trauma or by irradiation

Davidson and Waymouth (1943) have investigated the effects of growth promoting factors on the nucleoprotein content of fibroblasts growing *in vitro*. They experimentally established the fact that a thermostable embryo extract caused a definite rise in the nucleoprotein phosphorus of such cells. This activity could also be ascribed to crude ribonuclease preparations but not to crystalline ribonuclease or the anterior pituitary extracts.

Their review discusses the contentions advanced by several investigators that the so-called growth promoting effects of embryo extracts are due solely to a special assemblage of nutrient protein break-down products. We feel that the evidence, while still incomplete, tends to refute this stand, and rather that the embryo growth promoting substance, while perhaps supplying necessary nutrient substrate, also partakes possibly in a catalytic stimulation of mitosis; further, that some or all of these factors are necessary for the normal physiological synthesis of substrate protomorphogens into new cell protein, and nuclear chromatin material.

It is not our purpose to enter into a prolonged review of the various growth factors which have been identified by various investigators. We feel that the embryo and placenta have a complex group of growth factors (which we might call the "embryo group") which stimulate various phases of growth and mitosis. As the embryo differentiates, the various components of this complex are relegated to various differentiated cells such as the anterior pituitary, thymus, epithelial cells, spleen, etc. Single celled animals do not differentiate in this manner; this could explain their cultivation without the need for additional embryo hormone while the cultivation *in vitro* of adult tissue requires a separate "embryo" factor. Reference to Chapter 3 will supply evidence that these factors are also present in adult tissue, but their effectiveness is "masked" by the concentration of inhibitory protomorphogens also contained therein. Removal of the protomorphogens with suitable solvents results in the manifestation of greater growth promoting activity in adult extracts (Baker and Carrel, 1925).

We feel that future investigations may show that these isolated growth "factors" such as the Anterior Pituitary Growth Hormone, Sperti's wound factor, the growth factor in autolyzed adult tissue, etc., are differentiated components of the original "embryo group." Included in these might be the epithelial fibrinolysin which we have discussed in Chapter 5 under "elutogenic factors." This fibrinolysin has also been identified in embryo tissue.

It is significant, however, that these growth factors are found in the epithelial tissues of the mammal. The epithelial tissue group is that which continues to undergo mitosis after parturition (skin, secretory epithelium, osteoblasts, etc); the presence of embryo growth factor components in them is to be anticipated.

Of greater importance in our discussion is the necrotic influence of the intense concentrations of protomorphogens which may accumulate as a result of trauma or burns. Were it not for the embryo-like growth factors or injury hormones also released, excess concentration of protomorphogen would prevent cell division and thereby inhibit regeneration and repair.

In the presence of the embryo-like growth factors moderate amounts of released protomorphogens are not toxic but are used as substrate material for the synthesis of new cells. Without these growth factors the protomorphogens might be lethal and exert a necrotic influence. Without these growth factors the protomorphogens give rise to scar tissue rather than a healthy regenerated group of cells.

Let us investigate inflammation and trauma with the thought of identifying the lethal influence of any excess amount of protomorphogen which might be released.

Toxic Influence of Released Protomorphogen.—Menkin (1945) has carefully investigated the fractions of inflammatory tissue exudates which give rise to the phenomenon associated with inflammation. He has separated four fractions as follows: (1) leukotaxine, a heat stable diffusible peptide responsible for the increase in permeability and chemotaxis; (2) leucocytosis-promoting factor, a heat labile, non-diffusible protein promoting leucocytosis especially of granulocytic and megakaryocytic cells; (3) necrosin, a heat labile, non-diffusible euglobulin responsible for necrosis, tissue damage and coagulation or thrombosis, and (4) pyrexin, a heat stable, non-diffusible factor released from protein by necrosin and responsible for pyrogenesis. He has amply demonstrate that none of these factors is associated with, or depends upon, histamine for its specific action.

These factors are all derivatives of tissue which have undergone autolysis. The protomorphogen from that tissue is probably associated with necrosin since this factor produces tissue damage and is also thromboplastic, two attributes of protomorphogen. The thromboplastic activity of the protomorphogen released in inflammation or injury, we feel, is one of the basic factors in the production of scar tissue in the healing process. (In Chapter 5 we discussed the activity of protomorphogen as the thromboplastic substance precipitating fibrin and causing the deposition of connective tissue.) Drinker (1942) recognizes some such activity, for he states "A bad scar is an expression of lymph blockade resulting in excess fibroblastic growth due to the accumulation of substances ordinarily removed by the lymph." We feel that the substance referred to is protomorphogen. This influence is also indicated by the observation that applications of embryo substance or of growth factors such as Sperti's wound hormone prevent scar formation. This is probably due both to the fibrinolysin contained in such extracts and their catalytic action in stimulating tissue regeneration by utilizing the released protomorphogen as substrate material for new protein molecules.

Turck (1933) has demonstrated that autolyzed tissue substance releases protomorphogen (Turck calls it "cytost") which causes inflammation and severe tissue damage. The administration of this substance results in shock and in many cases the involution of an organ homologous to that from which the protomorphogen was prepared. This specific effect was demonstrated to be a characteristic, even of tissue ashed at temperatures up to 700 degrees C. It is difficult to reconcile this extreme thermostability of Turck's cytost (protomorphogen) with the labile nature of Menkin's necrosin; we shall not attempt to do so. It is sufficient to note that Turck's ash was made of whole tissue rather than of inflammatory exudates; and, although autolyzed tissue substance exerts the same effect, it is apparent that something happens to the protomorphogen molecule itself as a consequence of autolysis. It is obvious that ashing has a different effect on the protomorphogen molecule than enzymatic reduction. Menkin's comment that necrosin is only recovered from exudates with an acid pH indicates that it is a product which may be released from the tissue (or the protomorphogen) as a result of enzymatic activity rather than protomorphogen itself. Probably the enzymatic reduction concurrent with autolysis splits the morphogen molecule, changing its thermostability in this profound degree.

SHOCK

Turck's production of shock as a consequence of the administration of protomorphogen from autolyzed or ashed tissue leads to a discussion of the nature of traumatic shock.

There is more literature on the problem of traumatic shock than is possible for us to review within these pages. We shall, as with similar problems in this chapter, simply review a few pertinent comments which seem to link the problem with protomorphogen influences.

A Toxic Factor in Shock.—There is much evidence, pro and con, referring to the existence of a toxic factor in shock. We feel the evidence for its existence to be overwhelming and shall attempt to link it with morphogen influences without ignoring other conflicting shock hypotheses. Moon (1942) has reviewed the recent conclusions of the consensus of investigators that the toxic products resulting from tissue autolysis in the traumatized area are the most important causes of shock in wounded men.

Marsh (1940) has reviewed the circumstances surrounding traumatic shock and remarks that a toxic "H" substance secreted from the injured tissue is responsible for the generalized increase in capillary permeability.

Mirsky and Freis (1944) demonstrated that administration of crude trypsin to rabbits resulted in a shock-like syndrome. They consider that this syndrome is due to the release of some proteolytic substance which is responsible for the functional depressions of shock. A recent review discusses the experiments of Rapport, Guild and Canzanelli in which by cross-circulation it was demonstrated that a toxic shock producing factor may be transferred in the blood stream to another animal. This"H"factor is not histamine. Roller (1943) has found that an ultra-filtrate from histamine shock exerts an effect upon injection similar to nephritic serum which is not due to histamine *per se* but to some other toxic factor released by histamine. Valy Menkin has conclusively demonstrated that the substance in inflammatory tissue exudates which increases permeability is not histamine.

Turck (1933) was able to induce the shock syndrome in test animals by the injection of autolyzed tissue but not by administrations of saline extracts of fresh tissue. However, the injection of the ash of fresh homologous tissue was followed by immediate and profound shock symptoms leading to death. Apparently the shock producing tissue factors are protected in some manner in fresh tissue and only become effective after a period of autolysis. It is altogether possible that the autolyzed tissue releases necrosin as a protomorphogen component and this product is responsible in part for the shock syndrome. However, when fresh tissue is ashed the thermostable protomorphogen mineral links remain and can exert the toxic shock-producing influence. The latent period of autolysis following trauma is an important consideration in the production of the toxic shock factor which must not be overlooked. It is significant that shock symptoms appear after such a latent period. Guanidine may be an important toxic factor released from the nucleoproteins by this autolysis. This substance is a powerful denaturant of proteins and may play a significant part in the irritation of the nerve termini, an activity of importance in the shock syndrome. Guanidine is responsible for the shock-producing effects of certain reptile venoms, being released by hydrolysis of nucleic acids brought about through the medium of a specific enzyme (Sevag, 1945). Robertson (1924) reports the increase of urinary methyl-guanidine in anaphylactic shock.

1 "Transmission of a Shock Producing Factor." Editorial, *J.AM.A.* 128:813-814, 1945.

This leads us to suggest that the toxic "H" factor in traumatic shock, especially burns, is in reality protomorphogen which is suddenly released. If protomorphogen is a basic toxic factor in shock (as Turck's ash experiments indicate) then we see no reason why protomorphogen or necrotic associates of protomorphogen, i.e., necrosin, should not be considered as *the* toxic factor in shock. In Chapter I we commented upon the potassium content of protomorphogen. Cicardo (1944) suggests that the toxic agent in traumatic shock is associated with potassium. Zwemer and Scudder (1938) consider potassium an important "H" factor in shock. Recently Tabor and Rosenthal (1945) reported that animals which show the shock syndrome are decidedly sensitive to administrations of potassium that would not inconvenience a normal animal. Mylon and Winternitz (1946) experimentally demonstrated by means of cross circulation experiments that the toxic factor initiated at the site of injury escapes primarily through the tissue fluids and lymph rather than directly through the blood stream as distinct from potassium. On the other hand, Pen, Campbell and Manery (1944) report that extensive investigations indicated that there is no substance in different types of muscle extracts other than potassium which can be considered toxic in nature.

Shorr, Zweifach and Furchgott (1945) supply a stimulating concept with their contention that a vaso-depressor substance is released by the liver and muscle and a vaso-excitor principle by the kidney consequent to anoxia in these tissues. They are emphatic in their evidence that the vaso-depressor material (VDM) is produced only from liver and skeletal muscle, but on the other hand the indications are that aerobic incubation with normal liver slices destroyed the VDM. Such incubation with other tissues gave negative results.

It might be interesting to speculate on the linkage of phosphagen with the toxic factor in shock in view of the above information. It will be remembered that phosphagen is broken down in muscle consequent to over-work, autolysis or injury, and also that phosphagen seems to be synthesized mainly in the liver where it is stored in comparatively large quantities. Potassium, of course, is an important part of the phosphagen molecule. (The importance of this molecule in cell biochemistry is reviewed in Chapter 3 of this volume.) Further evidence in this direction is the release of sugar by tissues in an inflamed or traumatized area.[1] The carbohydrate component of phosphagen (potassium creatine hexose phosphate) is conceivably the source of this glucose.

Quite possibly toxic end products of phosphagen breakdown are responsible for the vaso-depressor influence in shock, its potassium and carbohydrate moieties giving rise to the other factors manifested. Phosphagen is located primarily in skeletal muscle and liver; it is significant that these tissues are the only sources of VDM. The liver might destroy the VDM by re-combining it as an innocuous phosphagen molecule. Quite possibly this vaso-depressor material or its precursors arising from the cells is transported in the tissue fluids rather than exuding directly into the blood stream accounting for the phenomenon described by Mylon and Winternitz (see above).

Challenging the tissue toxin hypothesis is the report of Prinzmetal, Freed and Kruger (1944) who concluded that the toxic factor released by crushed muscle is a product of bacterial action in the traumatized area, since shock did not occur when animals were treated by local or systemic use of certain antibacterial agents. Abraham and associates (1941) also subscribe to this conclusion as a result of their experiments, but admit that large amounts of sterile autolyzed tissue may still release factors which increase capillary permeability resulting in traumatic shock.

The experiments of Feigen and Deuel (1945) can be explained by the hypothesis that protomorphogen is the toxic factor in shock. They report that the injection of a cattle brain extract *before* severe scalding prolonged the survival period of mice. Its administration *after* scalding, however, seemed to intensify the toxic effect of the burn. The cattle brain product used was thromboplastic. We have previously discussed the observation that brain substance is high in protomorphogens which are well "sheathed" by the associated phosphatide substances. If this product is administered before the burn, the protomorphogen contained therein could be adequately removed and excreted releasing the phosphatide "wrappers" with their affinity for the protomorphogen liberated by the burn, hence the anti-toxic influence. When administered after the burn, however, the added

1 Menkin, V.: "The Effect of Necrosin on the Blood Sugar Level." *Am. J. Physiol.,* 147:379-383, 1946.

protomorphogen in the preparation would further overload the protective systems already overtaxed with protomorphogen and thus intensify the toxic symptoms. They report that soya bean lecithin had no effect, considering therefore that the cattle brain influence was due to the thromboplastin rather than the phosphatides. We do not consider this a satisfactory explanation, however, since there is no evidence that soya bean lecithin has the protective affinity for protomorphogens exhibited by brain phosphatides, particularly cephalin.

Nervous Theories of Shock.—Crile (1914, 1936) has emphasized the importance of the nervous system in the etiology and control of shock. His kinetic theory of shock (1914) holds that shock is the result of a conversion of potential into kinetic energy in the nerve centers of vital tissues (brain, suprarenals, liver) with a resultant exhaustion of these centers followed by the symptoms of shock. He presents the conclusion (1936) that the best preventative of surgical shock is spinal anesthesia which presumably prevents the transfer of traumatic stimuli to the balance of the nerve centers producing the exhaustion preceding shock. (He emphasizes that the fall in blood pressure must also be prevented.) This narcotic excludes nervous stimuli from the brain thyroid-adrenal-sympathetic systems and prevents them from exhausting the vital energy and discharging of the electrical potentials important to the maintenance of cell vitality. (See our discussion of electrical potentials and cell vitality in Chapter 3.)

This theory excludes the possibility of the humoral transfer of a toxic shock-factor as the exclusive cause of shock.

The production of the shock syndrome by nervous stimulation or exhaustion is well known and has received much attention under the study of "shell shock." Turck (1933) for instance, produced shock and death by discharging a blank cartridge in the cage of experimental animals. He considers the sound waves as stimuli of traumatic injuries preceding shock, although no such visible damage was reported. Smitten (1946) has reviewed a highly sensitive protoplasmic system of a liquid sol which under the influence of injury (possibly an abnormal nervous stimulus) passes instantly into a highly viscous gel associated with a distinct shift towards lower pH values.

Crile's theory that an exhaustion of the vasoconstrictor center is the primary etiological factor in shock has been supported and refuted by various investigators, the most popular opinion being that direct proof of this exhaustion is difficult to obtain but that nervous changes play a significant part in the shock syndrome.[1] Speransky (1943) has published an advanced exposition of the importance of the nervous system in disease. Regarding inflammatory processes his experimentally supported theory postulates that these reactions in tissues evoke dystrophic influences in the nervous system which react to effect various local changes in tissue. This creates a damaging cycle in which the original inflammation is enhanced by the generalized irritation of nervous structure. He states that ". . . the process produced by the immediate irritation of a particular point of the nervous system becomes the originator of dissimilar tissue changes of a biochemical character in various other points of the organism" and concludes as follows ". . . in chronic inflammation both the maintenance of the primary focus and the formation of secondary foci become nothing less than a new pathological function of the nervous system."

From the evidence herein presented it becomes apparent that no matter how strongly we may embrace the toxic theory of shock, it is obvious that the nervous system plays an exceedingly important, if not vital, part in both the etiology and pattern of the shock syndrome.

Decreased Fluids and Increased Permeability in Shock.—Wiggers (1942) has reviewed the shock problem and makes the comment that reduced quantity of circulating blood is one of the most fundamental phenomena in the shock syndrome. McMillan (1940) comments that the most important abnormality in shock is the drastic reduction in the effective volume of circulating blood. McDowell (1940) comments that in shock, just before death, there is low blood pressure and a weak heart. A disturbed fluid balance resulting in a defective blood volume, followed by failure of the circulatory system seems to be the immediate cause of death in shock. The recent widespread successful use of blood plasma in the treatment of shock is testimony to the accuracy of this conclusion, which (incidentally) was suggested by Crile in 1909.

---

1 "Anociassociaton and Treatment of Shock." *J.A.M.A..*, 128:773, 1945.

Anaphylactic Shock.—Vaughan, Vaughan and Vaughan (1913) have reviewed the problems of anaphylactic shock and presented an hypothesis with adequate experimental substantiation. They demonstrate the existence of a universal toxic constituent in all proteins; they suspect histamine. If it is suddenly released into the blood stream, this constituent is responsible for the genesis of the shock syndrome.

The recent reviews of anaphylactic shock (Feldberg, 1941) leave little doubt but that the primary toxic substance in this type of shock is histamine released from the protoplasm of cells. The administration of sensitizing amounts of foreign proteins results in the production of immune antibodies which are able to destroy the specific protein by lyric enzyme action. The destruction of the foreign protein following the first injection releases the "toxic factor" slowly and it is disposed of adequately without lethal effect. The second administration of protein (after the sensitization), however, is destroyed promptly by the blood antibodies, releasing the toxic factor so rapidly that the eliminative processes are not able to cope with it and the shock syndrome is produced.

Shwartzman (1937) reports that the injection of foreign protein into a localized tissue area, sensitizes this area so that subsequent administrations of the same foreign protein into remote parts of the organism, give rise to an inflammatory reaction at the site of the first local administration. In this case it is likely that the lyric antibodies are produced and held locally, consequent to the sensitizing injection. Subsequent administration of the foreign protein results in a lysis and release of toxic products, in lethal concentrations, only at the locale where the lyric antibodies are stored. Burnet (1941) has commented on evidence that antibodies may be stored in restricted locales, where they originate.

It is important to distinguish between the shock produced by the anaphylactic method and by the direct method of the administration of tissue autolysates. Anaphylactic shock is a result of the production of lyric antibodies to a foreign protein, consequent to a sensitizing administration of small amounts. It increases in severity as the source of the administered protein is chosen from more distant relatives of the recipient on the phylogenetic scale (the more foreign the protein, the more powerful the antibody) .Tissue autolysate shock, however, is due to the direct acting homologous protomorphogens present in the first injection and *decreases* in severity as the injected autolysate is chosen from animals further removed on the phylogenetic scale (Turck, 1933). (Non-homologous protomorphogen is not inhibitive to cell activities.)

Cyclic Nature of the Shock Syndrome.—All of the theories we have discussed are too well substantiated by experimental evidence to be dismissed. The only conclusion to make when presented with such an impasse is that all of these theories are linked together and can ultimately be reconciled. On this basis it is possible to establish a "shock cycle" which will embrace all of the experimental evidence. In suggesting this cyclic nature of the shock syndrome, we are accepting the obvious contention that this cycle may be started at any point and it may be interrupted at any point. Thus, it becomes evident that the primary etiology of shock may refer to a variety of stimuli and there is more than one effective treatment of shock possible.

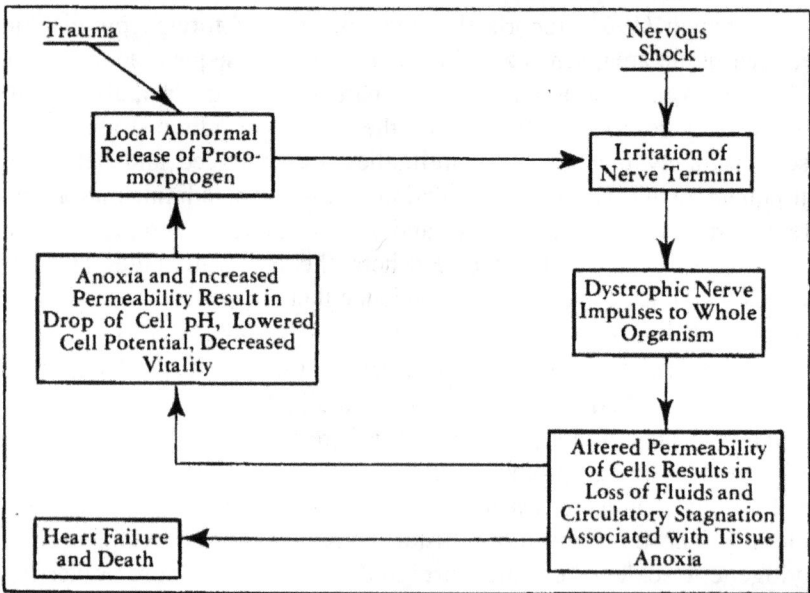

FIGURE 9
The Cyclic Nature of the Shock Syndrome.

259

1. For convenience we are assuming that we are dealing with traumatic shock. The cycle starts with the abnormal release of protomorphogens at the site of trauma due to autolysis possibly enhanced by bacterial metabolism. (Protomorphogens in this case are the traumatic toxic shock-factor.)

2. The released protomorphogens irritate the local nerve termini with consequent dystrophic influences on the entire organism (Speransky, Crile). It is important to note that normally the sympathetic nervous system exerts a trophic influence on tissue, preventing its degeneration (Asher).[1]

3. The dystrophic nervous influences lower the permeability of all the body cells. (Sympathetic nervous system influences the permeability of cells (Engel, 1941).)

4. The increased permeability interferes with the normal blood volume and the dystrophic nervous impulses cause capillary constriction and dilatation of the large blood vessels, draining the remaining blood into the internal vascular network.

5. Reduced circulating blood volume consequent to stagnation in the large internal blood vessels decrease the oxygen supply to the tissues, encouraging further the enervation of the cells.

6. The increased permeability and anoxia of tissue cells also permits abnormal diffusion of electrolytes with resultant reduction of potential differences and lowered vitality of the cell. The lowered pH following electrolyte diffusion shifts many reversible enzymes from the constructive to the destructive phase releasing more protomorphogens, completing the cycle.

7. The end-result of these processes is failure of the circulation and death.

This cycle is graphically illustrated in figure 9. It will be seen that traumatic injuries can start this cycle at the point of protomorphogen release. Also, nervous "shocks" may start this cycle at the point of nerve irritation. (Speransky states that this irritation may be chemical or purely biological.) Large losses of blood may initiate this cycle at the point of circulatory stagnation. Moon (1942) has published a comprehensive review of the shock problem and included a cycle similar to the one we describe but outlined in considerably more detail than we are able to do in this volume.

Similarly the various partially successful treatments of shock can inhibit

1 Asher, L.: "Trophic Function of the Sympathetic Nervous System." *J.A.M.A.,* 108:720-721, 1937.

this cycle at any point. For instance: (1) the administration of fluids or plasma may stop the cycle at the point of decreased blood volume; (2) local administration of phosphatides or "growth factors" such as Sperti's wound hormone may inhibit the local toxic influences of the released protomorphogen; (3) anesthetization of the nerve trunks may impair the transfer of the dystrophic impulse from the traumatic area; (4) pharmacological heart stimulants may prevent the exhaustion of the circulatory system; (5) the administration of adrenal cortex hormone may prevent the release of the potassium component of the protomorphogen toxins, and (6) the use of adrenal cortex hormone and of vitamin C may enhance the oxygen carrying capacity of the blood preventing the tissue anoxia.

Normally, shock producing influences are so slight that the tissue antibodies stimulate vital endocrine and protomorphogen eliminative functions, arresting the cycle and preventing death.

All of the above enumerated methods of treating shock have been reported in the literature with varying degrees of success. The fact that the various shock treatments and also the experimental evidence on the etiology of the shock syndrome is in harmony with the shock cycle we outline, leads us to present it without apologies for the minor inaccuracies it no doubt contains. Were we to attempt an evaluation of the relative importance of the various factors concerned in shock, it is apparent from the available evidence that the nervous system and its dystrophic influence is of prime importance irrespective of the manner in which this dystrophic influence is produced.

SENESCENCE

The problem of senescence and its causes is still the major riddle of medical science although competent and comprehensive reviews of the knowledge to date have been published (Cowdry, 1939, 1942). The paucity of careful investigations of the basic causes of ageing *per se* is probably a consequence of the viewpoint that senescence is not a disease but a natural process, and further, one that will never be altered in its nature. Studies have been more or less diverted to the pathological conditions associated with old age as illustrated by the science of geriatrics.

Senescence in Single Celled Organisms.—We have reviewed the morphogen cycles of single celled organisms in Chapter 3. If a colony of single celled individuals is kept in a restricted media, they will cease to divide, grow aged and eventually undergo autolysis. This phenomenon appears even though the culture is kept well supplied with available foodstuffs.

Senescence in single cells was postulated to be due to the accumulation of their protomorphogen in the media where it polymerizes under higher concentrations. This polymerization prevents the protomorphogens inside the cell from escaping into the media, either by causing the intra-cytoplasmic protomorphogen to polymerize so it is no longer diffusible, or by "clogging" the cell boundary.

The inability to excrete protomorphogens into the media as a part of the metabolic cycle results in their cytoplasmic accumulation. This affects the surface boundary and the electrical potential which exists between the nucleus and cytoplasm. As a consequence of this potential change (or vice versa), the pH of the protoplasm is lowered and the reversible enzymes of the cell drift more and more towards the destructive rather than constructive phase. A gradually lowering of cell vitality accompanies these changes which eventually result in dissolution of the cell.

This is a brief review of the hypothesis we have discussed in Chapter 3 which attempts to explain the basic causes of senescence in single cell organisms. The fundamental factor involved is the accumulation of protomorphogens in the media.

Senescence in the Metazoan Organism.—It is not unreasonable to suspect that the fundamental basis of senescence in metazoans is the same as in single celled organisms. Senescence in metazoan life is the ageing of the individual cells of the organism whose "medium" can be considered the tissue fluids and blood. We may consider the metazoan organism as a unit "culture" in which the protomorphogens discharged by the cells into the tissue fluids are eliminated by the specialized mechanisms discussed in Chapter 5.

We believe the primary index of senescence to be the degree of vitality and activity of the postmitotic cells, and the ability of the intermitotic and differentiating mitotic cells to quickly regenerate and repair damage.[1]

There is some indication that protomorphogens do, in fact, accumulate in the blood and tissue fluids of the mammal as age increases. Alexis Carrel (1924) has reported that the blood serum becomes more growth-inhibiting with advancing years and has even proposed that this characteristic be used as a basis of measuring biological age. McCay (1939) has reviewed other evidence for the existence of an inhibitor in aged tissue. We have discussed evidence (Chapter 3) that this inhibitor is protomorphogen. It is significant that the blood protein, total lipid and lipid phosphorus fractions increase with age while cholesterol decreases (Baker and Carrel, 1927). Phosphatides have been discussed in our hypothesis as "wrappers" which form a protective-association with protomorphogens and their increase in the blood very possibly parallels the increase of protomorphogens.

Ruzicka (1924) has called attention to the progressive condensation of colloids with age, a phenomenon which he terms "hysteresis."[2] This condensation is a result of a decrease of the particle charge reducing their dispersion and producing a polymerization of the colloidal molecules which is progressively less reversible. Linfoot (1944) has called attention to the particles in blood (increased in infectious conditions) of a thromboplastic nature and whose concentration varies with the health and vitality of the patient. It will be recalled that protomorphogens are assumed to undergo polymerization into larger molecules under the unfavorable environment of increased concentration.

Smitten (1946) reports that living protoplasm is a very liquid sol which passes instantly into the state of a highly viscous gel under the influence of injury. This is accompanied by a lowering of the intracellular pH. There is an implication that these reactions may be preceded by various pathological processes, in particular upsets of the nervous system. We have already discussed

---

1 Cowdry (1942) defines these cells as follows: (r) postmitotic cells, those which are most differentiated some of which can revert and regenerate such as liver, renal and thyroid cells, others which are on a higher plane of differentiation such as nerves, cardiac, skeletal, rod and cone cells, etc., whose life cycle is equal to that of the organism, and (z) intermitotic and differentiating mitotic cells, those which progressively undergo mitosis to renew epidermis and repair damage such as fibroblasts, marrow cells, epithelial cells and those differentiated further such as the erythroblasts, myelocytes, etc.

2 Apparently Ruzicka cliose this term to represent the progressive irreversibility of the ageing processes.

the influence of the nervous system over the tissue pH changes associated with traumatic shock. It becomes apparent that it is also concerned with the same changes associated with senescence.

The accumulation of polymerized protomorphogen in the blood and tissue fluids with advancing age would exert characteristic effects on the cells of the organism. Carlson[1] has listed the following progressive age changes which have not been shown to be due to specific diseases:

1. Gradual tissue dessication.

2. Gradual retardation of cell division, capacity of cell growth and tissue repair.

3. Gradual retardation in the rate of tissue oxidation (lowering of B. M. R.).

4. Cellular atrophy, degeneration, increased cell pigmentation and fatty infiltration.

5. Gradual decrease in tissue elasticity, and degenerative changes in the elastic connective tissue.

6. Decreased speed, strength and endurance of skeletal neuromuscular reactions.

7. Decreased strength of skeletal muscle.

8. Progressive degeneration and atrophy of the nervous system, impaired vision, hearing, attention, memory and mental endurance.

The following of these conditions listed by Carlson can be interpreted by means of the morphogen hypothesis as a direct effect of protomorphogen accumulation in the tissue fluids: (No. 2) retardation of cell division and capacity for repair; (No. 3) retardation of rate of tissue oxidation; (No. 4) cellular atrophy, degeneration, increased pigmentation and fatty infiltration, and (No. 5) degenerative changes in the elastic connective tissue (we believe due to the accumulation of protomorphogens adsorbed on this tissue (see Chapter 5)). These consequences of increased protomorphogen concentration in pericellular fluids are discussed in more detail in Chapter 3.

The balance of the progressive age changes enumerated above can all be interpreted as necessary corollaries to the changes we have listed as basically due to accumulating protomorphogens.

Further change of age that we consider directly due to the higher concentration of protomorphogen in the tissue fluids are to be found in the

1 Carlson, A.].: "The Physiology of Ageing." Private Publication, 1944.

literature. For instance, MacNider (1942) analyzed the increased susceptibility of aged animals to toxic substances and concludes that this is due to the lower intercellular oxidation found in older tissues. We have emphasized the reduced enzymatic activity and nuclear energy exchanges resulting from accumulation of excess protomorphogens in the media. Ruzicka (1922) reports that older cells have a pH nearer the isoelectric point. We have discussed this change in intercellular pH consequent to protomorphogen increase in the media.

The increase of fibrils and connective tissue elements as age progresses has been recognized for many years (Minot, 1908). Weidman (1939) reports that cutaneous ageing is associated with altered elastic tissue and probably with hyperplasia of fibrous tissue (local only). In Chapter 5 we reviewed evidence that the secretion of protomorphogens into the tissue fluids leads to a hyperplasia of fibrous tissue (it being precipitated following the thromboplastic activity of protomorphogen). We also analyzed reports that protomorphogen discharged into the tissue fluids was adsorbed on elastic and fibrous tissue. This adsorption probably would account for the ultimate degeneration of the elastic tissue. Blumenthal (1945) has analyzed the ageing processes in the endocrine glands and remarks that the stroma of these glands progressively changes from a loose fibrillar to a hyaline-fibrous tissue. Reference to Chapter 5 will review the evidence that excessive protomorphogen accumulations cause a hyalinization of local tissues.

Further indication that the degenerative changes associated with age are due to the accumulation of metabolic products (protomorphogens) is supplied by the observation that the retardation of growth by lowered calorific intake prolongs the life of rats by simply lengthening the various life cycles and lowering the metabolism. Lowry and associates (1942) conclude from these experiments that the life span is not dependent upon any definite time but rather upon a certain sum total quantity of metabolism.

To sum up the evidence which leads us to suggest that senescence in metazoan organisms is a natural consequence of the gradual accumulation of protomorphogens surrounding the tissues:

1. The primary index of senescence, i.e., the loss of the vitality of the cells and their ability to regenerate is experimentally demonstrated in culture

265

studies to be caused by the accumulation of protomorphogens in the media.

2. The primary index of senescence is the same in metazoan organisms and is likely caused by the same condition, i.e., accumulation of protomorphogens in the pericellular tissues and fluids surrounding cells.

3. There is evidence of the accumulation of growth-inhibiting substances in the blood and tissues of aged animals and of the "polymerization" of colloids, both characteristics of which are seen in the media of an ageing culture.

4. The altered enzymatic activity in ageing metazoan cells (as indicated by changes in the oxidative rate and intracellular pH) has been seen (Chapter 3) to be a characteristic of single cell cultures following an increased concentration of protomorphogen in the medium.

5. The degenerative changes in elastic tissue and hyperplasia of local areas of fibrous tissue have been presented (Chapter 5) as end results of protomorphogen activity.

The Cycle in Metazoan Senescence.—In Chapter 5 we discussed various physiological systems whose function appears to be the elimination of the protomorphogens released from the cells. It was not possible for us to establish the exact relationships between these various functions due to a lack of sufficient experimental data. It appears that all these eliminative systems cooperate in the excretion of protomorphogens. Their progressive impairment results in the changes which are associated with senescence.

It is unlikely that any one of the eliminative systems has a monopoly on this process. Nature has evolved multiple mechanisms in biology for the performance of the most important functions, and the impairment of one usually results in a compensatory activity on the part of another. As an example we discussed the compensatory elimination of protomorphogens by the kidney as a consequence of liver damage and impaired bile secretion.

The anterior pituitary, however, seems to assume a position of primary importance in the control of senescence. Sajous (1911) has called it a "test organ" because it seems to hold the reins which control the secretion of practically all the other endocrine organs by reacting to blood changes. Evans (1935) reports that administrations of anterior pituitary growth hormone impairs the development of the degenerative tissue changes associated with old age but strangely enough this does not represent a

266

lengthening life span, but quite the reverse. Apparently the depolymerizing influence of the growth hormone accelerates metabolic processes, and this activity is quite apart from its influence over the immune-biological system which, perhaps, is the most significant as far as senescence is concerned. It is significant that the anterior pituitary is involved in progeria, a condition of premature senescence. Also significant is the report that the anterior pituitary growth principle, while of benefit in stimulating regeneration and healing in adults, has no stimulating effect in children (Barker, 1922). If the anterior pituitary were concerned with protomorphogen disposal, it would naturally exert a greater influence in adults whose extracellular protomorphogen assumes a greater importance in the systemic economy as age progresses.

The anterior pituitary, through its master control of other endocrines and through the medium of the adrenotropic hormone (maintaining the natural tissue antibody (see Chapter 5)), may well be the key organ in the prevention of the degenerative changes associated with old age.

Although the degenerative changes of old age are, we believe. primarily caused by the accumulation of protomorphogens in the tissues, the life span of various species is too well defined to ascribe its determination to the chance accumulation of this substance. It is more likely that the life span is determined by some key organ which is "wound up like a clock" and gradually "runs down" during the metabolic activities associated with the life cycle. The position of the pituitary in the center of the endocrine system and its evident participation in protomorphogen metabolism strongly suggests that it is the master organ whose control of the life cycle is in tum determined by heredity.

Its medium of determination for the life cycle may revolve around its control over the systems which eliminate protomorphogens, particularly its maintenance of the immune mechanism. (See Chapter 5 discussion of adrenotropic hormone influence on natural tissue antibodies.) In this respect it is interesting to note the significant association of a relatively high percentage of neutrophile polymorpholeucocytes and high total leucocyte count in rats with the longest life spans.[1]

---

1 Reich, C.: and Dunning, W. F.: "Leucocyte Level and Longevity in Rats." *Science,* 93:429-430, 1941.

Death from "old age" is usually ascribed to an involution of one or more of the vital organs, heart, liver, kidney, etc. Accumulating protomorphogens cause degeneration of various organs until one or the other finally succumbs to this influence and death ensues. It is important to keep in mind, however, that our hypothesis assumes a master control over the protomorphogen eliminative systems (probably the anterior pituitary) and this master control thus determines the life cycle within the variations determined by environmental modifications.

Specifically, the master organ determines the life cycle by its control over protomorphogen elimination and consequent regulation of the rate of accumulation. The accumulated protomorphogen, however, is the basic cause of the degenerative changes of "old age" and the inevitable death of the organism.

There may be some interesting differences between the male and female protomorphogen eliminative mechanisms. Insurance statistics[1] indicate that women have a slightly longer life expectancy than men. However the differential between the mortality rates of married and unmarried individuals shows an advantage of the married of 2:1 for males and only about 1.2:1 for females. It is possible that the synthesis of spermatozoan chromosomes offers an important special avenue of protomorphogen disposal in the male. Corresponding disposal in the female represents a relatively insignificant volume of chromatin material.

We suggest that these statistics illustrate the importance of the spermatozoa as a protomorphogen eliminating avenue in the male, the altered sexual activity in the single or widowed male being responsible for a reduced protomorphogen excretion and consequently a hastening of the ageing processes and a lowering of tissue vitality. We have not overlooked the probable influence of better food and care which is characteristic of married life, but the female differential of 1.2:1 perhaps represents the proportionate influence of these factors probably present to the same extent in the male.

It is interesting to note that the course of phenomena consequent to accumulating protomorphogens in old age seems to follow quite closely the same cycle we have illustrated in traumatic shock. In shock the cycle is sudden, dramatic and pathological, but in old age it is gradual, prosaic and normal. The difference seems to be on, primarily, of rate rather than characteristics.

1 "Marriage and Long Life." *Statistical Bulletin of the Metropolitan Life Insurance Co., New York,* 18:7-10, 1937.

Slashed to its basic essentials, the "old age" cycle may be therefore outlined:

1. Protomorphogens accumulate surrounding the cells primarily because the eliminative systems (under master organ control) are gradually but progressively impaired.

2. In addition to the direct inhibitory effect on local cells, there is a progressive irritation of the nerve termini and gradual increase of dystrophic and decrease of trophic nervous activity.

3. There is, as a result of this dystrophic influence, a progressively altered permeability of all cells and increased tendency to vasoconstriction of the capillaries.

4. This altered permeability influences the electrical potential and pH of the cell fluids and protoplasm, resulting in loss of vitality and emphasis on the destructive rather than the constructive phases of intracellular enzyme activity.

5. This process continues with progressively decreasing cell vitality and changing morphology until some vital organ succumbs and death ensues.

Morphogen Concentrations in Senescence.—The accumulation of protomorphogens associated with senescence may be further differentiated from traumatic shock phenomena in the nature of its direct influence upon local cells. In shock the release of toxic protomorphogens in high concentrations by trauma (burns, etc.) normally exerts a highly lethal influence upon the cells in the immediate neighborhood, but the main channel of the shock cycle seems rather to follow other, more extensive lines, such as irritation of nerve termini and the subsequent pattern of collapse.

In cultures of single celled organisms the high media concentration of protomorphogen also exerts a highly lethal influence on the cells inhabiting the culture by preventing the excretion of intracellular protomorphogens. Senescence in these cultures can be traced directly to this influence. It may be easily removed and the culture rejuvenated by washing the cells and transferring them to a fresh media.

Unfortunately the situation is not quite so simple in the metazoan animal organism. Various types of "blood purification" schemes have been tried with moderate degrees of success in rejuvenation. In one case the blood of a senile dog was removed and washed, the serum being replaced with normal saline solution. The dog was remarkably rejuvenated according to the reports of the investigators but the change was only temporary and after a period of a few weeks the conditions of senescence had all returned.

It is evident that in such experiments the protomorphogens accumulated in the tissue fluids and blood serum have been considerably reduced. In consequence, the deleterious influence of tissue fluid protomorphogens has been removed. From the temporary nature of such a stimulus it is evident that the concentration of protomorphogens which has accumulated during the lifetime of the individual has not been materially reduced.

The crux of this problem is manifest in the tremendous adsorbing power of connective tissue for protomorphogens excreted from cells. This property, which we have extensively discussed elsewhere, makes the connective tissue a vast storehouse of accumulated protomorphogens. This increases progressively with age.

Normally, elutogenic factors (i.e., sex hormones) remove the protomorphogens from the connective tissue storehouse into the tissue fluids or blood serum where they are either utilized for tissue repair or excreted from the organism by the mechanisms outlined in Chapter 5. As age increases the organs responsible for elutogenic action (gonads, thyroid) and those responsible for excretion (kidneys, liver, spleen, etc.) progressively regress and the whole eliminative cycle is thrown out of balance. (These organs in themselves regress primarily due to protomorphogen accumulations in their locale-for instance MacNider (1945) has shown that the stainable lipoid in the kidney progressively increases with age-this stainable lipoid probably representing protomorphogen accumulations.)

There is thus a general damming up of protomorphogen throughout the organism which impairs the vitality of all tissues. Of more importance, however, is the disordered balance in the eliminative cycle, resulting in overconcentration in some areas and underconcentration in others. In many cases for instance, the senile individual is lacking primarily the elutogenic substances which remove protomorphogens from the connective tissue storehouses.

In such an individual there may actually be a deficiency of protomorphogens in the tissue fluids to allow regenerative and healing processes.

In such cases the administration of homologous protomorphogens or elutogens may be beneficial provided the other healing promoting factors are available. The experiments of Reissner 1941, 1942) reviewed earlier in this chapter are possibly a case m point.

(We might insert at this point some further speculations on the elutogenic aspects of the sex hormones. The evidence seems to rule out the possibility that progesterone is an elutogen. The female elutogens can probably be associated solely with the estrogenic fractions and the malt elutogens with the testosterone fractions. We feel that the influence of these two groups of hormones is manifested primarily by reason of their elutogenic activity but this influence is *over specific tissues.* We know that the estrogens are always present in the male and the male principle in the female. It may be that both these principles are necessary for a balanced elutogenic activity. The irreducible minimum necessary to prevent premature senility can probably be estimated as the amount of female principle normally in the male, or male principle normally in the female; there is in this a suggestion that such amounts are present in the castrate. In clinical practice therefore, there may be some substantiation for the suggestion that administration of testosterone should be accompanied by a modicum of estrogens, and vice versa.)

On the other hand, if the eliminative mechanisms such as the kidney, liver and reticulo-endothelial defenses are inadequate, the addition of elutogenic factors may be exceedingly toxic since they simply raise the blood and tissue fluid concentration further above the optimum. Excessive administration of sex hormones and thyroid in particular may be disastrous in the elderly patient for this reason.

The protomorphogen eliminative route in the metazoan animal as reviewed in Chapter 5 is a complicated and nicely balanced system which may compensate for minor inadequacies but which seems to be particularly sensitive to major imbalances which may be promoted inadvertently by various therapeutic measures.

It is unfortunate that diagnostic measures are not available which clearly indicate the protomorphogen levels in the blood and tissue fluids since these would enable the practitioner to more accurately assuage the condition of the senile patient. Perhaps the level of blood cholesterol esters may be significant since these are a part of the lipid sheathed protomorphogen molecule in the blood stream. Considerably more experimental evidence is necessary before this diagnostic information can be assembled and adequately appraised.

PREGNANCY

Pregnancy is a condition in which there is enhanced mitotic activity in the developing embryo. Such enhanced mitotic activity must necessarily result in the production of abnormal amounts of protomorphogen; these must be eliminated, otherwise the effect on the fetus and the mother would be disastrous. We shall briefly mention the indications that the protomorphogen eliminative mechanism overcompensates as a result of this additional demand and analyze other interesting associated phenomena in the light of the morphogen hypothesis.

The presence of urea (an elutogenic factor) in the allantoic and amniotic fluids, the intense thromboplastic activity of placental substance, as well as the fibrinolysin of embryo substance (see Chapter 5), all indicate that the protomorphogens produced by the dividing embryo are prevented from remaining in the embryonic locale and are, instead, passed through the placenta into the maternal circulation where they must be eliminated. The increased thromboplastic activity of blood in late pregnancy (Winternitz et al, 1941; Pickering et al, 1932) also indicates that fetal protomorphogens are excreted via the maternal circulation.

Nausea of Pregnancy.—The onset of pregnancy is often accompanied by nausea, the so-called "morning sickness." We feel that this may be caused by the additional protomorphogens eliminated by the embryo into the maternal circulation. Nausea is often associated with the symptoms of traumatic shock and is a characteristic influence of free protomorphogens. This effect is quite possibly produced as a result of the irritation of the nerve termini. It is quite likely that various emetic drugs produce their effects through protomorphogen release. (Turck reported that mustard has this influence.) Possibly the emesis is a natural defensive reaction to toxic concentrations of free proto-morphogens as protein poisons in foods.

After a period of a month or more the nausea normally disappears. This we believe is due to the appearance of a natural tissue antibody whose activity is that of promoting the elimination of the embryo protomorphogens from the maternal circulation. We have received clinical reports that the administration of vitamins, particularly vitamin C, is helpful in some cases in overcoming this nausea. This is to be anticipated inasmuch as vitamin C is a cooperator in the function of the adrenal cortex, this organ stimulating the immune-biological system (see Chapter 5).

Eclampsia of Pregnancy.—"The etiology of eclampsia constitutes one of the major mysteries of medicine" (Bodansky and Bodansky, 1940). There are many clinical observations in eclampsia; various investigators emphasize different symptoms such as the vascular spasms, pyelitis, hypertension and the tetany-like spasms. We cannot completely discuss the various observations of this condition and include all the phenomena reported, such as changes in the blood chemistry and endocrine metabolism. Such discussions are adequately presented elsewhere (Bodansky and Bodansky, 1940) . We shall confine ourselves to a discussion of eclampsia relative to the morphogen hypothesis[1].

Robertson (1924) states that the application of creatine to the motor areas of the cortex throw an animal into convulsions, although it is devoid of stimulating effect on nerve fibers. He and others have commented upon the sharp rise in urinary creatine excretion which precedes eclamptic convulsions. Creatine is methyl-guanido-acetic acid and a normal constituent of muscle tissue. The above effect is probably due to the guanidine released from creatine by contact with cerebral enzymes. We have mentioned that guanidine is probably an end product of nucleoprotein (protomorphogen) hydrolysis. Guanidine and methyl-guanidine have a powerful effect upon neuromuscular tissue, producing intense twitching and tetanic spasms.

The increased protomorphogen in the maternal blood in late pregnancy should be carefully investigated in view of its possible relationship to blood guanidine or methyl-guanidine, with the thought that these products may be

1 Recently Schneider (1947) has demonstrated forcibly that the factor responsible for toxemia of pregnancy is thromboplastin. Inasmuch as thromboplastin is a special form of protomorphogen the intimate relationship between toxemias of pregnancy and the morphogen hypothesis is self-evident.

of primary importance m the production of tetanic spasms in eclampsia. In this case we are dealing with toxicity from protomorphogen split products rather than protomorphogens themselves as in nausea of pregnancy.

The relationship of guanidine to creatine offers some interesting speculations in diverse fields of physiology. It has been reported that the tetany following parathyroid removal is minimized if preceded by gonadectomy.

We believe the gonadal hormones promote the release of protomorphogens into the blood (elutogenic effect, see Chapter 5) to be carried to the gonadal workshop and elaborated into germ cell chromosomes. It is likely that there is a certain degree of breakdown of the protomorphogen thus carried in the blood into guanidine. Gonadectomy results in a lesser release of the protomorphogens from the connective tissue reserves; in the castrate these attached protomorphogens are probably only eliminated by being attacked *in situ* as it were, by the normal protomorphogen antibody. (Note other remarks on prostate hormone action as a protector of protomorphogen in transit.)

Beard (1943) has reviewed evidence that the parathyroid hormone increases the creatine phosphate content of muscle, and comments upon the relationship of the parathyroids to creatine phosphate metabolism. We feel that an increase in guanidine or methyl-guanidine is possibly responsible for the convulsions and tetany consequent to parathyroid removal. This whole picture suggests that the parathyroids normally fix guanidine or methyl-guanidine into creatine, thereby preventing the spasm-producing effects of these products. The conversion of guanidine into methyl-guanidine would off er a valid basis for the beneficial influences of methionine and betaine (methyl donors). Their activity has been carefully studied relative to fat metabolism and creatine production. Evidently methyl donors cooperate with the parathyroids in the control of guanidine toxicity. Recently Dr. Paul Gyorgy of the University of Pennsylvania Medical School suggested at the November, 1946 meeting of the New York Academy of Sciences that a lack of choline may be an etiologic factor in eclamptic convulsions. Choline is intimately associated with the methyl donors in that the latter are precursors for it in respect to their influence over fat metabolism.

This would explain the reduced tetanic influence of parathyroid extirpation after gonadectomy, assuming that the gonads maintain the blood guanidine or methyl-guanidine as a consequence of their elutogenic influence. By the same token administrations of sex hormones or of other elutogenic factors would probably be more toxic in the absence of the parathyroid principle or deficiency of methyl donors.

Summarizing these suggestions: (1) the eclamptic convulsions of lat pregnancy may be caused by the increase in blood guanidine consequent to the overloading of the protomorphogen eliminative mechanisms, guanidine being a derivative of nucleoproteins and thereby associated with protomorphogens; (2) parathyroid hormone catalyzes the synthesis of creatine phosphate, utilizing blood methyl-guanidine, and (3) this whole physiological system indicates another necessity for methyl groups.

These postulations suggest many experimental investigations: (1) blood guanidine should be checked in eclamptic and tetanic convulsions; (2) the influence of the administration of elutogenic factors, such as sex hormones and thyroid, should be checked in parathyroidectomized animals, and (3) the influence of parathyroid hormone and methyl donors on eclampsia should be investigated for possible beneficial action.

The "Rejuvenation" of Pregnancy.—Pregnancy has often been reported by various clinicians to sometimes result in a "rejuvenation" of the mother. A more careful study of these observations might uncover some interesting data. It would be interesting to find, for instance, the relative mortality rates of child bearing and non-child bearing married women, or the relative mortality of those women who complete successful pregnancies in later years.

The morphogen hypothesis establishes theoretical bases for these miscellaneous clinical observations. In our discussion of senescence we emphasized as the basic etiological factor the rate at which protomorphogens accumulate in the tissues. Although we realize that this is primarily under the control of a master organ which establishes a fairly constant "life cycle" for the various species, minor variations in the rate of accumulation due to alterations in the excretory systems can probably vary the rate of the ageing process within certain limits.

We suspect that the "rejuvenating" influence of pregnancy is a direct reflection of the enhancement of the protomorphogen eliminating systems consequent to the pressure of fetal protomorphogen. We can list several systems of protomorphogen elimination which might easily be stimulated during pregnancy.

1. The natural tissue antibody: the apparent arresting of nausea of pregnancy after a preliminary latent period we believe is due to enhanced activity of the natural tissue antibody, stimulated by the antigenic activity of embryonic protomorphogens in the maternal circulation.

2. The elutogenic factors are greatly increased during pregnancy (sex hormones, guanidine, urea, etc.).

3. Large amounts of "embryo growth principles" are delivered to the mother, these factors containing fibrinolysins (elutogens) and other protective influences which tend to keep the protomorphogens from exerting toxic effects and aid in their elimination.

The enhanced protomorphogen excretion as a result of the stimulation of these systems during pregnancy probably continues for awhile after parturition (particularly the augmented natural tissue antibody) and the inhibition of protomorphogen accumulation during this period might well exert a "rejuvenating" influence.

Varicosities Accompanying Pregnancy.—Pregnancy is often accompanied by varicosities, especially hemorrhoids, whose etiology is usually ascribed to "pressure." While mechanical pressure may be an important consideration in the etiology of these conditions, we believe other factors should also be considered.

We have received some clinical evidence that hemorrhoids are adversely affected by the administration of bile salts. Constipation has been considered a cause of hemorrhoids, and it is significant that it may of ten be accompanied by intestinal toxemias which as we have previously noted may result in the reabsorption of bile toxins into the blood. Bile secretion is increased in pregnancy.

We have previously considered evidence that bile is an important avenue of protomorphogen excretion. In pregnancy the blood is overloaded with protomorphogens consequent to fetal excretion into the maternal circulation. Pregnant blood has been shown to be highly susceptible to coagulation, and its platelet count is high.

The existence of a high protomorphogen (thromboplastic) content in the blood and the observation of its hypercoagulability is suggestion enough that the blood in the capillaries under local pressure and faced with stagnation would exert a powerful dystrophic influence o these vessels.

CANCER

Cancer is a disease which has probably received more attention from investigators than any other malady which affects mankind. Volumes have been published even in the past few years which would require a lifetime to competently study and analyze. The exact etiology of this condition remains an enigma and consequently no really effective therapeutic measures have been forthcoming (except radiation and surgery, both drastic procedures). There are as many suggested methods of treatment and prevention as there are hypotheses of cancer and each method of treatment has met with some measure of success.

It is far from our purpose to attempt even a cursory analysis of the cancer problem. There are many excellent reviews on both the whole picture and on various angles of the question available in the literature. No purpose can be served by adding another in this volume. However, any hypothesis, as comprehensive as we claim the morphogen hypothesis to be, would necessarily consider a disease in which the primary observation is that of disordered cell morphology and metabolism. We shall, therefore, review only those parts of the cancer problem which we believe are concerned with the morphogen hypothesis, leaving the irrelevant material to more competent reviewers.

Local Protomorphogen Concentration in Cancer Areas.—Burrows (1927) concludes that "Cancer develops from anything which crowds the cells and allows them to develop and maintain a high concentration of the archusia independent of the blood stream." (Reference to previous chapters will recall evidence that Burrow's archusia is protomorphogen.) He has also commented that the hyaline degeneration and precipitation preceding cancer is a consequence of an increased local concentration of archusia (protomorphogen). Burrows' conclusions and investigations lend support to the hypothesis that cancer is preceded and accompanied by an intense local concentration of what we term protomorphogen.

We should call attention to Burrows' (1927) investigations of the lecithin-like "wrapping" of the protomorphogen molecule which he terms "ergusia."[1] This substance is of a fatty nature produced by and intimately associated with archusia. Burrows makes the statement that "Cancer, as has been pointed out in previous papers, is the result of anything which acts to remove the excess of the ergusia from a local area of tissues." The reader will recall our discussion in Chapter 5 in which we consider archusia to be protomorphogen and ergusia to be the lipoid "wrapper" which associates itself with protomorphogen, reducing its untoward influence on neighboring cells.

Further Evidence of Anti-Cancer Influence of Phosphatides.— Burrows' conclusions on the potency of his ergusia in preventing the carcinogenic stimulus of intense local concentrations of archusia (protomorphogen) have received support from investigations indicating the anti-cancer and anti-protomorphogen influence of the phosphatides.

Robertson (1924), author of the allelocatalyst conception that has been such an important link in the morphogen hypothesis, has investigated the influence of phosphatides and cholesterol on his allelocatalyst. (The reader will recall our contention that Robertson's allelocatalyst, Burrows' archusia and Turck's cytost are all different names for what we are referring to as protomorphogen.) He reports that the administration of lecithin retards development at the early stages of embryonic life while it enhances growth and development in later stages. He concludes, therefore, that the lecithin exerts a "solvent" action upon protomorphogen. During the early growth stages in which the protomorphogen is in low concentration, its beneficial effects are adversely affected in that it is withdrawn into an external lipoidal phase. Later, when the concentration of protomorphogen passes the critical state and becomes inhibitory to growth, its removal by lecithin tends to restore the natural balance and promote growth.

---

1 Burrows considered "ergusia" as a substance complete and distinct from "archusia." In Chapter 5 we suggested that his "archusia" is "raw" protomorphogen and "ergusia" is protomorphogen associated with a lipoid "wrapper." In this discussion of cancer we use the word "ergusia" rather loosely to refer to the lipoid "wrapper" free of protomorphogen. (NOTE: We have included much bibliographic material of Drs. Burrows and Jorstad not specifically referred to in the text with the thought that the reader might desire to more extensively study this work.)

Robertson also comments that the influence of cholesterol is exactly opposite to that of lecithin, possibly by reason of its solvent effect on the natural lecithin "wrapper" material. He also reports that injections of lecithin have retarded the growth of carcinoma while cholesterol administration stimulates cancerous growths. Many investigations have since been published which lend credence to Robertson's conclusions. Haven and Levy (1942) report disorders in the phospholipides of tumor cells which indicate that lecithin becomes concentrated in the nuclei. This suggests that cancer cells are not able to supply the lecithin (ergusia) to the media necessary to prevent the untoward influence of the protomorphogens in the tissue fluids. Beard (1935) comments that lipoid and especially cholesterol is increased in malignant tumors. According to Robertson's conclusions this cholesterol would exert a solvent influence on the lecithin, preventing it from "wrapping" excess intercellular protomorphogens. Stern (1941) emphasizes the tumor inhibiting influence of lecithin as opposed to the stimulating influence of cholesterol. Oberling (1944) has also reviewed the cancer stimulating influence of cholesterol and inhibiting ability of lecithins. We feel that this "antagonism" is also exhibited as a consequence of the supposition that cholesterol adsorbs protomorphogen, forming a fixed monolayer: This "activates" the protomorphogen because of the surface exposed. Lecithin, by sheathing this monolayer, "inhibits" the protomorphogen.

Burrows and Jorstad (1926) concluded that the lecithin-like material which combines with protomorphogen (ergusia) is identical or similar to vitamin A; this similarity is discussed in more detail in Chapter 5 of this volume. The anti-cancer influence of vitamin A is too well established to warrant a detailed consideration here. Burrows emphasizes that the hyalinization and keratinization of tissue that precedes cancer is also observed in vitamin A deficiency. Jorstad (1925) was one of the first to report that carcinogenic tars exert a more pronounced influence in vitamin A deficient animals.

Rosenberg (1942) has stated that there is an increase in purines in growing vitamin A depleted tissue after the administration of vitamin A. He further comments that all the primary and secondary symptoms of vitamin A deficiency may be explained by reason of this influence. Purines, as derivatives of nucleoproteins, may represent a link with protomorphogen molecules and it is our opinion that the vitamin A is associated with them in the nature of promoting the protective association with the phospholipide sheath.

Vitamin E has also been reported by Davidson (1937) to retard the development of cancer. Many other investigators have reported the same. In view of the influence of vitamin E in the processes of cell maturation and differentiation (Rosenberg, 1942) as distinguished strictly from mitosis, it is likely that this principle also is concerned with the sheathing of the morphogen molecules. Bomskov and v. Kaulla (1941) have reviewed the vitamin E picture and assumed that vitamin E is concerned primarily with embryonic development. Rosenberg reviews a significant experiment in which it is concluded that in vitamin E deficiency there appears a lysis of the germinal chromatin and morbid morphology of the germinal cells. It is apparent, therefore, that both vitamin A and E are involved in the protective association of protomorphogens with lipid sheaths, and both also are effective in retarding the development of experimental cancer.

We should also mention Hanson's (1930) reports that thymus administration was beneficial in several cancer cases. The thymus is known as the center of the lymph system and has been discussed in Chapter 5 with reference to protomorphogen sheathing metabolism.

Of singular interest is the recent announcement by Drs. Salmon and Copeland of the Alabama Agricultural Experiment Station that a specific dietary deficiency of choline results in a high percentage of cancer in experimental animals. Choline is a vitally important dietary precursor of phosphatides and, as such, is most essential for the normal sheathing processes by which protomorphogens are rendered innocuous.

To review, the original contention of Burrows that ergusia is a lecithin-like substance which prevents local concentrations of protomorphogen from becoming carcinogenic seems to be well supported by subsequent investigations which demonstrate the anticancer influence of lecithin and vitamin A and the cancer stimulating effects of cholesterol.

Carcinogens and the Irritation Hypothesis.—Burrows (1927) calls attention to the observation that the so-called carcinogenic influences all can be considered as factors which tend to eliminate ergusia from tissue fluids (ergusia being the lipoid-like "wrapper" for protomorphogen.) Coal tar lipoid solvent carcinogens are all solvents for the lipoid wrapper of protomorphogen Burrows, Jorstad and Ernst (1927) have demonstrated that carcinogenic x-rays protect against other symptoms of vitamin A deficiency, and their toxic effect is prevented by vitamin feeding. The intimation from these experiments is that the carcinogenic x-rays remove vitamin A from the tissues where it may be preventing protomorphogen toxicity and makes it available for other metabolic purposes.

Shabad (1946) has reviewed Soviet research in cancer and commented upon the fact that it is not necessarily the inflammation at the local point of administration which gives rise to the blastomogenic action, but rather some inherent activity of the chemical carcinogen itself that promotes the formation of neoplasms in various locales.

It is significant that the carcinogenic hydrocarbons are among the most powerful inductors of embryonic development (Refer to discussion in Chapter 4 of this volume). Needham (1942) reviews this problem in detail and evidence is available that these carcinogenic hydrocarbons induce embryonic development by means of facilitating the release of protomorphogen from a "bound" or "combined" form. Burrows' hypothesis that the carcinogenic substances remove the lecithin-like barrier releasing a local concentration of active protomorphogens seems to have a sound experimental basis.

The irritation concept of cancer etiology thus assumes new importance in our thinking. This concept and the experimental evidence accumulated in its support should receive a careful review. This re-investigation might very probably bring to light more evidence that all carcinogenic irritation reacts simply to reduce the tissue content of the lecithin-like (ergusia)"wrapper"which prevents the carcinogenic influence of concentrated local accumulations of protomorphogen from becoming manifest; such irritation also may produce local inflammation with increased free protomorphogen activity.

Immune Theory of Cancer Defense.—It is of greatest significance that an organism's susceptibility to transferable cancer is in ratio to its susceptibility to heterologous transplants. There is a natural immune system which prevents the intrusion of foreign protein molecules and also disposes of the tissue fragments and protomorphogen of the animal's own tissue. Whether the collapse of this mechanism is basically responsible for cancer or not, it is obvious hat its function must be impaired for cancer to gain a foothold in an organism.

Lumsden (1931) emphasizes the importance of the natural tissue antibody (discussed in Chapter 5) and the immune system in general in the protection against cancer invasion.* He concludes that there are two protective immune factors, one in the serum and one associated with the leucocytes. Jacolsen (1934) believes that cancer is a result of chronic "irritation" of the reticulo-endothelial system. Jaffe (1927) reports the conclusions of Erdmann that "irritation" of the reticulo-endothelial system makes possible a transplantation of tumors by killed tumor cells or filtrates. Arons and Sokoloff (1939) have concluded that the resistance to transplantable tumors and malignant disease in general is broken down by blocking the reticulo-endothelial system.

Russ and Scott (1945) have demonstrated that cell-free tumor fluid stimulates tumor growth in normal rats; but if extracted from a shrinking tumor it produces the opposite effect, indicating a local immune response responsible for the regression of the tumor. Murphy, Maisin and Sturm (1923) found that moderate doses of x-ray inhibited cancer but not as a result of their direct effect on the cancer cells. Murphy (1935) comments that this effect is probably associated with the activity of the lymph cells.

We have discussed the activity of the reticulo-endothelial system, lymph and immune-biological system and their relationship to removal of protomorphogens in detail in Chapter 5. *Stimulation of protomorphogen removal is beneficial in cancer, and inhibition of this removal is detrimental.* (The reticulo-endothelial system has a general anti-toxic function. How many of these toxins may be protomorphogens and their split products, and how many from other sources we are not prepared to say.)

* Recently Dr. Shapiro of Brooklyn has reminded investigators of the importance of Lumsden's significant work which involves the autoimmunization and formation of cytotoxic antitumor antibodies. (*J. A. M . A .*, 134:1042, 1947.)

Depolymerizing Influences in Cancer.—Attempts to induce immunity to specific cancer strains or to produce immune bodies to the same have been singularly disappointing. Recently, however, Gamaleia (1946) has reported some success in this direction in his review of Soviet work in this field. Clinical immunization seems to have been successful when fresh metastatic centers of the tumor are wrapped in a piece of the omentum and sutured under the skin or in the abdomen to act as an autovaccine.

We wish to speculate that the antigen in this case may be separated by diffusion from a substance which impairs its antigenic activity.

There are three attributes of the cancer cell which are manifest in tissue cultures and are of singular importance in this respect: (1) cancer cells exhibit no lag period when transferred; (2) an isolated sarcoma cell will initiate mitosis and grow into a colony while a normal cell will not, and (3) cancer tissue contains an intensely active growth stimulator which does not require ether extraction to exhibit its properties as does normal adult tissue. From these facts we can deduce two important concepts: (1) cancer tissue contains a powerful depolymerizer which prevents the accumulation of polymerized protomorphogen, and (2) cancer protomorphogen is normally not sheathed in lipoids.

The second concept we have already discussed relative to the suggestions advanced by Burrows. The first may be integrated into a hypothesis with startling ramifications.

There is other evidence that there is a depolymerizing factor in cancer tissue. Orekhovich (1944) reports that the muscle of sarcomatous rats was split more easily by tumor enzymes than that of normal animals after about 19 days from the transplant. This may represent the time necessary for the cancer depolymerizing factor to become available in all parts of the organism and exert its catalytic effect in cooperation with tumor enzymes. On the other hand the protein of the skin and blood of sarcomatous animals became more resistant to enzymatic destruction than that of normal animals. Note that these tissues are those in which mitosis continues throughout life and is not restricted after parturition as in muscle.

Apparently there is a cooperative factor in the epithelial cells which prevents the cancer depolymerizing substance from exerting a destructive influence.

Let us postulate the existence of a powerful depolymerizing factor present in cancer and also present in all embryonic tissue, but in embryonic tissue it is balanced by a cooperative synergist which prevents the depolymerizer from breaking the tissue down to the point of "competence" to receive new morphogens. (Review the significance of this word "competence" in Chapter 4.) In short, the cooperative factor prevents the complete depolymerizing of the morphogen determinants in the tissue. It is interesting to note the analysis of Kline and Rusch (1944) in which they suggest that the initial phase in tumor formation is a "period of induction" which we may compare with a similar period in the embryo during which *competent* embryonic cells receive their morphogen determinants.

Under the spell of this hypothetical depolymerizing substance, and in the absence of the natural embryonic balancer, a normal cell could conceivably be deprived of its adult morphogen determinants, regress back to its primordial low-organization state, and multiply as a neoplasm of this type of cell. Many investigators have commented on the appearance of this type of normal, but low-organization, cell in tumors. This type of cell could conceivably regain its embryonic competence and thus under the influence of mutation promoting substances such as viruses develop into a new type which could then reproduce as a tumor or malignant metastasis.

If such a depolymerizing substance were present, it could conceivably so degrade morphogens that they would no longer be adequate antigens; thus it is unlikely that in its presence a natural tissue antibody to the cancer cell could be produced, restricting its growth. We are speculating that in Gamaleia's experiments this depolymerizer has been separated from the cancer morphogen allowing the latter to function successfully as an antigen.

Antibodies do not appear in the blood streams of children until some time after parturition; even the natural tissue antibody does not appear for several weeks. This gives rise to the possibility that in the embryo this depolymerizer, although associated with a cooperating factor, nevertheless prevents the formation of antibodies. In view of our contention that the primary mode of estrogen activity is elutogenic activity and consequent release of protomorphogens it is interesting to note that multiple dosages of estrogens have no influence in animals a few days from parturition.[1]

Admittedly, these thoughts represent, perhaps, the more unrestricted type of speculation but we feel that if there be any germ of thought in them giving rise to some experimental investigation along new lines, certainly no harm will come from making them available to scientific investigators in this field.

Mutation Theory of Cancer.—The fact that antibodies to cancer tissue may exist is proof of the appearance in some cancer tissue of an antigenically new substance. Kidd (1944), for instance, has demonstrated the suppression of the Brown-Pearce carcinoma by a specific antibody. This antibody reacts with a macromolecular sedimentable constituent of the cancer tissue which he feels may play a part in the proliferative activities of the tumor cell. Whether the new antigen is an external parasite (virus hypothesis) or a "mutated" gene (mutation hypothesis) is the subject of an intensive biological investigation. In this brief discussion we shall consider it as a "mutated" gene and present later a conception which envisions it as also a transmittable virus.

Murphy (1935) has suggested the existence of what he terms a "transmittable mutagen," a thermolabile carcinogenic substance which is inhibited by a more thermostable"inhibitor."The "transmittable mutagen" exhibits no antigenic activity and is considered as a substance that induces the changes in cells we know as cancerous. He considers the carcinogenic activity of a tissue extract to be in ratio to the relative amounts of the "mutagen" and its "inhibitor."

Haddow (1944), on the other hand, advances the theory that a cytoplasmic determinant may transfer malignancy to susceptible cells and result in the continuous production of the altered character including its hereditary maintenance. He reviews the evidence suggesting the existence of such a cytoplasmic determinant consisting of nucleoproteins associated with phospholipides. He recognizes the fact that many carcinogens produce a genetic change in the tissues which does not require the presence of the carcinogen for its maintenance. He suggests that tumor virus may alter the genetic constitution of the host's cytoplas-

1 Dempsey, E. W.: "Metabolic Functions of the Endocrine Glands." *Ann. Rev. Physiol.,* 8:451-466, 1946.

mic determinant, initiating a cancerous condition which is self-supporting. Shabad (1946) reports that the carcinogenic factor from filtrable chicken sarcoma is a cytogenic protein rather than an exogenous viable component, lending some support to this contention.

Many of the phenomena associated with this problem are analogous to those met in a study of embryonic differentiation and the observations of experiments with embryonic transplants. Indeed, the differentiation or de-differentiation of normal cells into tumor types no doubt follows the same outline of biological activity as those differentiating in the embryo under experimental conditions of transplants, inductors, etc.

In our review of the embryonic problem and its relationship to the morphogen hypothesis we emphasized that inductor substances may be placed in two general classifications: (1) those of a determinant nature which in themselves cause morphogenic differentiations and transformations in *competent* cells, and (2) those of a non-determinant nature (this includes carcinogens) which unmask the determinant morphogens already present, making possible their active influence over morphology. It seems that the same two general classifications may apply to the cancer problem.

Murphy (1935) outlines the problem and lists both the chemical type of carcinogen and also a cell-free extract of an osteosarcoma which induces the growth and differentiation of local cells. Meissel (1944) describes the new heritable forms which appear following prolonged cultivation *in vitro* in the presence of chemical type carcinogens. Wachtel (1946) describes lipid extracts from cancer which induce cancer when injected into other animals. Meissel is probably describing the influence of a chemical carcinogen in making possible the activity of a determinant morphogen and Wachtel may be describing a morphogen determinant principle itself.

In order for embryonic transplants to become effective in inducing differentiation in the host of the nature of the cells of the donor the following conditions must be met (it seems to follow that the same circumstances must be present in order for a cancerous differentiation or de-differentiation to arise) (see Chapter 4): (1) the graft m6rphogens (normal in the case of the embryo and pathological "mutagens" or determinant viruses in the case of cancer) must be stronger and more effective than the morphogens present in the host's tissue, and (2) the host's tissue must be of such nature that it is competent to receive and act upon the directions of the new morphogen.

The natural tissue antibody, specific antibodies, sheathing lipoids, adequate circulation to carry away foreign morphogenic influences, all tend to prevent the first situation, i.e., they prevent the donor morphogens from becoming more effective than the host's. All of these factors are experimentally demonstrated to retard the development of cancer. Their relative strength probably accounts for the percentage of "takes" in transplanted cancer.

The second condition, i.e., that of competent host's cells, we have already discussed under the sub-title of a general depolymerizing substance associated with cancer.

There is considerable evidence that mutations may occur in biological systems, especially in cancer. Regaud, Gricouroff and Villela (1933) report that the precancerous metaplasia of the epithelium of the cervical canal is preceded by the establishment of a new type "hybrid" cell of mixed characters. Potter (1945) suggests that cancer protein may arise spontaneously as the result of a mutation or introduction of a virus. Manwaring (1934) has reviewed the environmental mutations occurring in bacteria which are initiated as a consequence of unfavorable culture conditions, especially stagnation. A stagnant culture is one in which there is a lethal excess of protomorphogen in the medium. The fact that such an environment is conducive to mutation is significant in view of Burrows' comments that cancer is preceded by a local intense concentration of protomorphogen unprotected by a lecithin "wrapping." We feel that the evidence is sufficiently complete to suggest the following hypothesis: (1) a cancer cell has a new genetic system due to a "mutation" of the cytomorphogen in the cell, or the introduction of a new abnormal morphogen; (2) the cytomorphogen "mutation" may be initiated by either an intense local concentration of free protomorphogens not protected by lecithin-like "wrappers" or induced by a specific virus, and (3) irritation, carcinogenic hydrocarbons and x-ray induce cancer because they either cause an intense local concentration of protomorphogens or they remove the lecithin-like "wrappers" which prevent the untoward effects of free protomorphogens.

Oberling (1944) in a masterly review and exposition of the virus theory of cancer objects to the mutation hypothesis on the following basis: (1) he explains the activity of carcinogens by suggesting that they prepare the cells for the advent of viruses (2) he says there is no need for the mutation conception since tumors can be explained by viruses without recourse to mutation, and (3) the fact that the onset of cancer is slow while mutations occur suddenly.

These objections may be met as follows: (1) we have outlined good grounds for the suggestion that carcinogens remove the lecithin-like "protectors" from local protomorphogen inducing precancerous conditions; (2) it is difficult to explain Burrows' observations that intense local concentrations of protomorphogen lead to cancerous conditions by means of the virus theory alone, and (3) we feel that the protomorphogen concentrations preceding cancerous conditions are latent in their development and may prepare the way for the advent of the virus which will stimulate the mutation, or the activity of the virus itself in stimulating mutation might be a complicated and time-consuming mechanism.

There is even a possibility that the virus in cancer is the cytomorphogen or organizer itself that has undergone mutation. Oberling objects to this on the following grounds: (1) organizers act only upon embryonic "competent" tissue while cancer virus acts upon adult tissue; (2) organizers differentiate cells while cancer viruses de-differentiate them; (3) organizers are not supposed to multiply in cells while cancer viruses apparently do; (4) organizers are not destroyed by boiling while sarcomatogenic agents are destroyed at 65 degrees C., and (5) several factors can induce organization (see Chapter 3) while there is only one virus for each specific cancer.

In the light of the morphogen hypothesis these objections can all be discussed as follows: (1) it is quite possible, indeed likely, that precancerous tissue approaches the "competency" of embryonic tissue (under the influence of the cancer depolymerizing substance we have postulated); (2) although organizers specifically differentiate cells, they are under the chromosome control in doing so, and quite likely would "run wild" in producing the specific and perhaps regressed types of cells found in cancer; (3) we have presented evidence in Chapter 3 that organizers (morphogens) reproduced themselves in the nucleus of the cell whose morphology they determine, indeed, this is the only place they do reproduce; (4) the heat stability of organizers or morphogens cannot be used as a reliable guide to their identification, as we have shown, in that it varies exceedingly, depending upon the complexity of the organizer molecule, and (5) we have emphasized that the complexity of factors inducing embryonic organization is due to the fact that while there is only one specific morphogen or "organizer" *per se* for each cell type, many factors, i.e., the carcinogenic hydrocarbons, may stimulate their appearance and activity.

Metabolic Changes in the Cancer Cell.—The biochemistry of the cancer cell is a subject embracing too much material for us to adequately discuss it here. We simply wish to present enough data to indicate that the morphogen metabolic cycle and some of the standard controls of morphogen elimination presented in this volume are considerably changed in the cancer cell.

First, it becomes necessary to clarify Burrows' original contention that cancer may develop as a consequence of an intense local concentration of "raw" protomorphogen (archusia). Active cancer tissue is, of course, in the process of intense mitotic activity. How, then, are we to reconcile the existence of this intense mitotic activity with the high extracellular concentration of protomorphogens which we know inhibits the mitotic index?

There are three avenues of explanation for this apparent paradox. First, if the cytomorphogens of the cancer cell have undergone "mutation," as we suggest, it is possible that the laws of normal protomorphogen growth control no longer are applicable, secondly, the morphological structure of an active cancer very often exhibits a necrotic center, the mitosis occurring at the boundaries, and third, factors quite possibly exist in cancer tissue which prevent the normal polymerization of protomorphogen. (We must remember the manner in which extracellular protomorphogen inhibits mitosis. The inhibition *per se* is a result of an inability of the cell to secrete protomorphogen from its protoplasm and the direct toxic influence of this protoplasmic concentration. This inability to secrete is a consequence of polymerization self-stimulated by concentrated protomorphogen in the pericellular fluids.)

We shall briefly discuss these possibilities. The change in protomorphogen controls as a result of "mutation" is suggested by the report of Cadness and Wolf (1934) that the serum fibrinolysin of normal

289

persons is different in specificity from that of cancer patients. The difference in genetic structure is indicated by Biesele's (1945) observation of a chromosomal enlargement in rabbit uterine adenocarcinoma. Also, cells of certain neoplasms show characteristic mineral disposition quite unlike the same type in the normal tissue (Scott, 1943).

The altered morphogen control in cancer is suggested by the observations of Crile, Telkes and Rowland (1931) that lipoids from cancer protein were not effective in forming normal autosynthetic cells but gave, rather, bizarre structures with many fatty droplets. The reader will recall our discussion of these experiments in Chapter 3 in which it was indicated that the fat component of the auto-synthetic cell substrate carried the protomorphogen or "organizer" which made possible the assemblage of a "synthetic" cell. Obviously the protomorphogen from cancer tissue is not found in the fatty residue as is normal protomorphogen, indicating a disordered protomorphogen control. Doljanski and Hoffmann (1940) have reported that cancer cells exhibit an extremely short or no lag period when explanted in vitro. In our discussion of lag period in Chapter 2 we reviewed evidence that it is that latent period in which excess intracellular protomorphogen can "leak" into the new media establishing a "common denominator" ratio allowing mitosis to proceed. Doljanski and Hoffmann's experiments indicate that the protomorphogen controls of cancer are altered since no latent period is observed.

The nature of the mitogenetic radiation from cancer tissue also indicates a disordered metabolism. Gurwitsch and Salkind (1928) for instance, reported that the blood of cancerous animals had lost its normal power of emanation of mitogenetic rays. Injection of a tumor extract into normal animals caused a temporary loss of this power. In our discussion of -mitogenetic rays in Chapter 3 we have reviewed their link with protomorphogen which indicates that protomorphogen is depolymerized by these emanations. The loss of mitogenetic activity in cancerous blood could conceivably be due to its abnormal absorption by the excess protomorphogen concentrations accompanying cancer.

This leads us to the discussion of the possibility of factors preventing the polymerization of the protomorphogen accumulating in cancer.

Cowdry (1924) has reviewed evidence that the growth stimulator from adult tissue is manifested only as a result of autolysis but the same (or similar) factor from cancer tissue is effective without this procedure. This suggests that this factor is "masked" by the inhibitory protomorphogen present in adult tissue but no such concentration is present in cancer tissue. This "masking" by protomorphogen is indicated by the reports reviewed in Chapter 3 that extraction of adult tissue with fat solvents enhances its growth promoting effects. We have extended the suggestion of a depolymerizing substance a few pages previous in our cancer discussion.

This suggestion of depolymerized protomorphogens in cancer tissue would seem to be invalidated by the observations of MacAlister (1936) that tumors are inhibited by local administrations of allantoin, probably a depolymerizer. (See Chapter 5 discussion of allantoin.) This does not necessarily follow. We suggest that there are factors or influences in cancer tissue which prevent the self-polymerization of intracellular protomorphogen, normally elicited by high intercellular accumulations. The intercellular accumulations are intimately concerned with the etiology of cancer (Burrows, 1926, 1927) and their removal might exert an inhibitory effect.

The possibility that the protomorphogen is secreted into the center of the tumor resulting in the necrosis often observed in that locale, but not preventing uncontrolled mitosis of the border cells needs no further discussion. It is a self-evident speculation and needs further experimental investigation and analysis.

It seems not illogical to assume therefore that there are metabolic changes in cancer tissue which allow the intracellular protomorphogen to be excreted into the substrate fluids at a normal rate, and these changes prevent the extracellular protomorphogens from effecting a polymerization of the intracellular morphogens. We have suggested various causes of these phenomena, all of which deserve further experimental consideration.

Assuming the correctness of the "mutation theory," the cancer cell morphogens can be considered antigenically different from normal morphogens. The intense local concentration of normal protomorphogen may be a constant stimulus to mutation but not inhibit the growth of the heterologous cancer cells.

Replica Hypothesis.—A significant observation on living protein is supplied by Beck (1944) who states that "In general it can be stated that the protoplasm produces and contains substances of specific asymmetry which cannot be elaborated in any non-living system." This asymmetry is of a stereochemical nature. The stereochemical specificity of normal tissue protein, molecules has long been recognized. It is manifested by the power to rotate polarized light.

Some years ago it was suggested that the proteins of cancer tissue were the stereochemical isomers of normal tissue. This has since been confirmed and denied by other investigators.[1] This has been compared to the stereochemical isomers in the proteins of the early embryo before the placental stage and the early stages of embryonic development in marsupials. These data suggest that a disorder in the stereochemical configuration of tissue is necessary for neoplastic mitosis and growth to occur.

Donovan and Woodhouse (1943) have presented what may be called a "replica hypothesis" of cellular duplication. They present evidence that mitosis is accompanied by the formation of mirror images of the nucleic acids of the chromosomes and a cleavage into stereochemical isomers. This odd stereotype then forms the normal as another mirror image, thus exactly duplicating the original. This conception is discussed by Gulland, Barker and Jordan (1944) in two respects in particular. First, they object to the emphasis upon nucleoprotein as a basic consideration in mitosis. And secondly, they maintain that no enantiomorphic nucleic acids have been recorded. To discuss these objections, first, we interpret an emphasis of nucleoprotein as an emphasis upon the known constituents of the chromosome and such discussion as referring to the whole chromosome rather than exclusively to nucleoproteins. Secondly, enantiomorphic forms, being the "pattern making molecules" would be so few in number compared to their normal isomers that it is not surprising they have not been recognized.

The replica hypothesis is also suggested in the comments of Astbury (1945) who discusses the x-ray impression that fibrous

1 Kogl:, F., and Erxleben, H.: "Chemistry of Tumors. Etiology of Malignant Tumors." *Z. physiol. Chem.,* 258:57-95, 1939. Chibnall, A. C., Reese, M. W., Tristam, G. R., Williams, E. F., and Boyland, E.: "The Glutamic Acid of Tumor Proteins. *Nature,* 144:71-72, 1939. White, J., and White, F. R.: "The Nature of the Glutamic Acid of Malignant Tumor Tissue." *J. Biol. Chem.,* 130:435-436, 1939.

proteins are constructed as a result of successive levels of organization based upon the "molecular template" action of primary determinant components.

Wilson (1928) reviews evidence on the duality (quardripartite chromosomes) of the chromosomes in the telophase and even the anaphase. There seems to be reliable evidence on the existence of this phenomenon but conflicting interpretations of it. It can be interpreted as an exhibition of the "replica hypothesis" in action during mitosis. The normal chromosomes create stereochemical isomeric "mirror images" which, in turn, create "mirror images" resulting in an exact duplication of the original. During this period the enantiomorphic "pattern making" chromosomes apparently appear, resulting in a quardripartite chromosome structure.

Sevag (1945) has discussed phenomena which seem to be linked with the "replica hypothesis." The suggestion has been made that d-antigen catalyzes the synthesis of l-antibody and l-antigen that of d-antibody. Lettre (1937) sets forth the proposition that the serologically active prosthetic group of the antibody is the mirror image of the naturally occurring l-active groups of the globulin molecule.

The "replica hypothesis" promises some interesting new conceptions of the cancer problem. An interference with the production and control of the enantiomorphic "pattern making" molecule could easily result in the uncontrolled production of determinant morphogens and the wild cell-growth associated with cancer. The proposition that stereochemical isomers are found in cancer tissue would then be interpreted as a demonstration of uncontrolled production of enantiomorphic "pattern making" genes or cytomorphogens.

Review of Pertinent Morphogen Links with Cancer.—To review our discussion of the links between cancer and the morphogen hypothesis: (1) cancer seems to be associated with an extraordinary concentration of protomorphogens in the local tissue fluids; (2) the lipoidal "wrappers" tend to prevent these accumulations from becoming carcinogenic; (3) irritation assists in the local accumulation of protomorphogens and thus may lead to cancer; (4) x-rays and the carcinogenic hydrocarbons have been shown to destroy or dissolve away the "sheathes" which protect against these protomorphogens, and thus can lead to cancer; (5) the immune-biological system is important as a protector against cancer probably because it assists in the removal of intensely concentrated protomorphogen in local areas. This is a function of the natural tissue antibody and is linked with the protection against foreign proteins; (6) there is a powerful uncontrolled depolymerizing influence in cancer tissue which tends to de-differentiate cells to the point of embryonic competence and which prevents the cancer cell morphogens from acting as antigens; (7) it seems as if the determinant cytomorphogens in cancer tissue have undergone a "mutation" giving rise to disordered genetic structure associated with neoplasms; (8) the existence of a virus in cancer is not questioned, but we suggest that it either causes the cytomorphogen "mutation" or the "mutated" cytomorphogen itself is the virus (We have previously associated the morphogens with viruses in Chapter 1); (9) the intense local concentration of protomorphogen in cancerous tissue fluids does not inhibit mitosis because: (a) cancer metabolism is considerably different than normal metabolism and protomorphogen controls are consequently altered in their nature, (b) the necrosis in the center of cancer tissue is a result of this protomorphogen accumulation which, however, does not prevent the mitosis of bordering cells, (c) there are factors present in cancer which depolymerize intracellular protomorphogen, thus preventing its accumulation in the protoplasm despite a high extracellular concentration or (d) cancer cell protomorphogen is heterologous to normal protomorphogen and thus cancer cells are not inhibited by it, and (10) the "replica hypothesis" suggests that mitosis is associated with the production of a "pattern making" stereotype of the chromosome which, in turn, produces another "mirror image" stereotype or exact duplicate of the original. In cancer the controls of the"pattern making"stereotypes are altered resulting in uncontrolled mitosis. This also indicates an altered morphogen metabolism.

OTHER DISEASES POSSIBLY ASSOCIATED WITH DERANGED MORPHOGEN METABOLISM

The scope of activity which we have envisioned in the morphogen hypothesis indicates that deranged morphogen metabolism probably is a concomitant phenomenon in many disease syndromes. We are limiting our discussion, however, to those in which there is some satisfactory experimental evidence of its link and those which suggest a possibility that deranged morphogen metabolism may be a primary etiologic factor.

Anemia.—We have reported the hypothesis that the platelet content of the blood is regulated by an autocatalytic response of the reticulo-endothelial cells in the production of a substance "thrombocytopen" which, by its influence over megakaryocytic protomorphogen, controls the activity of these platelet-forming cells.

The possibility of such an autocatalytic regulation of the erythropoietic centers should not be overlooked. Its derangement would be a fundamental factor in the etiology of anemia.

Hurwitz (1922) has suggested that a lipoid substance derived from hemolyzed erythrocytes stimulates the production of red blood cells by the bone marrow. (Erythrocytic protomorphogen would be assumed to appear in association with a lipoid substance in autolyzed red blood cells.)

Spleen, liver and marrow extracts have been reported by many investigators to stimulate erythropoiesis. Leake and Leake (1924) consider that spleen and marrow act in synergism in this respect, stimulating the rate of erythrocyte production and causing a hyperplasia of the bone marrow. Downs and Eddy (1922) have demonstrated the influence of protein free spleen extracts in stimulating the number of circulating red blood corpuscles. Leake and Bacon (1924) report that the spleen-marrow product is thermostable, water soluble and inactivated by alcohol and ether. They do not deny that phosphatides may be an important constituent.

Leake (1924) has also demonstrated an initial moderate anemia following the extirpation of the spleen. Recently White and Dougherty (1945) have reported the erythropoietic influence of adrenotropic hormone and commented upon the trend towards anemia in adrenalectomized animals.

The anti-anemia factor from liver has been extensively studied. It is significant to note that Prinzmetal, Hechter, Margoles and Feigen (1944) have reported an anti-burn-shock factor from liver. (This factor may intercept the shock cycle by preventing the toxic irritation of protomorphogen released by the burn.) They comment, however, that this fraction is not the anti-pernicious-anemia factor.

These comments suggest the possibility that the reticulo-endothelial system, and possibly the liver, secrete protective substances (possibly under autocatalytic control) which control the blood serum content of erythrocytic protomorphogen. The blood serum erythrocytic protomorphogen concentration would be the basic erythropoietic regulator in this system. The stimulation of this activity by the adrenal cortex would be a product of the immune-biological regulation of that organ.

Maegraith, Martin and Findlay (1943) have studied a species specific heat labile agent from tissue which exerts a lytic influence on homologous erythrocytes. There is a possibility that this factor is associated with erythrocytic protomorphogen for it is inhibited by a factor in the blood serum whose concentration rises and decreases in various hemolytic diseases.

Dr. F. Fenger of the Armour Laboratories comments[1] that the liver anti-anemic factor raises the blood platelet count before its effect on the erythrocyte count is observed. This activity can be interpreted as the elimination of erythrocytic protomorphogen and its conversion into platelets thus stimulating erythropoiesis by lowered protomorphogen concentrations.

The observation that extracts from hemolyzed erythrocytes stimulate red blood cell production might seem to disprove the suggestion that erythropoiesis is controlled by regulating protomorphogen concentration. This report should be reinvestigated with the object of determining whether the erythropoietic action is direct or a consequence of the stimulation of a reticulo-endothelial immune response which, in turn, lowers the blood erythrocyte protomorphogen concentration.

These ideas are not presented as a "theory" of anemia. Rather we have hesitantly suggested the possibility of a mechanism whereby erythrocyte production is regulated by the concentration of its protomorphogen which in turn is controlled by an immune autocatalytic response in the reticulo-endothelial system. The suggestion seems to have enough merit to warrant further experimental consideration.

1 Personal communication.

Arthritis.—Turck (1933) has supplied significant data illustrating the importance of deranged protomorphogen metabolism in the etiology of arthritis. He reviews comments that arthritis deformans has been considered an infectious condition because of the local inflammatory reactions, but such inflammatory reactions may be produced by trauma or numerous other agents. He concludes that this disease is not fundamentally a result of infection. He reports the experiments of Harding (1921) in which blood cultures from 300 cases of arthritis were negative.

Turck experimentally produced stiff and sensitive joints and arthritic lesions in test animals by injecting homologous tissue ash into the bone marrow. Heterologous tissue ash caused no untoward effects. The injection of homologous tissue ash intramuscularly near the sciatic or ulnar nerve caused marked sensitivity and lameness similar to a neuritic condition in humans. Microscopic examination of the inflamed area showed congestion and extension of the vessels surrounding the nerve in the nerve sheath.

Barrett[1] experimentally produced arthritic lesions by injection of anterior pituitary growth hormone into vitamin A deficient animals. In this experiment the active cell proliferation induced by the anterior pituitary hormone naturally led to the production of augmented amounts of protomorphogen by the osteoblasts. We have reviewed evidence (Chapter 5) that vitamin A is a protector against toxic protomorphogen activity by reason of its catalyzing the protective-association between protomorphogen and lipids. The absence of this stimulus in anterior pituitary treated rats apparently allowed the excess "raw" protomorphogens to adversely influence local elements with the production of inflammatory arthritic lesions.

In our discussion of the cancer problem we reviewed Rosenberg's comments (1942) on the relationship between vitamin A and purine metabolism. We interpreted this, along with Burrow's evidence, as indicating that vitamin A is a necessary catalyst in the protective association of protomorphogens with lipid sheathing substance. In our discussion of senescence we indicated that the anterior pituitary growth hormone is associated with the protomorphogen disposal system. There are indications that by its promotion of the rate of nuclear metabolism, increased amounts of protomorphogen and its toxic associated nucleoprotein protomorphogen and its toxic associated

1 Personal communication.

nucleoprotein degradation products are produced. Adequate vitamin A reserves apparently are necessary in order to efficiently sheath these toxic products and prevent local inflammation. This is further indicated by the reports of Ershoff and Deuel (1945) that administration of anterior pituitary growth hormone to vitamin A deficient rats resulted in an increased mortality rate and precipitated the acute symptoms of vitamin A deficiency in a more pronounced form than saline treated controls.

Howell, Han and Ittner (1941) have reported arthritic inflammatory and degenerative lesions in horses subjected to prolonged vitamin A deficiency. They report on the appearance of arthritis in Army horses in Panama which were fed a low vitamin diet. Inclusion of vitamin A in their diet (in the form of green alfalfa) later corrected this abnormality.

Recently we have received encouraging clinical reports on the effectiveness of betaine and choline in alleviating arthritic pains. This is significant in view of the importance of both these substances as methyl donors in the metabolic cycle of the protomorphogen sheathing substance as reviewed in Chapter 5.

Small amounts of potassium (in the form of potassium bicarbonate) seem to cooperate in this action. In our discussion of the biochemistry of the sheathing molecule we also discussed the formation of dipotassium creatine hexose phosphate (phosphagen) as the normal avenue of disposal for the toxic nucleoprotein degradation product, guanidine. The production of guanidine at the site of arthritic lesions consequent to the degradation of the raw protomorphogens present may be one of the etiologic factors in arthritic pains. The influence of methyl donors, and of potassium, in alleviating these pains become significant in view of their position as necessary precursors to phosphagen. In a deficiency of these products the formation of phosphagen and consequent detoxification of guanidine may be impaired.

Turck reports experiments conducted in the Russian Red Cross hospital in Bulgaria. At Turck's suggestion "anti-cytost"[1] treatments were instigated. In one group 79 middle-aged and elderly sufferers from arthritis chronica deformans and polyarthritis chronica responded to this treatment no only in that the "affected joints were cured, but their very essence of life, their vitality, was increased in a very considerable degree."

1 For details on the "anti-cytost" therapy see following discussion of "Cytotoxins."

Other Diseases.—Turck reports that similar treatment has been tried with varying degrees of success in arteriosclerosis, asthma bronchialis, neurasthenia and other degenerative conditions associated with the progressive changes of old age.

One of the most universal characteristics of the degenerative diseases of old age is the deposition of cholesterol at various critical sites such as the aorta and blood vessels, leading to high blood pressure, coronary occlusion and related conditions. Faber (1946) reports that the deposition of cholesterol in such conditions depends upon the amount of cholesterol in the blood stream *and the influence of a tissue factor responsible for the localization of the deposits.* This cholesterol deposition seems to occur at points of injury and the tissue substance is of a metachromatic nature. In our discussion of the sheathing molecule normally associated with protomorphogen we reviewed the close association of cholesterol with protomorphogen and commented upon the powerful adsorptive influences of the latter, this molecule forming a monolayer activating thromboplastic protomorphogen.

It is apparent that the protomorphogen concentrations concomitant with old age may form an attraction for the localization of cholesterol and formation of atheromatous lesions. Therapy aimed at promoting the elimination of protomorphogens might very well achieve a measure of success in atheromatous diathesis associated with senescence.

Cytotoxins

This leads us to a brief discussion of one of the most promising fields of therapy and medical research that exists today. This is the field of "cytotoxins" or the therapeutic use of antibodies to homologous protomorphogens. This field of research has been developed by Soviet scientists and constitutes a spectacular chapter in medical and biological history.

Turck's anti-cytost treatment consisted of sub-lethal injections of homologous autolyzed tissue preparations and the consequent active immunization of the animal. We have discussed the existence of a natural tissue antibody and Turk's protomorphogen immunization experiments seems to cause simply an augmented production of this protective substance.

Antibody was also obtained and injected along with minute amounts of active protomorphogen (autolyzed homologous tissue extract) at varying intervals. Treatment consisted of from 1 to 5 cc. of immune serum with 0.4 to 0.5 cc. of active protomorphogen in intervals of 10 days. Total treatment consisted of from 10 to 50 cc. of immune serum and 2 to 3 cc. of active homologous protomorphogen.

As an alternative Turck employed the intramuscular administration of sub-lethal doses (0.25 cc.) of chloroform with the purpose of creating points of focal necrosis and protomorphogen concentration which would antigenically stimulate the animal's immune-biological system, increasing the content of natural tissue antibody. Prof. K. Victorov of the Timiryazew Agricultural Academy, Moscow, has published an interesting monograph reviewing the development of the science of cytotoxins and presenting much original material from his laboratory. He has been kind enough to send us an English review of his monograph which we have published in full in the appendix.

Cytotoxins are the immune antibodies developed from the antigenic action of injected tissue extracts. The administration of this homologous immune serum to animals is followed by various manifestations of its activity. The first work emphasized the extreme toxic effect of these antibodies on the tissue homologous to the antigen. Guyer (1920) (1921) demonstrated the influence of immune serum to lens substance when administered to a pregnant animal. The development of lens tissue in the fetus was adversely affected.

The use of cytotoxins in small dosages to stimulate the vitality of organs has been developed in the soviet during the past 20 years particularly under the guidance of Academician Bogomolets and co-workers. We have been unable to find evidence of other participation in this development with the exception of Dr. Fenton B. Turck who was apparently unaware of similar activities in the Soviet Union.

Bogomolets and his co-workers have reported spectacular effects using spleen and bone marrow cells as antigens. The immune serum, which they term "Anti-Reticular Cytotoxic Serum" (ACS), is prepared from human organ extracts, injecting the antigen so obtained into rabbits. Small amounts of the immune serum so obtained stimulate the activity of the reciculo-endothelial system, larger amounts causing its involution.

The therapeutic influences of this serum, which may be attributed to a "rejuvenation" of the reticulo-endothelial system are spectacular. Many pathological states are reported to respond to this treatment including infections and degenerative conditions, senile manifestations and experimental malignant tumors.

Good English reviews of the work of Bogomolets and his collaborators have recently appeared in the American Review of Soviet Medicine.[1] Leake (1946), participating in an evaluation of ACS serum has published an excellent review of its possibilities and the general relationship of the reticulo-endothelial system to resistance to disease.

Medvedev (1940) has reported on the production of protective enzymes against various tissues consequent to the injection of spleen extracts. These extracts stimulate autocatalytically the functions of the reticulo-endothelial system, increasing hydrophilic colloids, augmenting the production of natural tissue antibody, stimulating phagocytosis and elevation of the opsonin titre. Note that the spleen extracts are used actively and not as antigens to produce ACS serum. This work emphasizes the vitality building influence of a stimulation of the reticulo-endothelial tissue similar to that obtained with Bogomolets' ACS serum.

Prof. Sakharov is another Soviet scientist who has extensively investigated the cytotoxins, reporting particularly satisfactory results in the treatment of schizophrenia and diabetes with cytotoxic serums. The schizophrenic immune serum was prepared using nervous tissue as an antigen, and the diabetic immune serum by using pancreas substance as an antigen.

Prof. Victorov states that the immune sera thus prepared, while exhibiting a varied species specificity, is nevertheless exactly specific for different tissues. His experiments demonstrate a strict tissue specificity and absence of species specificity of the extracts. This is a phenomenon which we have reported to be characteristic of protomorphogens.

1 Bogomolets, A. A.: "Anti-Reticular Cytotoxic Serum as a Means of Pathogenetic Therapy." *Am. Rev. Sov. Med.,* 1:101-112, 1943. Marchuk, P. D.: "A Method of Preparing and Preserving Anti-Reticular Cytotoxic Serum," *Ibid,* 1:113-123, 1943. Linberg, B. E.: "Anti-Reticular Cytotoxic Serotherapy of Frostbite and War Wounds."*Ibid,* 1:124-129, 1943.

Specific immune sera prepared by using various organs as antigens are effective in small amounts in stimulating the organs of the specific type from which the antigen is prepared. Obviously this opens up the wide vista of a new therapeutic approach to many stubborn conditions, an opportunity which is not being ignored by this group of investigators.

We feel that the stimulating activity of cytotoxins is in the reduction of inhibitory protomorphogen concentrations in specific areas, allowing regeneration and repair to manifest themselves. Larger doses of these sera no doubt induce lysis of the cells themselves accounting for the adverse effects of excessive dosage. The reader will recall our hypothesis that excess protomorphogen accumulations in the areas surrounding tissues inhibits their regeneration and is a basic cause of the progressive degeneration with age. Cytotoxin therapy offers a rational method of lowering these protomorphogen concentrations in specific locales. How much influence may be expected over protomorphogen adsorbed on connective tissue, however, is still questionable.

Pomerat and Anigstein (1944, 1945) are American investigators who are experimentally studying the ACS serum. They have demonstrated the inhibitory influence of ACS on spleen and other tissue *in vitro* which changed to stimulation when the concentration of ACS was greatly reduced. The inhibitory influence of larger doses *in vivo* has also been demonstrated by these investigators.

Cytotoxin therapy therefore offers a spectacular therapeutic measure which should receive wide attention among investigators and clinicians. Its experimental and clinical basis has been firmly established in 25 years of careful study. We feel that the morphogen hypothesis offers a rational theoretical approach to the phenomenon. Its use in pregnancy, however, should be accompanied with caution since the cytotoxic immune sera can influence embryonic differentiation unfavorably (Guyer, 1920, 1921) and it is not impossible that it may so reduce the blood content of certain specific protomorphogens as to adversely affect the organization of the chromosomes in the gonadal germ cells. (This has been reported by various investigators.)

We have been struck at times by the close identification of the properties of several drugs with the phenomena characteristic of certain specific classes of protomorphogen metabolism.

Thiosinamin (Waugh and Abbott, 1905), for instance, is a mustard derivative which has a valuable effect in softening scar tissue, dissolving pathological fibrous tissues and promoting the absorption of generalized cicatricial adhesions. This would seem to indicate that it is a powerful elutogen or fibrinolysin. This postulation is further supported by the results of overdosage: burning at site of injection, headache and debility, fever and nausea. These are all clearly recognizable symptoms of increased concentration of raw protomorphogen in the general system. Possibly the concomitant usage of sheathing promoters, such as methyl donors, lecithin or vitamin F would alleviate some of these symptoms to a degree.

The use of nitrogen mustards has recently enjoyed considerable interest. Goodman and co-workers (1946) have analyzed their utility in the treatment of various neoplastic conditions: In our analysis of neoplasms we emphasize the evidence that intense concentrations of protomorphogens predispose to cancer. That the nitrogen mustards assist in their removal is indicated also by the nausea accompanying injection.

Ipecac is another drug which can be classified on the basis of its influence over protomorphogen. Potter (1913) reports that it is useful in stopping hemorrhagic tendencies and increases bile flow. The former can be interpreted as due to the promotion of thromboplastin formation from protomorphogen and the latter as a compensatory excretory reaction to increased blood protomorphogen. These interpretations are supported by the report that this drug causes lung congestion in overdosage. We have reported that lung congestion and inflammation is one of the most striking evidences of a sudden increase of raw protomorphogens.

Iris versicolor (lridin) (Ellingwood, 1915) is recommended in nausea, chronic liver disease, enlarged spleen, neuralgia, wasting muscles and other atrophies. These are all conditions which we would expect to be associated with a general increased concentration of protomorphogen in the tissues. Apparently iridin promotes the elimination of protomor-phogens, a postulation which does not sound too absurd when we re-read

Ellingwood's comment that it promotes the elimination of toxic and effete material from the blood. It is not within the scope of this work to exhaustively analyze the pharmacological ramifications of the morphogen hypothesis. It seems that a re-analysis of the effects of various drugs in the light of the morphogen hypothesis might supply the necessary theoretical background that a drug must enjoy today in order to be recognized. It is entirely likely that many discarded drugs might be accepted again if their empirical reputation could be interpreted on a sound theoretical basis.

Such a classification must group these drugs into those whose effect is wholly elutogenic, depolymerizers, cholagogues, promoters of sheathing protection, general eliminators, etc. The effect of eluting protomorphogens from connective tissue must be delineated from the influence of degrading the molecule once it has been eluted. Some drugs will no doubt be found to be effective in each of these activities, others in both. The utility of such therapeutic adjuncts will depend upon their specific action, which may be desirable in one disease but not in another.

Review of Morphogens and Pathology

Inflammation.—Protomorphogens released in trauma or injury inhibit the regeneration and repair of tissue unless accompanied by "embryonic" growth factors or wound-hormones which catalyze the synthesis of new tissue, using the protomorphogens as substrate material. Necrosin, the toxic necrotic factor in inflammatory exudates is likely a derivative of protomorphogen.

Shock.—We have suggested a shock cycle in which the release of protomorphogens subsequent to trauma or injury irritates the nerve termini resulting in dystrophic impulses to the whole system. This cycle is illustrated in figure 9.

Senescence.—Although the life cycle of various species is controlled by the hereditary influence over a master organ (probably the anterior pituitary) the direct cause of tissue ageing is the accumulation of protomorphogens surrounding the tissues. This accumulation is progressive with age and depends upon the gradual incapacitation of the elimination systems under master organ control.

Pregnancy.—We have suggested that nausea of pregnancy is due to excess protomorphogens in the maternal blood stream as a consequence of enhanced mitotic activity in the embryo. Eclampsia is likely a manifestation of similar phenomena complicated by the seriousness of guanidine derivatives in the blood resulting from excess protomorphogen breakdown. We have reconciled parathyroid tetany and eclampsia, showing them to be similar phenomena but not necessarily subject to the same treatment. The so-called "rejuvenation" of pregnancy we have attributed to the stimulation of the protomorphogen removal systems compensatory to their overload during pregnancy.

Cancer.—This condition has been reviewed as a mutation of cytomorphogens or genes resulting from virus activity or caused by intense local concentrations of protomorphogens which Burrows has shown to precede and predispose to cancer. We have also speculated on the existence of a depolymerizing substance in cancer which prevents the accumulation of excess protomorphogens from inhibiting growth and which may de-differentiate the cells into the low-organization competent embryonic type. We have reviewed the "replica hypothesis" which envisions the mirror image duplication of genic material into a "pattern making" stereoisomer which in turn produces a mirror image product identical with the original. We have suggested that the controls of the "pattern making" stereoisomer are lacking in cancer allowing uncontrolled mitosis, a suggestion with some experimental backing.

Anemia.—We have presented a few suggestions leading to the possibility that the erythrocytes are under the autocatalytic control of their protomorphogen, the concentration of which is regulated by an immune reaction of the reticulo-endothelial system.

Arthritis.—We have reviewed Turck's experiments and comments which seem to describe arthritis as inflammation caused by local concentrations of protomorphogens.

Cytotoxins.—The phenomenon of specific tissue immune sera and its influence in stimulating or inhibiting tissue vitality is reviewed and explained on the basis of altering the protomorphogen concentration in specific organs.

The aspects of the morphogen hypothesis when applied to pathological conditions may seem too far-reaching to the average. investigator or clinician. But when we remember that we are dealing with one of the fundamental aspects of control of cell vitality and life it would not be surprising to find all branches of physiology and pathology eventually linked with the morphogen concept.

BIBLIOGRAPHY

ABRAHAM, E. P., BROWN, G. M., CHAIN, E., FLOREY, H. W., GARDNER, A. D., and SANDERS, A. G.: "Tissue Autolysis and Shock." *Quart. J. Exp. Physiol.,* 31:79-97, 1941.

ARONS, I., and SOKOLOFF, B.: "Role of Reticulo-Endothelial System in Cancer with Reference to Congo Red Therapy in Roentgen Sickness and Anemia." *Am. J. Roentgenol.,* 41:834-849, 1939.

ASTBURY, W. T.: "The Structural Proteins of the Cell." *Biochem. J.,* 39:lvi-lvii, 1945.

BAKER, L. E., and CARREL, A.: "Lipoids as the Growth-Inhibiting Factor in Serum." *J. Exp. Med.,* 42:143-154, 1925.

———— "Effect of Age on Serum Lipoids and Proteins." *Ibid,* 45:305-318, 1927.

BARKER, L. F.: *Endocrinology and Metabolism.* Vol. I. D. Appleton & Co., New York. 1922.

BEARD, H. H.: "Cancer as a Problem in Metabolism." *Arch. Int. Med.,* 56:1143-1170, 1935.

———— *Creatine and Creatinine Metabolism.* Chemical Publ. Co., Brooklyn, N. Y. 1943.

BECK, W. A.: "The Organization of Protoplasm." *Growth,* 8:259-309, 1944.

BIESELE, J. J.: "Chromosomal Enlargement in Neoplastic Rabbit Tissues." *Cancer Research,* 5:179-182, 1945.

BLUMENTHAL, H. T.: "Ageing Processes in the Endocrine Glands of the Guinea Pig: The Influence of Age, Sex and Pregnancy on the Mitotic Activity and the Histologic Structure of the Thyroid, Parathyroid and Adrenal Glands." *Arch. Path.,* 40:264-269, 1945.

BODANSKY, M., and BODANSKY, O.: *Biochemistry of Disease.* Macmillan Co., New York. 1940.

BOGOMOLETS, A. A.: "Anti-Reticular Cytotoxic Serum as a Means of Pathogenic Therapy." *Am. Rev. Sov. Med.,* 1: 101-112, 1943.

———— "Blood Transfusion in the Treatment of Internal Disease." *Ibid,* 2:196-198, 1945.

BOMSKOV, C., and V. KAULLA, K. N.: "The Mode of Action of Vitamin E." *Klin. Wochschr.,* 20:334-340, 1941.

BURNET, F. M.: *Production of Antibodies.* Macmillan Co., Ltd., London. 1941.

BURROWS, M. T.: "The Mechanism of Cancer Metastasis" *Arch. Int. Med.,* 37:453-472, 1926.

———— "Energy Production and Transformation in Protoplasm as Seen Through a Study of the Mechanism of Migration and Growth of Body Cells." *Am. J. Anat.,* 37:289-349, 1926.

———— "The Nature of Atrophy and Hyalinization of Cells and Tissues." *J. Cancer Research,* 11:72-85, 1927.

———— "Malignant .Sarcoma in Embryonic Tissue Produced by Berkefeld Filtrates of Jensen Sarcoma of Rats." *Arch. Path. & Lab. Med.,* 4:168-192, 1927.

——— "The Production of Malignant Ulcers in the Skin of Rats, with Notes on Cause of Cancer in the Uterus, Breast and Liver." *Ibid,* 4:380-395, 1927.

——— "The Mechanism of Cell Division." *Am. J. Anat.,* 39:83-134, 1927.

BURROWS, M. T.: and JORSTAD, L. H.: "On the Source of Vitamin A in Nature." *Am. J. Physiol.,* 77:38-50, 1926.

——— "On the Source of Vitamin B in Nature." *Ibid,* 77:24-37, 1926.

BURROWS, M. T., JORSTAD, L. H., and ERNST, E. C.: "Cancer, Vitamin Imbalance, and Roentgen Ray Activity." *J.A.M.A.,* 87:86-89, 1926.

——— "Further Studies of the Effect of Graded Doses of X-rays on Animals Fed on Various Dietaries, with a Note on the Effect of Dietary in the Treatment of Patients with X-rays." A report appearing in Radiology and reviewed by BURROWS, M. T., in his article: "The Nature of Atrophy and Hyalinization of Cells and Tissue," *J. Cancer Research,* 11:72-85, 1927.

CADNESS, B. H. E., and WOLF, C. G. L.: "Fuchs Reaction for the Serodiagnosis of Carcinoma." *Biochem. Ztschr.,* 238:287-306, 1931.

CARREL, A.: "Tissue Culture and Cell Physiology." *Physiol. Rev.,* 4:1-17, 1924.

OCARDO, V. H.: "Volumen Globular del Perro Durante el Choque Traumatico.". *Rev. Soc. Argentina Biol.,* 20:339-350, 1944.

COWDRY, E. V.: *General Cytology.* University of Chicago Press, Chicago, Ill. 1924.

——— *Problems of Ageing.* Williams & Wilkins Co., Baltimore, Md. 1939.

——— *Problems of Ageing.* 2nd Edition, Williams & Wilkins Co., Baltimore, Md. 1942.

CRILE, G.W.: *The Phenomena of Life.* W. W. Norton & Co., New York. 1936.

CRILE, G, W., and LOWER, W. E.: *Anoci-Association.* W. B. Saunders Co., Philadelphia, Pa. 1914.

CRILE, G. W., TELKES, M., and ROWLAND, A. F.: "The Nature of Living Cells." *Arch. Surg.,* 23:703-714, 1931.

DAVIDSON, J. R.: "Attempt to Inhibit Development of Tar-Carcinoma in Mice; Effects of Vitamins on Tumor Threshold." *Canadian M.A.J.,* 37:434-440, 1937.

DAVIDSON, J. N., and WAYMOUTH, C.: "Factors Influencing the Nucleoprotein Content of Fibroblasts Growing In Vitro." *Biochem. J.,* 37:271-277, 1943.

DOLJANSKY, L., and HOFFMANN, R. S.: "Cancer Cell and Growth-Regulating System of the Body." *Nature,* 145:857-858, 1940.

DONOVAN, H., and WOODHOUSE, D. L.: "A Speculation on the Nature of the Chemical Structure Which is the Essence of the Malignant Cell." *Nature,* 152:509-510, 1943.

DOWNS, A. W., and EDDY, N. B.: "Further Observations on the Effects of the Subcutaneous Injection of Splenic Extract." *Am. J. Physiol.,* 62:242, 1922.

DRINKER, C. K.: *The Lymphatic System.* Lane Medical Lectures, Stanford University Press, Stanford University, California. 1942.

ELLINGWOOD, F.: *Am. Materia Medica.* Ellingwood's Therapeutisa, Chicago. 1915.

ENGEL, D.: "The Influence of the Sympathetic Nervous System on Capillary Permeability." *J. Physiol.,* 99:161-181, 1941.

ERSHOFF, B. H., and DEUEL, H. J., JR.: "The Effect of Growth Hormone on the Vitamin A-Deficient Rat." *Endocrin.,* 36:280-282, 1945.

EVANS, H. M.: "The Growth Hormone of the Anterior Pituitary." *J.A.M.A.,*104:1232-1237, 1935.

FABER, M.: "Cholesterol Deposit in Organism, with Special Regard to Conditions m Atheromatosis." *Nordisk Medicin,* Gothenburg, 30:1239-1245, 1946.

307

FEIGEN, G. A., and DEUEL, H. J., JR.: "Comparative Effects of Thromboplastin, Lecithin. and Saline Solution on Mortality of Mice Following Experimental Burn Shock." *Proc. Soc. Exp. Biol. & Med.,* 58:81-84, 1945.

FELDBERG, W.: "Histamine and Anaphylaxis." *Ann. Rev. Physiol.,* 3:671-694, 1941.

FISCHER, A.: "Nature of the Growth-Promoting Substances in the Embryonic Tissue Juice. A Review of the Author's Investigations." *Acta. Physiol. Scand.,* 3:54-70, 1940.

GAMALEIA, N.: *U.S.S.R. Information Bulletin,* Published by the Soviet Embassy, Washington, D. C. October 23, 1946.

GOODMAN, L. S., WINTROBE, M. M., DAMESHEK, W., GOODMAN, M. J., GILMAN, A., and MCLENNAN, M. T.: "Nitrogen Mustard Therapy." *J.A.M.A.,* 132:126-132, 1946.

GULLAND, J. M., BARKER, G. R., and JORDAN, D. O.: "Structure of Nucleic Acid in the Dividing Cell." *Nature,* 153:20, 1944.

GURWITSCH, L., and SALKIND, S.: "Mitogenetic Behavior of the Blood of Carcinoma Patients." *Biochem. Zeitschr.,* 211:362-372, 1928.

GUYER, M. F.: "Immune Sera and Certain Biological Problems." *Am. Naturalist,* 55:97-115, 1921.

GUYER, M. F., and SMITH, E. A.: "Studies on Cytolysins. II. Transmission of Induced Eye Defects." *J. Exp. Zool.,* 31:171-215, 1920.

HADDOW, A.: "Transformation of Cells and Viruses." *Nature,* 154:194-199, 1944.

HANSON, A. M.: "A Report of Four Cases of Inoperable Carcinoma Treated with Intramuscular Injections of Karkinolysin." *Minnesota Medicine,* 13:65-73, 1930.

HARDING, M. C.: "A Group Study of 300 Cases of Arthritis." *Calif. State J. Med.,* 19:26-28, 1921.

HAVEN, F. L., and LEVY, S. R.: "Phospholipids of Tumor Cells and Nuclei." *Cancer Research,* 2:797-798, 1942.

HOWELL, C. E., HART, G. H., and ITTNER, N. R.: "Vitamin A Deficiency in Horses." *Am. J. Vet. Research,* 2:60-74, 1941.

HURWITZ, S. H.: "Pathological Metabolism of the Blood and Blood-Forming Organs." *Endocrinology and Metabolism,* Vol. 4. Ed. by L. F. Barker, D. Appleton & Co., New York. 1922.

JACOLSEN, C.: *Arch. Dermatol. u. Syph.* 169:562, 1934.

JAFFE, R. H.: "The Reticulo-Endothelial System. Its Role in Pathologic Conditions in Man." *Arch. Path. & Lab. Med.,* 4:45-76, 1927.

JORSTAD, L. H.: "The Relation of the Vitamins to the Reaction Induced by Coal Tar in the Tissues of Animals." *J. Exp. Med.,* 42:221-230, 1925.

KIDD, J. G.: "Suppression of Growth of the Brown-Pearce Tumor by a Specific Antibody." *Science,* 99:348-350, 1944.

KLINE, B. E., and RUSCH, H. P.: "Some Factors That Influence the Growth of Neoplastic Cells." *Cancer Research,* 4:762-767, 1944.

LEAKE, C. D.: "Leukocytic Reactions to Red Bone Marrow and Spleen Extracts." *J. Pharm. & Exp. Therap.,* 22:109-115, 1924.

——— "The Reticulo-Endothelial System and Resistance to Disease." *Hawaii Med. J.,* 5:251-256, 1946.

LEAKE, C. D., and BACON, F. J.: "A Preliminary Note on the Properties of an Alleged Erythropoietic Hormone." *J. Pharm. & Exp. Therap.,* 23:353-363, 1924.

LEAKE, C. D., and LEAKE, E. W.: "The Erythropoietic Action of Red Bone M arrow and Splenic Extracts." *J. Pharm. & Exp. Therap.,* 22:75-88, 1924.

LETTRE, H.: *Ang. Chemie,* 50:581-592, 1937.

LINFOOT, S.: "The Common Cold." *Med. Press & Circular No. 5467*, 107-112, 1944.

LOOFBOUROW, J. R., COOK, E. S., DWYER, C. M., and HART, M. J.: "Production of Intercellular Hormones by Mechanical Injury." *Nature*, 144:553-554, 1939.

LOWRY, O. H., MCCAY, C. M., HASTINGS, A. B., and BROWN, A. N.: "Histochemical Changes Associated with Ageing. III. The Effects of Retardation of Growth on Skeletal Muscle." *J. Biol. Chem.*, 143:281-284, 1942.

LUMSDEN, T.: "Tumour Immunity." *Am. J. Cancer*, 15:563-640, 1931.

MACALISTER, C. J.: *Investigation Concerning an Ancient Medicinal Remedy and Its Modern Utilities.* John Bale, Sons & Danielsson, Ltd., London. 1936.

MCCAY, C. M.: *Problems of Ageing.* Ed. by E. V. Cowdry, Williams & Wilkins Co., Baltimore. 1939.

MCDOWELL, R. J. S.: "The Circulation in Relation to Shock." *Brit. Med. J.,*, 1:919-924, 1940.

MCMILLAN, W. O.: "Shock." *Industrial Med.*, 9:567-569, 1940.

MACNIDER, W. DEB.: "The Aging Processes and Tissue Resistance." *Sci. Month.*, 54:149-154, 1942.

―――― "Occurrence of Stainable Lipoid Material in Renal Epithelium of Animals Falling in Different Age Segments." *Proc. Soc. Exp. Biol. & Med.*, 58:326-328, 1945.

MAEGRAITH, B. G., MARTIN, N. H., and FINDLAY; G. M.: "The Mechanism of Red-Blood-Cell Destruction." *Brit. J. Exp. Path.*, 24:58-65, 1943.

―――― "Lyric Agent and Inhibitory Factors in Human Tissue and Serums." *Lancet*, 1:573-575, 1943.

MANWARING, W. H.: "Environmental Transformation of Bacteria." *Science*, 79:466-470, 1934.

MARSH, F. B.: "Disturbed Fluid Balance." *Industrial Med.*, 9:546-551, 1940.

MARSHAK, A., and WALKER, A. C.: "Effect of Liver Fractions on Mitosis in Regenerating Liver." *Am. J. Physiol.*, 143:226-234, 1945.

―――― "Effect of Chromatin Derivative on the Healing of Skin Wounds." *Proc. Soc. Exp. Biol. & Med.*, 58:62-63, 1945.

―――― "Mitosis in Regenerating Liver." *Science*, 101:94-95, 1945.

―――― "Transfer of P^{32} From Intravenous Chromatin to Hepatic Nuclei." *Am. J. Physiol.*, 143:235-237, 1945.

MEDVEDEV, N.: "Sur la Regulation Autocatalytique des Fonctions du Systeme Physiologique du Tissu Conjonctif." *J. Med. de L'Acad. des Sci. de la RSS D'Ukraine*, 10:1447-1450, 1940.

MEISSEL, M. N.: "Effect of Cancerogenic Substances Upon Microbic Cells." *Compt. rend. (Doklady) Acad. Sci. URSS*, 42:399-400, 1944.

MENKIN, V.: "Role of Inflammation in Immunity." *Physiol. Rev.*, 18:366-418, 1938.

―――― "Chemical Basis of Fever." *Science*, 100:337-338, 1944.

―――― "The Significance of Biochemical Units in Inflammatory Exudates." *Ibid*, 101 :422-425, 1945.

―――― "Chemical Basis of Fever with Inflammation." *Arch. Path.*, 39:28-36, 1945.

―――― "Further Studies on the Mechanisms of Injury and Fever with Inflammation." *Fed. Proc.*, 4:149, 1945.

MINOT, C. S.: *The Problem of Age, Growth, and Death.* B. P. Putnam's Sons, The Knickerbocker Press, New York. 1908.

MIRSKY, I. A., and FREIS, E. D.: "Renal and Hepatic Injury in Trypsin 'Shock'." *Proc. Exp. Biol. & Med.*, 57:278-279, 1944.

MOON, V. H.: *Shock. Its Dynamics, Occurrence and Management.* Lea & Febiger, Philadelphia, Pa. 1942.

MURPHY, J. B.: "Experimental Approach to the Cancer Problem. I. Four Important Phases of Cancer Research. II. Avian Tumors in Relation to the General Problem of Malignancy." *Bull.]. Hopkins Hosp.,* 56:1-31, 1935.

MURPHY, J. B., MAISIN, J., and STURM, E.: "Local Resistance to Spontaneous Mouse Cancer Induced by X-Rays." *J. Exp. Med.,* 38:645-653, 1923.

MYLON, E., and WINTERNITZ, M. C.: "The Differentiation of K From Other Agents Associated with the Toxic Effects of Tourniquet Shock." *Am. J. Physiol.,* 146:254-261, 1946.

NEEDHAM, J.: *Biochemistry and Morphogenesis.* Cambridge University Press, London. 1942.

OBERLING, C.: *The Riddle of Cancer.* Yale University Press. New Haven, Conn. 1944.

OREKHOVICH, V. N.: "A Possible Relationship in Animals Between Tumor Susceptibility and Stability of Tissue Proteins." *Am. Rev. Sov. Med.,* 1:517-531, 1944.

PEN, D. F., CAMPBELL, J., and MANERY, J. F.: "Toxic Substances From Muscle." *Am. J. Physiol.,* 141:262-269, 1944.

PICKERING, J. W., MATHUR, S. N., and ALLAHAB, M. B.: "The Role of Tissue Juices in Thrombosis." *Lancet,* 2:387-388, 1932.

POMERAT, C. M., and ANIGSTEIN, L.: "Anti-Reticular Immune Serum: Its Action Demonstrated by Tissue Culture Technique." *Science,* 100:456, 1944.

———— "Reticulo-Endothelial Immune Serum (R.E.I.S.): I. Its Action on Spleen In Vitro." *Texas Repts. on Biol. & Med.,* 3:122-141, 1945.

———— "The Effect of Reticulo-Endothelial Immune Serum (R.E.I.S.) On Heart Fragments in Tissue Culture." *Fed. Proc.,* 4:56, 1945.

POTTER, S.: *Therapeutic Materia Medica & Pharmacy,* 12th Ed. G. Blankenton & Sons, Philadelphia, Pa. 1913.

POTTER, V. R.: "The Genetic Aspects of the Enzyme-Virus Theory of Cancer." *Science,* 101:609-610, 1945.

PRINZMETAL, M., FREED, S. C., and KRUGER, H. E.: "Pathogenesis and Treatment of Shock Resulting From Crushing of Muscle." *War Med.,* 5:74-79, 1944.

PRINZMETAL, M., HECHTER, O., MARGOLES, C., and FEIGEN, G.: "A Principle From Liver Effective Against Shock Due to Burns." *J. Clin. Invest.,* 23:795-806, 1944.

REGAUD, C., GRICOUROFF, G., and VILLELA, E. U. D.: "Hybrid Cell Mutation in Uterus Preceding Cancer.".*Compt. rend. Acad. d. sci.,* 197:537-539, 1933.

REISSNER, A.: "Dental Organotherapy." *Dental Items of Interest,* May, 1941. See *Dental Reflector* for October 15, 1941 and January 1, 1942. See "Dental Organotherapy" by P. Edelman in *Dental Items of Interest,* March, 1942. (These publications give extensive foreign bibliographical material.)

ROBERTSON, T. B.: *Principles of Biochemistry.* Lea & Febiger, Philadelphia, Pa. 1924.

ROLLER, D.: "The Occurrence of Permeability-Increasing Substances in Histamine Shock." *Klin. Wochschr.,* 22:704, 1943.

ROSENBERG, H. R.: *Chemistry and Physiology of the Vitamins.* InterScience Publishers, Inc., New York. 1942.

RUSS, S., and SCOTT, G. M.: "Resistance to Tumor Grafts Produced by Cell Free Tumor Extract." *Brit. J. Radiology,* 18:173-175, 1945.

RUZICKA, V.: "Ueber protoplasma-hysteresis und eine methodf zur direkten Bestimmung derselben." *Arch. Ges. Physiol.,* 194:135-148, 1922.

———— "Beitrage zum Studium der Protoplasmahysteretischen Vorgange." *Arch. f. mikr. Anat.,* 101:459-503, 1924.

SAJOUS, C. E. DEM.: *The Internal Secretions and the Principles of Medicine.* Vol. I, 4th Ed., F. A. Davis Co., Philadelphia, Pa. 1911.

SCHNEIDER, C. L.: "The Active Principle of Placental Toxin: Thromboplastin; Its Inactivator in Blood: Antithromboplastin." *Am. J. Physiol.,* 149:123-129, 1947.

SCOTT, G. H.: "Mineral Distribution in the Cytoplasm." *Biol. Symp.,* 10:277-289, 1943.

SEVAG, M. G.: *Immuno-Catalysis.* Charles C. Thomas, Springfield, Ill. 1945.

SHABAD, L. M.: "Trends in Cancer Research in the USSR." *Am. Rev. Sov. Med.,* 4:166-174, 1946.

SHORR, E., ZWEIFACH, B. W., and FURCHGOTT, R. F.: "On the Occurrence, Sites and Modes of Origin and Destruction, of Principles Affecting the Compensatory Vascular Mechanisms in Experimental Shock." *Science,* 102:489-498, 1945.

SHWARTZMAN, G.: *Phenomenon of Local Tissue Reactivity.* Paul B. Hoeber, New York. 1937.

SMITTEN, N. A.:"Vital Studies of the Neuroplasm."*Am. Rev. Sov. Med.,* 3:414-425, 1946.

SPERANSKY, A. D.: *A Basis for the Theory of Medicine.* International Publishers, New York. 1943.

SPERTI, G. S., LOOFBOUROW, J. R., and LANE, M. M.: "Effects on Tissue Cultures of Intercellular Hormones From Injured Cells." *Science,* 86:611, 1937.

STERN, K.: "Aspects of the Relationship Between Malignant Tumors and Reticulo-Endothelial System." *New Intern. Clin.,* 3:111-131, 1941.

TABOR, H., and ROSENTHAL, S. M.: "Experimental Chemotherapy of Burns and Shock; Effects of Potassium Administration, of Sodium Loss, and Fluid Loss in Tourniquet Shock." *Public Health Rep.,* 60:373-381, 1945.

———— "Experimental Chemotherapy of Burns and Shock; Electrolyte Changes in Tourniquet Shock." *Ibid,* 60:401-419, 1945.

TURCK, F. B.: *The Action of the Living Cell.* Macmillan Co., New York, 1933.

VAUGHAN, V. C., VAUGHAN, V. C., JR., and VAUGHAN, J. W.: *Protein Split Products In Relation to Immunity and Disease.* Lea & Febiger, Philadelphia, Pa. 1913.

VICTOROV, K. R.: Personal Communication.

WACHTEL, H. K.: "Carcinogenic Substance From Human Cancer." *Nature,* 158:98, 1946.

WAUGH, W. F., and ABBOTT, W. C.: *A Textbook of Alkaloidal Therapeutics.* Clinic Publishing Company, Chicago. 1905.

WEIDMAN, F. D.: *Problems of Ageing.* Ed. by E. V. Cowdry, Williams & Wilkins, Baltimore, Md. 1939.

WHITE, A., and DOUGHERTY, T. F.: "Effect of Prolonged Stimulation of the Adrenal Cortex and of Adrenalectomy on the Numbers of Circulating Erythrocytes and Lymphocytes." *Endocrin.,* 36:16-23, 1945.

WIGGERS, C. J.: "Present Status of Shock Problem." *Physiol. Rev.,* 22:74-123, 1942.

WILSON, E. B.: *The Cell in Development and Heredity.* 3rd Ed., Macmillan Co., New York. 1928.

WINTERNITZ, M. C., MYLON, E., and KATZENSTEIN, R.: "Studies on the Relation of the Kidney to Cardiovascular Disease." *Yale J. Biol. & Med.,* 13:595-622, 1940-41.

ZWEMER, R. L., and SCUDDER, J.: "Blood Potassium During Experimental Shock." *Surg.,* 4:510-527, 1938.

CHAPTER 7

SUMMARY OF THE MORPHOGEN HYPOTHESIS

WE HAVE PRESENTED the morphogen hypothesis fully aware of its suggestive nature. It will not be possible to review in detail all the evidence or present again in detail the varied hypotheses we have suggested in this volume. Rather, we shall attempt an extremely brief outline of the salient points.

Fundamental Scientific Concepts.—It appears that the form and characteristics of both inorganic and simple organic molecules are influenced by their environment, i.e., temperature, atmosphere, pressure, etc. The determining factor in the organization of these simpler molecules is the natural chemical affinities and valences of the component elements. The development of inorganic and organic molecules can be considered "evolutionary" in the sense that it has been under the influence of environmental factors.

True organic evolution, we believe, begins at the point where molecular complexity reaches the critical point beyond which it cannot proceed without the introduction of new means of organization other than the natural atomic valences and affinities.

The new means of organization are met by the basic biological determinant, protomorphogen, whose function is to serve as a "pattern-making" matrix of specific linkages for the assemblage of biologically specific proteins.

Basic Biological Determinants.—The basic biological determinants may be grouped as those specific factors whose function is the organization of living form and the component living molecules. These determinants include, in inverse order of their complexity, chromosomes, genes, cytomorphogens and protomorphogens. The *chromosome* determines the characteristics of the species and the individual. It is an organized assemblage of *genes* which determine the separate characteristics of the individual. Genes are composed of an organized group of *cytomorphogens* which are the determinants of individual cell morphology. These are, in turn, composed of an organized assemblage of *protomorphogens* which are determinants for the biological protein molecules of living cells.

Nature of Protomorphogen and Its Determinant Influences.— The protomorphogen structure is fundamentally a spatial array of mineral molecules. Their influence may be demonstrated antigenically even when all organic constituents are removed by ashing. In biological systems, however, they are associated (as are cytomorphogens) with nucleoproteins and may even represent a particular class of viruses.

It appears that the determinant influence of protomorphogen is due in large part to the organized groups of mineral linkages which catalyze the formation of specific protein molecules. While the specificity of a protein molecule is a result of an active surface pattern, we think that this surf ace pattern is organized with the assistance of protomorphogen links. The antigenic effect of ashed protomorphogen is no doubt due to a catalytic influence over the synthesis of new proteins which are antigenic, rather than a specific antigenic activity *per se.*

Protomorphogen, therefore, is not necessarily the specific component of, or the original specific structure preceding, the biological protein, but rather a means by which biological proteins are produced and maintained under the influence of the living organism.

Dynamic State of Living Matter.—All living molecules are in a constant state of dynamic equilibrium and have a limited life period. This dynamic state requires a constant energy input for its maintenance and is a characteristic difference between living and nonliving protein molecules. Protomorphogens, as the determinants for biological proteins, therefore must also be in a state of dynamic metabolism.

Chemical Nature of Protomorphogens.—The most characteristic phenomenon exhibited by protomorphogen active extracts is *thermostability.* This, however, varies considerably with the complexity of the protomorphogen and the degree to which the extract has been subjected to enzymatic hydrolysis. Extracts containing active protomorphogens range in this nature from the 300 degree tissue ash described by Turck to

314

many products from cells which are injured by prolonged treatment at 60-100 degrees C.

The *specificity* of protomorphogen extracts also varies with the degree of complexity, but in general they exhibit a limited degree of species specificity and a marked degree of tissue specificity. Certain of their activities, i.e., substrate material for biological syntheses, are marked by a lack of specificity. In these cases perhaps only fragments of the protomorphogen molecules are useful.

The *molecular size* of protomorphogens seems to vary considerably, although in general they are associated with macromolecular sedimentable constituents. At certain stages of their metabolism they appear to be diffusible and at other stages highly non-diffusible. This variation no doubt depends upon the degree of polymerization of the associated nucleic acids which may exhibit molecular weights from 1500 to over a million.

In biological systems protomorphogens are always found associated with lipoid substances for which they have a peculiar affinity. We postulate that this phenomenon is in the nature of a protective association by means of which many of the toxic activities of the protomorphogens are inhibited.

The fact that protomorphogens, as found in living systems, are highly *adsorbable* on connective tissues or charcoal is a most important characteristic intimately connected with their biological activity.

The association of protomorphogens with nucleoproteins probably accounts for the solubility of the former in *saline solution* which can be used for the extraction of nucleoproteins. Similarly their association with lipoid substances explains the fact that they seem to be removed with and activated by *acetone, alcohol-ether* and other *fat solvents* including the *carcinogenic hydrocarbons*.

Factors Influencing the Rate of Cell Division, the Allelocatalyst Theory.—The allelocatalyst theory holds that an autocatalytic substance (protomorphogen) is excreted from cells especially during mitosis, its concentration in the medium increasing as a result.

Small amounts of protomorphogen in the media stimulate the division rate of cells and higher concentrations inhibit. The stimulation is exerted by both homologous and heterologous protomorphogens, but the inhibition is exhibited only by homologous protomorphogens.

There is a reciprocal relationship between the protomorphogens in the protoplasm and those in the media. If this balance is optimum cell division will occur. This optimum condition for initiating cell division we have tentatively called a balance which all cell transfers must reach (by loss of protomorphogen from the protoplasm) before cell division begins.

As the protomorphogens accumulate in the media this optimum ratio is changed and cell division is progressively inhibited. It can only be restored by lowering the media protomorphogen concentration, thus allowing a loss of protomorphogen from the protoplasm, re-establishing the optimum balance.

Similarly, cell division will not proceed if the media volume is so large that sufficient protomorphogen (excreted by the transferred cell) cannot accumulate in it to stimulate growth. Cells transferred into too large a volume of media will not develop.

Lag Period.—The transfer of cells into a new culture media is followed by a latent period before cell division begins. This is the period necessary for protomorphogen concentrations in the protoplasm and in the media to reach the optimum ratio. In the case of aged cells transferred into a fresh media sufficient protomorphogen must diffuse out of the protoplasm into the media to lower the intra- and raise the extracellular concentrations to the optimum points. In the case of young cells transferred into a fresh media sufficient protomorphogen must still leak out to raise the extracellular concentrations to an optimum point; the metabolic activities of the cell must proceed without mitosis for enough time to produce more intracellular protomorphogens. Various original ratios of intra- and extracellular protomorphogens suggest a variety of possibilities for varying lag period and culture phenomena discussed in the text of Chapters 2 and 3.

Nucleus As the Seat of Cell Division and Cell Vitality.—We feel that nuclear metabolic activities are the seat of mitotic activity and their increase must precede cell division. The nucleus, we believe, is also the seat of the metabolic activities which maintain cell vitality.

Electrical Potentials in the Cell.—The electrical potential between the cytoplasm and the media, and that between the nucleus and cytoplasm, varies directly with the health and vitality of the cell. These potentials are a function of the integrity of the cell and nuclear " membranes."

Cell Permeability.—An increase in the permeability of these surface areas results in the dispersion of ions which "short-circuit" the system and decrease the electrical potential. The health and vitality of the cell, therefore, is secondary to the integrity and health of the cell surface boundaries.

Intracellular pH.—The potential hydrogen ion concentration is also lowered in the protoplasm as a consequence of the increased permeability of the cell surface boundaries. This decrease and the decrease in cell boundary potentials are inevitably associated with ageing processes and innervation of cells.

Reversability of Cell Enzymes.—Most enzyme reactions are reversible, the predominant direction of the reaction depending upon substrate conditions. A lowered pH emphasizes the destructive and inhibits the constructive phases of intracellular enzyme systems. In this manner a decrease in cell vitality and impairment of morphological integrity is produced by changes in pH.

Influence of Protomorphogens Over Cell Vitality.—An increase in the protomorphogen content of the cell medium alters the integrity of the cytoplasmic surface boundary, resulting in: (1) the increased concentration of cytoplasmic protomorphogens, and (2) an increased diffusion of electrolytes which lower the cell potential and hydrogen ion concentration. The increased cytoplasmic protomorphogen concentration exerts a similar influence on the nuclear membrane. The end result is a gradual lowering of the cell potentials (including pH) and emphasis on destructive enzymatic activity which inhibits mitosis and eventually results in cell dissolution.

Nuclear Energy Systems.—The synthesis and destruction of *phosphagen* by nuclear *phosphatase* is an important nuclear energy system which may be associated with chromosome (and consequently protomorphogen) activity. We speculate that *ribonucleic acid* is synthesized at the nuclear boundary and converted into *desoxyribonucleic*

317

acid, becoming a part of *chromatin nucleoprotein.* Chromatin metabolic activities break this down again into the *ribonucleic acid* form and it is secreted into the cytoplasm in the nucleoprotein constituent of chromatin granules which serve as determinants for cell structure. The morphogen moiety of the chromatin granule is used as a cytoplasmic determinant, releasing ribonucleic acid for further chromatin synthesis.

The *nucleoprotein chromatin* is the only self-duplicating molecule in living organisms and, as such, is the key to cell division, regeneration and growth. The above cycle is connected with the synthesis of new protomorphogens and assumes a position of key importance in the protein moiety of chromatin nucleoprotein.

Mechanism of Protomorphogen Excretion.—We have seen that new protomorphogens are synthesized only in the chromosome as a part of new nucleoprotein chromatin material. This cycle of chromatin synthesis and discharge is greatly augmented during, and primarily responsible for, cell division. It, however, occurs regularly throughout cell life as an energy mechanism and an expression of the dynamic state of living molecules.

The waste protomorphogen components of chromatin nucleoprotein accumulate in the cytoplasm under the protection of a fatty "envelope." This accounts for the appearance of fatty degeneration and cytoplasmic fat droplets in ageing cells.

This "split" or "waste" protomorphogen is further lost into the media, and gradually accumulates as an expression of the constant dynamic metabolic activity of chromatin.

Use of Media Protomorphogen in Cell Growth.—The "split" protomorphogen in the media is utilized by the cell as substrate material for the synthesis of new cytoplasmic protein molecules at the cytoplasmic boundaries.

This cytoplasmic protein is in turn used as substrate material for synthesis at the nuclear boundaries. In this manner protomorphogen "fragments" appearing in the media are utilized for the nuclear energy mechanisms and consequently their presence stimulates growth and mitosis. These "fragments" may not necessarily be homologous since they are utilized as fragmentary protomorphogen components and not as complete specific units. Homologous protomorphogens, however, are used for this purpose and are probably preferred by the cell.

Inhibition of Mitosis By Media Protomorphogen Concentrations.—As the protomorphogen accumulates in the media it polymerizes into larger molecules, polymerization and formation of chain molecular systems being a peculiar characteristic of morphogen substance. These polymerized molecules either "clog" the cytoplasmic boundary preventing further excretion of cytoplasmic protomorphogens, or they influence the cytoplasmic protomorphogens to polymerize.

The accumulation of polymerized protomorphogens in the cytoplasm lowers the cell potentials and pH, inhibiting the constructive enzyme phases resulting in cessation of mitosis and eventual cell dissolution.

This polymerizing influence is a specific effect and heterologous protomorphogens cannot induce it. It is for this reason that homologous protomorphogens alone can inhibit cell division by reason of their concentration in the media.

Mitogenetic Rays.—Ultra-violet radiations are given off by actively dividing cells. These radiations are called mitogenetic rays because they stimulate the mitotic index of neighboring cells. These radiations are produced by gylcolysis, oxidation and proteolytic reactions in the cell nucleus. They seem to be associated with nucleoproteins and absorbed by them. These mitogenetic ray-producing reactions are associated with the metabolic cycle of chromatin.

We suggest that the beneficial influences of mitogenetic radiation is due to its influence over the polymeric state of nucleoprotein and protomorphogen, eliminating the adverse effects of excessive polymerization of the latter in the cytoplasm.

Mineral Distribution in the Dividing Cell.—Microincineration experiments at different phases of mitosis indicate that the mineral constituents are concentrated in the chromatin material (which we would suppose, since protomorphogen has an important mineral framework). During the mitotic phases the mineral distribution follows a pattern linking it with chromatin activity.

Differentiation in the Developing Embryo.—We suggest that the chromosomes map "fields" in the blastoderm and deliver genic groups of cytomorphogens to these areas. The pattern is determined by the electric field surrounding the chromosomes which directs the mosaic of thread-like molecules emerging from the chromosome. These thread-like molecules are fibrin, precipitated by means of thromboplastic protomorphogen. The genes are "sent out" along these fibers to their locale of determinant activity by the chromosome.

When the genes have been "deposited" in the areas they are to organize, they begin to"unwind"and release cytomorphogens into the receptive embryonic cells in these areas; this results in differentiation. Up to the point where the cytomorphogens are released to receptive cells the cells, in general, are low-organization cells incapable of differentiation (without their determinants). This period of low-organization is a period known as "competence" during which transfer of determinants may influence or alter the nature of the subsequent differentiation.

Determinants As a Virus System.—We have previously held that the nucleoprotein chromatin is the only self-duplicating molecule in the living organism. Viruses are submicroscopic macromolecular nucleoprotein particles which can exist outside a living organism but can only reproduce in a living host. The determinants, from the complex chromosome to the simpler component protomorphogen, may constitute a specialized virus system.

Experimental Transfer of Determinants.—Determinant material may be transferred to receptive "competent" embryonic areas of another individual stimulating abnormal differentiation in locales which would normally differentiate in a different manner.

Inductors and Organizers.—Some transfers do not specifically cause the differentiation of morphology which they would normally do in the donor embryo. Such transfers represent, we believe, an inductor mechanism distinct from the organizer system. The organizer is the determinant *per se,* which confers to "competent" cells a differentiation stimulus of a specific nature. An inductor, however, is a substance which either releases, alters the nature of, or stimulates the activity of the host's own organizer material. Inductors, therefore, may consist of hydrocarbons, fat solvents for the lipoidal "wrapper" of real organizers, or organizers themselves may exert an "inductor" influence by increasing the permeability of the organizer cells. Any mechanical or chemical irritation may, through this means, exert an "inductor" effect.

Whether the chromosomes of individual cells only contain the genic material necessary for the organization of specific locales, or whether they contain a full genic complement much of which is "masked" is not critically pertinent to the morphogen hypothesis.

Determinant Morphogen Cycle.—At mitosis the chromosome discharges a significant amount of chromatin into the cytoplasm which organizes the morphology of the cytoplasm and of the cell. This is a part of the determinant cycle of morphogens which is specifically concerned only with the organization of the cell morphology.

Metabolic Morphogen Cycle.—We have reviewed the constant discharge of protomorphogen into the cytoplasm and cytoplasmic excretion into the media that is a pan of the energy mechanisms in the nucleus. This synthesis and excretion is a function of the dynamic state of nuclear constituents and probably occurs independently of the determinant cycle since protomorphogens continue to accumulate in the media even after cell division ceases. The possibility that the determinant cycle is simply an augmented manifestation of the metabolic cycle is recognized, however.

Chromosome Synthesis.—Chromosome synthesis occurs in the germinal cells of the metazoan animal. We postulate that a heterochromatin framework which organizes the spatial relationships of genic material and which is responsible for the general determinant effects, is passed from parent to offspring as the primary hereditary factor.

We believe that morphogen determinant material, however is synthesized only in the somatic cell nucleus. It is transferred to he germ cells through the blood stream and serves as substrate material for complete chromosome assembly. In this manner, some environmental modifications may be transferred as recessive hereditary characteristics since deficiencies of specific cytomorphogen fragments can impair the nature of the chromosome assembly of that particular determinant gene.

Environmental Modifications of Structure.—Environmental modification of the structure of cells *in vitro* has been demonstrated, but seems to be restricted to the influence of morbid protomorphogen in the media over cytoplasmic protein synthesis. This may produce recessive changes and these may be "bred out" of the strain by a series of transfers under optimum culture conditions. Environmental "mutations" of cells *in vitro* into entirely new and unrelated species, however, is a phenomenon often reported, especially in the field of bacteriology. These "mutations" apparently are stimulated by extraordinary concentrations of protomorphogens in a stagnant media or by other unfavorable media characteristics. This observation has considerable importance in relation to the cancer problem as we shall shortly see.

Maintenance of Morphological Integrity.—The bisexual reproductive method of metazoan animals is a key factor in the maintenance of species integrity. In protozoan life, however, it seems that the chromatin reorganizations concomitant with conjugation or endomixis have this responsibility.

Morphogens in Various Forms of Living Organisms.—The morphogen hypothesis applies to all living cells, although the details of the metabolic cycles and the information regarding the lowering of cell potentials by accumulating protomorphogens probably does not strictly apply to plant as well as animal cells. Among the metazoan organisms the methods of control over the morphogens differ in plants, reptiles and mammals. We have discussed only the probable methods of control in mammals and pointed out significant differences between these and those of the cold blooded animals.

Thromboplastic Properties of Morphogens.—The morphogens are thromboplastic (stimulate the precipitation of fibrin). Their presence is responsible for the thromboplastic activity of platelets and tissue extracts, especially those from brain, lung, kidney, testicle and placenta. They are often associated with lipoid substances in these tissues which "sheath" them, preventing the toxic influences of "raw" protomorphogens.

Morphogens and Connective Tissue.-Connective tissue arises from the precipitation of fibrin under the influence of the thromboplastic activity of protomorphogens excreted from the body cells. Protomorphogen is then adsorbed on these fibers giving rise to what we know as white connective tissue. Yellow elastic tissue also has an affinity for free protomorphogens excreted by the cells. These two fibrous tissues constitute a great storehouse of adsorbed protomorphogens in the organism.

Excretion of Morphogens from the Organism. —The protomorphogen discharged by the cells into the pericellular fluids is adsorbed on and stored in *connective tissue*. It is released from this adsorbed state by *elutogenic factors* which include *sex hormones, epithelial fibrinolysins, urea and allantoin, thyroid and guanidine.*

Some of the intact protomorphogens thus released are enveloped in lipoid "wrappers" and prevented from further lysis by the prostate secretion. They are transferred in the blood to the germinal centers where they are attached to the chromatin network to form "active" chromosomes. This is the "determinant" morphogen cycle in the metazoan organism.

Vitamin A catalyzes the protective association of protomorphogen with lipoids. The hyaline degeneration associated with vitamin A deficiency is an expression of the connective tissue forming activity of free protomorphogens.

The elimination cycle begins with splitting of the protomorphogen released by elutogenic factors under the hydrolytic influence of *blood trypsin* and *kidney enzymes*. The *diffusible fractions* are excreted by the *kidney* in the *urine.*

The non-diffusible colloidal fractions remaining after enzymatic hydrolysis react with the *natural tissue antibody,* which agglutinates them with the assistance of *alexin* so they are sensitized and ready for phagocytic removal. In the *phygocytes* they may undergo further enzymatic destruction with excretion of the diffusible constituents in the *urine* and the non-diffusible in the *bile.*

Those protomorphogen particles picked up by the *megakaryocytes* form the *thromboplastic* substance in the platelets. They are then disposed either through the blood coagulation mechanism or are destroyed and excreted in the *bile.*

Tissue Inflammation.—Free protomorphogens are intensely irritating to neighboring cells. This irritation is probably normally prevented by various protective functions, especially the protective association with lipoids, catalyzed by vitamin A. Under hydrolysis the inflamed tissue releases a series of inflammation producing products, of which necrosin is probably a protomorphogen derivative.

Tissue Regeneration and Repair.—Protomorphogens in an injured area function as substrate material for the synthesis of new protomorphogens during repair. In order for them to be available as such, embryonic or epithelial "growth" factors must be present to stimulate the synthesis of new tissue. Epithelial fibrinolysin probably keeps them in depolymerized form so they may be available for their determinative action in this synthesis.

Traumatic Shock.—We have presented a cycle of the shock syndrome (Fig. 9). The local release of protomorphogens by trauma irritates the nerve termini in that area resulting in a dystrophic stimulus to the whole organism. This dystrophic stimulus increases permeability, lowering the potential differences and pH, resulting in a lowered cell vitality and release of more protomorphogens, completing the cycle. The dystrophic influence also constricts the capillaries and dilates the large internal blood vessels into which the blood drains with resulting circulatory stagnation and anoxia further lowering the vitality of all the cells.

The immediate cause of death in shock is the circulatory failure attendant upon these influences. Nervous influences can also precipitate this cycle by reason of the dystrophic stimuli thus generated.

Senescence.—The primary *index* of senescence is lowered cell vitality and inability to undergo the mitosis necessary to repair damage. This lowered vitality is a direct consequence of the accumulation of discharged protomorphogens in the connective tissue and fluids surrounding the cells.

The primary *cause* of senescence, however, is the progressive impairment of the protomorphogen elimination systems which results in the gradual accumulation of these toxic products in the cell environment.

This gradual impairment of the eliminative functions is under the control of a master organ which determines the life cycle of the species. This is fundamentally an expression of hereditary influences. This master organ is probably the anterior pituitary which operates through its position as a "test organ" controlling vital functions. One of the most important of these is probably its regulation of the immune-biological system through its influence over the adrenal cortex.

We have called attention to the fact that the cycle in senescence is similar to the cycle in traumatic shock except that it is progressive, slow and lacking the dramatic nature of the latter.

Nausea of Pregnancy.—We have suggested that emesis is a consequence of toxic protomorphogen concentrations, probably an automatic protective mechanism against the occasional toxic protomorphogens contained in foods. Probably most emetic drugs (mustard, for instance) produce their effects through release of toxic protomorphogens.

The nausea occurring in the first months of pregnancy is an expression of the increased protomorphogen in the maternal circulation as a result of embryonic mitotic activity. This is brought under control after a latent period by the increased activity of the natural tissue antibody stimulated by the antigenic influence of this protomorphogen increase.

Eclampsia.—The eclampsia occurring later in pregnancy we feel is due to the excess production of protomorphogen decomposition products, i.e., guanidine, resulting from the overactivity of the protomorphogen hydrolytic systems, overloading the eliminative functions. It is probable that the parathyroid normally converts this guanidine into creatine-phosphate with the assistance of methyl groups which form methyl-guanidine.

"Rejuvenation" of Pregnancy.—The "rejuvenation" often reported to follow pregnancy is, we believe, a consequence of enhanced protomorphogen eliminative systems, i.e., natural tissue antibody, as a consequence of the overcompensation during pregnancy.

Varicosities in Pregnancy.—We have suggested that the increased thromboplastic content of pregnant blood assists the pressure processes m the formation of varicosities such as hemorrhoids due to the stagnant blood circulation in the affected area.

Cancer.—We have outlined the relationships of cancer to the morphogen hypothesis. High local concentrations of protomorphogens seem to lead to the appearance of cancer probably either by causing "mutations" directly or preparing the way for the advent of viruses which will stimulate "mutations." This high local concentration is prevented from exerting carcinogenic influences if it is combined in a protective-association with lipoids, this combination being catalyzed by vitamin A. This explains why vitamin A deficiency predisposes to cancer. It also offers a modus operandi for carcinogenic hydrocarbons and x-rays, both of which remove the lipoids from the protective-association releasing intense local concentrations of "raw" protomorphogens.

Replica Hypothesis.—The Replica Hypothesis envisions the possibility of the production of "pattern making" enantiomorphic mirror-images of normal chromatin material which, in turn, produces another mirror-image of itself, thus exactly duplicating the original molecule. This stereoisomeric mirror-image reproduction method is likely a normal chromosome mechanism in physiological mitosis. It is also linked with the antigenic catalysis of antibodies.

During the initial stages of embryonic development and during cancer, this replica method is greatly stimulated resulting in the appearance of enantiomorphic "pattern making" stereoisomers. In cancer, it is possible that the controls of the enantiomorphic "pattern making" molecules are impaired, resulting in their pathological increase and resultant uncontrolled cell division.

Anemia.—There is a possibility that both the blood platelet and erythrocyte counts are regulated by the concentration of their homologous protomorphogen in the blood stream. This concentration is, in turn, controlled by an immune response involving the reticulo-endothelial system.

Arthritis.—Some arthritic lesions are apparently an expression of the accumulation of toxic protomorphogens in the osteoblast areas, resulting in inflammatory and degenerative processes in the bones and joints. This accumulation is prevented from exerting inflammatory influences by vitamin A, which catalyzes the protective-association with lipoids.

Cytotoxins.—The antigenic stimulation of specific immune antibodies to specific organ protomorphogen and the use of small amounts of the antibody to reduce the protomorphogen accumulations in specific tissues is known as Cytotoxin therapy. The reduced local protomorphogen accumulations, so obtained, are followed by an increase in the vitality of the organ concerned.

Protomorphogens are specific for organs but only relatively specific for tissues. Therefore it is practical to produce immune sera against the protomorphogens of specific organs and use it to promote their vitality. Bogomolets has extensively studied immune serum against reticuloendothelial protomorphogen and by the use of small amounts succeeded in stimulating the activity of this system. As would be anticipated, spectacular therapeutic results have been reported.

The use of specific immune sera (cytotoxins) in excess or during pregnancy should be cautioned against. In excess, the immune sera causes an involution of the tissues in the organ itself, instead of simply reducing the surrounding protomorphogen to a more favorable balance. In pregnancy the immune sera can impair the transfer of protomorphogens to the germinal cells and a chromosome deficiency in certain determinants may result.

The Morphogen Hypothesis.—Finally, the morphogen hypothesis basically contends that the chromosome material is probably a virus system which exhibits a constant dynamic metabolism and consequently secretes fractions into the medium surrounding the cell. In a depolymerized form these fractions (protomorphogens) are used as substrate material for protein synthesis at the cytoplasm boundary, stimulating growth and mitosis. In a concentrated and polymerized form, however, they prevent further excretion with the result that these "waste" products (protomorphogens) accumulate in the cell. These accumulations lower the cell potential preventing mitosis, decreasing vitality and eventually causing death.

The morphogens, therefore, in addition to serving as the determinants for cell morphology regulate the life and vitality of all cells. The balance of the morphogen hypothesis is concerned with their specific influence in disease and the manner of controlling these influences in the metazoan organism.

APPENDIX

On the Importance of Cytotoxines In Zootechnics,
Veterinary Science and Medicine
PROFESSOR K. R. VICTOROV
Timiryazew Agricultural Academy
Moscow, U. S. S. R.

APPENDIX

On the Importance of Cytotoxines in Zootechnics,
Veterinary Science and Medicine
PROFESSOR K. R. VICTOROV

Timiryazew Agricultural Academy Moscow,
U. S. S. R.

CYTOTOXINES BELONG to immune substances which are most interesting from the theoretical point of view and are of great practical value. Their discovery in 1898 by Borde stimulated numerous investigations examining their action on the organism from various standpoints. The study of cytotoxines may be divided into two periods. The first period includes the first ten years after their discovery and continues till about 1913. During this period a very important property of cytotoxines has been studied, viz. their strong virulence. The introduction of large dosages of cytotoxines has caused different pathological processes, from inflammation to necrosis in the corresponding organs. The Mechnicov point of view that small dosages of cytotoxines may and ought to produce the opposite effect, namely stimulation, which could be utilized in the treatment of various diseases, has not attracted attention and has almost been forgotten.

Only beginning with about 1925 has science begun a close study of the effect of small dosages of cytotoxines. This second period lasts till now. Between these two periods there is an interval when hardly any papers were published on cytotoxines.

The second period began with the works of Academ. Bogomoletz and his collaborators who until this very day are intensively studying the stimulating effect of small dosages of specific cytotoxines increasing the function of the elements of the active mesenchyme of the animal organism. Owing to this, the protective factors rise considerably in connection with this most important function of the mesenchyme in the organism. As a result of this Bogomoletz continues to treat with great success various diseases of man especially infectious ones, as, for instance, scarlet fever, measles, various kinds of typhus. Most hopeful results have been obtained in the treatment of malignant tumors in test animals.

The second institution, where small dosages of cytotoxines have been applied with great success, is the laboratory of Prof. Sacharov, whose work is connected with that of the clinics. This laboratory has attained splendid results in the treatment of schizophrenia by means of neurotoxic sera, and particularly, in the treatment of diabetes by means of pancreatoxic sera.

We started studying these problems in 1932 under the influence of the success of Prof. Sacharov. During this period my collaborators and I have solved several important problems in the study of cytotoxines, beside working on the application of cytotoxines in zootechnics, veterinary science and medicine.

A few words should be said about the method applied by us. We prepared cytotoxic sera from the blood of rabbits who were immunized with emulsions of cells from various organs or of extracts from dessicated organs. Several injections produce a sufficient amount of cytotoxines in the organism whose concentration may be titrated after the reaction of complement binding. The problem of the size of the small dosages we solved by conditionally accepting the size of the titre for the quantity of units of cytotoxines in 1 ccm. and finally, in several experiments we found that a small dosage, i.e. the dosage which produces a stimulating effect, fluctuates near 0.2 of a titre unit per kilogram of body weight.

First of all we verified the property of the cytotoxines' double action. In several investigations small dosages invariably stimulated and caused an increase of the function of corresponding organs, but at the same time they accelerated and increased the growth processes and the development of the tissues. The tests of large dosages showed the intense action of specific cytotoxines. For instance, a splendidly laying hen received a large dosage of ovariotoxic serum; the very next day the hen stopped laying eggs, lost her appetite and ruled her feathers; however, four days later the hen was already well; twelve days after the injection the hen was killed; autopsy showed complete absence of ovaries which had been resolved. We had made no systematic experiments with large dosages, as we were interested only in small dosages for stimulation. The following problem which we had to solve was that of the specificity of the action of cytotoxines. The literature gives no positive answer to this question. Some writers consider that cytotoxines are endowed with a strict tissue-specificity and, therefore, act only on that organ or tissue which served as antigens during immunization of the animal. Other writers considered that the cytotoxines affect not only the given organs, but others as well, and some investigators add that their action is feebler than on the homologous organs. Thus, only relative specificity was attributed to cytotoxines. As regards species-specificity, there is also no agreement among the investigators. Some writers consider that cytotoxines affect the organ and tissues of only those species of animals whose organ served as antigens. Other writers have come to the conclusion that species-specificity is feebly manifested or is completely absent in cytotoxines and that owing to this cytotoxines affect the organs of other species of animals as well.

It seemed to me that this problem is especially important for the practical application of cytotoxines, since the fact of species-specificity and the absence of tissue-specificity would inhibit the preparation, of cytotoxic sera and their practical application. In our investigations and observations, we invariably came to the conclusion that cytotoxines possess strict tissue-specificity and are almost devoid of species-specificity. Since the solution of this problem is of great importance, I recommended that one of my collaborators carry out a special investigation. We carefully isolated the corpus lutea from cow ovaries and immunized rabbits with the corpus lutea; the luteotoxic serum was applied to female mice; as a result, we observed pronounced growth stimulation of the corpus lutea and sometimes to such an extent that the ovary turned into an organ consisting entirely of corpus lutea and interstitial tissue. This investigation showed marked tissue-specificity and absence of species-specificity (Chushkin). However, in order to solve the problem of the presence or absence of species-specificity in cytotoxines, we carried a special investigation with ovariotoxines prepared from pig and cow ovaries and treated pigs with both kinds of sera; as indices of action served the clinical manifestations of oestrus and of their duration; it was found that the "pig ovariotoxines" are more effective than the "cow ovariotoxines." Thus we have the right to affirm that cytotoxines are endowed with very feeble species-specificity.

As already mentioned, Academ. Bogomoletz and Prof. Sacharov use cytotoxines for increasing the action of feeble sick organs, viz. they apply cytotoxines for treatment. We also obtained very good results when we applied cytotoxic treatment. We treated some diseases of animals and man with cytotoxines as, for instance, impotence of stallions, bulls, hogs and man was treated with a specific testiculatoxic serum (Riabov, Beirakh). Hypophyseal plus ovarial-toxic sera also gave good results in treating sterility of mares (Kasakov). We called forth sufficient lactation during feeble milk secretion in pigs and thus saved the young from death (Morosov). Unexpected good results were obtained during the treatment of such complicated processes as chronic prostatitis, especially on gonorrhoeal basis (Beirakh). Of no less interest proved the results of treating wounds of horses with reticulotoxic sera as the wounds healed much quicker (Oserov).

However, it seemed to us of no less importance to investigate in the interests of zootechnics the problem of the possibility of stimulating healthy organs, in the prime of their physiological equilibrium. In this case we had the aim of increasing the productivity of farm animals.

Our investigation of this problem gave positive results. The experiments with treating cows with mammotoxines (Morosov) made it possible to increase the milk yield till 60 to 90 per cent as compared with the initial milk yield before the injections. Of great interest are the experiments of Professor Pirogov in this respect who obtained with this preparation profitable milk increase in koumiss mares. The experiments of Ivanov on hens showed that there are possibilities of increasing egg laying.

If it is to be admitted that the stimulation of normal organs is possible and is practically produced, this problem becomes still more important when it is applied to the organ of internal secretion. The stimulation of the function of the endocrine organs may affect such general functions of the organism as, for instance, growth, metabolism, propagation. We obtained (Birikh) a visible growth increase in mice and rabbits after treatment with hypophyseotoxic sera. The work of Averin carried on in the farm "Achkassovo" on young pigs showed that by means of hypophyseal plus thymotoxic sera it is possible to increase the growth of sucking pigs till 120 per cent at slaughter period as compared with 100 per cent of the control pigs.

The experiments with stimulation of development in mice by means of thyreotoxic sera (Arkhangelskaja) were less effective. However, in these cases we also have decided outlooks for carrying out more thorough and larger experiments since the development in the mice was stimulated by four to five days.

I can not consent to the objection that the stimulation of the normal organs with the aim of increasing normal productivity may be dangerous as the overworked organs may become emaciated. We know that the organism of animals function on a certain average and that animals possess considerable reserves for enlarging their activity within the limits of the physiological norms. The function of kidneys, salivary glands, muscles, heart and vessels, of the nervous system, etc., exhibits some times such degrees of activity increase in responding to various requirements of the organism that there is no doubt that even during such moments of overwork the organism remains within its limits of health and norm. It is of course, evident that, for instance, lactation requires corresponding increase of feeding of the animal.

Before concluding I must concentrate on the problem of the mechanism of the action of cytotoxines. We find only one explanation in the literature: Academ: Bogomoletz assumes that during the action of the cytotoxines the tissues perish and lysates are produced which are responsible for the action of the cytotoxines. Owing to the fact that Bogomoletz does not explain the primary action of the cytotoxines, and that so far we have no explanation of the activity of lysates, this explanation is unsuccessful.

It has always seemed to me that the explanation of such facts should be looked for in the histological pictures of the stimulated organs. Therefore, all our investigations were accompanied by thorough microscopic examinations, which showed that the cytotoxines irritate all the elements of the corresponding organs, viz. parenchyme, interstitium, vessels. A hyperaemia is invariably observed here, swellings, increase of mesenchymal elements, somewhat juicy parenchyme at those cells passing over into an increase of secretory processes, cellular multiplication passing over into the development of the active parts of the organs, in the glands into the. development of the alveoles. I do not doubt that introceptive innervation takes part in this irritation of the organ, that this process in its tum involves the action of the central nervous system which thereupon directs the stimulation process. Possibly, this latter circumstance may explain the duration of the action of the cytotoxines.

GLOSSARY OF NEW TERMS

This glossary contains those common words to which have been imparted special meanings and a few terms the authors have been forced to coin in order to satisfactorily express some new concepts.

ALLELOCATALYST—Robertson's name for an autocatalytic growth factor which inhibits growth in higher concentrations. (Protomorphogen is one of its physiological forms.)

ARCHUSIA—Burrows' name for protomorphogen in one of its physiological forms.

BIOLOGICAL PROTEIN—Any protein requiring an input of energy to maintain its dynamic state and exhibiting antigenic specificity.

BOUND PROTOMORPHOGEN—Protomorphogen in a relatively inactive form because of its association with other cell components.

COMPETENCE—(of tissue), a temporary state of receptiveness to morphogenetic influence.

CYTOST—Turck's name for antigenic tissue ash. (A protomorphogen end product.)

CYTOTOXIN—An artificial, exogenous form of natural tissue antibody, which may be used to reinforce the action of the natural endogenous form.

DEPOLYMERIZER—Any factor which reduces the molecular size of colloidal structure, usually with an increase in biological activity.

DETERMINANT—Any cell component which organizes living structure.

ELUTOGEN—Any endogenous or exogenous factor that is able to clear the tissues of stores of adsorbed or combined protomorphogen.

ERGUSIA—Burrows' name for a cell component secreted under certain circumstances. (Protomorphogen carried in a lipoid protector.)

INDUCTOR—Any cell component which induces differentiation. May or may not also organize structure. Some inductors may be of extracellular source, such as carcinogens.

LIVING PROTEIN—See BIOLOGICAL PROTEIN.

MORPHOGEN—The chromosome or any of its components which is responsible for morphogenesis.

MORPHOGEN HYPOTHESIS—A hypothesis that links the control of growth and primary biological phenomena with the chromosome factors that determine morphology.

NATURAL TISSUE ANTIBODY—That group of antibodies developed by the immune mechanism towards the organism's own specific tissue proteins.

ORGANIZER—Synonym for determinant.

RAW PROTOMORPHOGEN—Protomorphogen dissociated from lipoid or protein (its most toxic form).

REPLICA HYPOTHESIS—A concept of molecular reproduction by means of template patterns in which the template is the stereoisomer of the reproduced unit.

SHEATHED PROTOMORPHOGEN—Protomorphogen with its chemical affinities masked by lipoid layers, usually monomolecular.

SHEATHING SUBSTANCE—A specialized lipoid complex that tends to form layers around protomorphogen molecules.

TEMPLATE—A stereoisomer of a functional biological unit, serving as a pattern for reproduction, just as a photographic negative serves as a pattern for a print.

UNWINDING OF CHROMOSOMES—The orderly and patterned release of morphogens to competent embryonic tissue.

WRAPPED PROTOMORPHOGEN—See SHEATHED PROTOMORPHOGEN.

\VRAPPER SUBSTANCE—See SHEATHING SUBSTANCE.

AUTHOR INDEX

347

Taurog, A. (co-workers), 183
Taurog, A. (Entenman, Chaikoff), 204
Tayeau, F., 198, 108
Taylor, C. V. (Barker), 54
Taylor, C. V. (Strickland), 53
Taylor, G. W. (Richards), 106
Telkes, M., 72, 75-77
Telkes, M. (Crile, Rowland, 72, 75, 76, 86, 290
Ten Broeck, C., 186
Tennant, R. (Liebow), 20
Tennant, R. (Liebow, Stem), 96
Thiery, J. P. (Maignan), 189
Thorell, B., 91, 110
Thorell, B. (Caspersson), 91
Tipson, R. S., 105
Tittler, I. A., 102
Tocantins, L. M., 218
Tompkins, E. H., 199, 222
Torrioli, M. (Puddu), 229
Tracy, M. M. (Cohn, Brues), 32
Tristam, G. R. (Williams, Boyland, Reese, Chibnall), 292
Troescher, F. M. (Greenberg), 224
Troland, C. E. (Lee), 129
Tserling, V. V. (Bobko), 16
Turck, F. B., 9-15; 22, 33-35, 59, 62-65, 83, 86, 99, 112, 150, 160, 161, 172, 173, 195, 209, 210, 225, 251-254, 256, 259, 272, 278, 297-300, 305

Underwood, E. J., 17
Ungar, G., 170

Vail, V. C. (Huggins), 215, 216
Van Camp, G., 96
Van Herwerden, M.A., 92
Vaughan, J. W. (Vaughan, Vaughan), 258
Vaughan, V. C. (co-workers), 231
Vaughan, V. C. (Vaughan, Vaughan), 258
Vaughan, V. C., Jr. (Vaughan, Vaughan), 258
Victorov, K., 300, 301
Vigneaud, V. du (Chandler, Moyer, Keppel), 204
Vigneaud, V. du (co-workers), 204

Villela, E. U. D. (Regaud, Gricouroff), 287
Von Haam, E. (Cappel), 185

Waddington, C. H. (Needham, Needham), 134
Wadleigh, C. H. (Shive), 16
Waksman, S. A. (Davison), 82
Walker, A. C. (Marshak), 170, 248 Walsh, J. H., 181
Warthin, A. S., 181
Watchel, H. K., 286
Waugh, W. F. (Abbott), 303
Waymouth, C. (Davidson), 91, 93; 145, 249
Wehmeier, E. (Spemann, Fischer), 134
Weidman, F. D., 265
Weinglass, A. R. (Tagnon), 186, 187
Weismann, A., 9, 140, 141, 155, 179, 182
Weiss, P., 123-116, 118, 131-133, 135, 138, 139
Wells, J. A. (Dragstedt, Cooper, Morris), 187
Wells, J. A. (Dragstedt, Rocha e Silva), 188
Werner, A. A., 180
Werner, H., 35, 56, 95, 101
Werner, H. (Carrel), 9
West, R. (Chargaff), 223
Whipple, G. W. (Hooper), 228
White, A. (Chase, Dougherty), 217, 219
White, A. (Dougherty), 295
White, F. R. (White), 292
White, J. (White), 292
White, M. J. D. (Pontecorvo), 151
Widenbauer, F. (Reichel), 172
Wiggers, C. J., 257
Wildiers, E., 37
Williams, E. F. (Boyland, Reese, Chibnall, Tristam), 292
Williams, T. L. (Rehfuss), 225, 226
Willmer, E. N., 85
Wilson, E. B., 87, 292
Winnek, P. S. (Smith), 17
Winternitz, M. C., 175
Winternitz, M. C. (et al), 272
Winternitz, M. C. (Mylon), 254, 255

SUBJECT INDEX